R. Wiesendanger H.-J. Güntherodt (Eds.)

Scanning Tunneling Microscopy III

Theory of STM
and Related Scanning Probe Methods

With 199 Figures

Springer-Verlag

Berlin Heidelberg New York
London Paris Tokyo
Hong Kong Barcelona
Budapest

Professor Dr. Roland Wiesendanger

Institute of Applied Physics, University of Hamburg, Jungiusstr. 11,
D-20355 Hamburg, Germany

Professor Dr. Hans-Joachim Güntherodt

Department of Physics, University of Basel, Klingelbergstrasse 82,
CH-4056 Basel, Switzerland

Series Editors

Professor Dr. Gerhard Ertl

Fritz-Haber-Institut der Max-Planck-Gesellschaft, Faradayweg 4–6,
D-14195 Berlin, Germany

Professor Robert Gomer, Ph.D.

The James Franck Institute, The University of Chicago, 5640 Ellis Avenue,
Chicago, IL 60637, USA

Professor Douglas L. Mills, Ph.D.

Department of Physics, University of California,
Irvine, CA 92717, USA

Managing Editor: Dr. Helmut K. V. Lotsch

Springer-Verlag, Tiergartenstrasse 17,
D-69121 Heidelberg, Germany

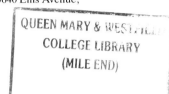

ISBN 3-540-56317-2 Springer-Verlag Berlin Heidelberg New York
ISBN 0-387-56317-2 Springer-Verlag New York Berlin Heidelberg

Library of Congress Cataloging-in-Publication Data. Scanning tunneling microscopy III : theory of STM and related scanning probe methods / R. Wiesendanger, H.-J. Güntherodt (eds.). p. cm. -- (Springer series in surface sciences ; 29) Includes bibliographical references and index. ISBN 3-540-56317-2 (Berlin : acid-free paper). -- ISBN 0-387-56317-2 (New York : acid-free paper) 1. Scanning tunneling microscopy. 2. Scanning probe microscopy. I. Wiesendanger, R. (Roland), 1961- . II. Güntherodt, H.-J. (Hans-Joachim), 1939- III. Title: Scanning tunneling microscopy 3. IV. Title: Scanning tunneling microscopy three. V. Series. QH212.S35S253 1993 502'.8'2--dc20   92-46679

Production Editor: A. Kübler

Typesetting: Macmillan India Ltd., India

54/3140-5 4 3 2 1 0 – Printed on acid-free paper

Preface

While the first two volumes on Scanning Tunneling Microscopy (STM) and its related scanning probe (SXM) methods have mainly concentrated on introducing the experimental techniques, as well as their various applications in different research fields, this third volume is exclusively devoted to the theory of STM and related SXM methods. As the experimental techniques including the reproducibility of the experimental results have advanced, more and more theorists have become attracted to focus on issues related to STM and SXM. The increasing effort in the development of theoretical concepts for STM/SXM has led to considerable improvements in understanding the contrast mechanism as well as the experimental conditions necessary to obtain reliable data. Therefore, this third volume on STM/SXM is not written by theorists for theorists, but rather for every scientist who is not satisfied by just obtaining real-space images of surface structures by STM/SXM.

After a brief introduction (Chap. 1), N.D. Lang first concentrates on theoretical concepts developed for understanding the STM image contrast for single-atom adsorbates on metals (Chap. 2). A scattering-theoretical approach to the STM is described by G. Doyen (Chap. 3). In Chap. 4, C. Noguera concentrates on the spectroscopic information obtained by STM, whereas the role of the tip atomic and electronic structure in STM/STS is examined more closely by M. Tsukada et al. in Chap. 5. The tunneling time problem is still of great topical interest, not only in conjunction with STM, and will therefore be focused on in a separate Chap. 6 by C.R. Leavens and G.C. Aers. A unified perturbation theory for STM and SFM is described by C.J. Chen in Chap. 7. The important issue of tip–sample interaction in STM and related SXM methods is addressed in two separate chapters (Chap. 8 by S. Ciraci and Chap. 9 by U. Landman). Chapter 10 by G. Overney is devoted to contact-force microscopy on elastic media, whereas theoretical concepts of atomic scale friction are described by D. Tománek (Chap. 11). Finally, U. Hartmann concentrates on the theory of non-contact force microscopy (Chap. 12).

We would like to thank all the contributors who have contributed to this third volume on STM, as well as Springer-Verlag for the continuous pleasant collaboration. Hopefully, this third volume on the theory of STM/SXM will help many experimentalists to better understand their data, and will stimulate even more theorists to concentrate on still unsolved issues in the exciting and challenging field of STM and related scanning probe methods.

Hamburg, March 1993
R. Wiesendanger
H.-J. Güntherodt

Contents

Contributors

G.C. Aers

Institute for Microstructural Sciences,
National Research Council of Canada, Ottawa, Canada,
K1A 0R6

C.J. Chen

IBM Research Division, T.J. Watson Research Center,
P.O. Box 218, Yorktown Heights, NY 10598, USA

S. Ciraci

Department of Physics, Bilkent University, Bilkent 06533, Ankara,
Turkey

G. Doyen

Fritz-Haber-Institut der Max-Planck-Gesellschaft, Faradayweg 4-6,
D-14195 Berlin, Germany

H.-J. Güntherodt

Department of Physics, University of Basel, Klingelbergstrasse 82,
CH-4056 Basel, Switzerland

U. Hartmann

Institute of Thin Film and Ion Technology, KFA-Jülich,
P.O. Box 1913, D-52425 Jülich, Germany

N. Isshiki

Institute for Knowledge and Intelligence Science, kao Corporation,
Bunka 2-1-3, Sumida-ku, Tokyo 131, Japan

H. Kageshima

NTT LSI Laboratories, 3-1 Morinosato-Wakamiya, Atsugi-shi,
Kanagawa-ken, Japan

K. Kobayashi

Department of Physics, Faculty of Science, University of Tokyo,
Hongo 7-3-1, Bunkyo-ku, Tokyo 113, Japan

U. Landman

School of Physics, Georgia Institute of Technology, Atlanta,
GA 30332, USA

N.D. Lang

IBM Research Division, T.J. Watson Research Center,
Yorktown Heights, NY 10598, USA

C.R. Leavens

Institute for Microstructural Sciences,
National Research Council of Canada, Ottawa, Canada,
K1A 0R6

W.D. Luedtke

School of Physics, Georgia Institute of Technology, Atlanta,
GA 30332, USA

C. Noguera

Laboratoire de Physique des Solides, Associé au CNRS,
Université de Paris Sud, F-91405 Orsay, France

G. Overney

Department of Physics and Astronomy, Michigan State University,
East Lansing, MI 48824-1116, USA

T. Schimizu

Department of Physics, Faculty of Science, University of Tokyo,
Hongo 7-3-1, Bunkyo-ku, Tokyo 113, Japan

D. Tománek

Department of Physics and Astronomy and Center for
Fundamental Materials Research, Michigan State University,
East Lansing, MI 48824-1116, USA

M. Tsukada

Department of Physics, Faculty of Science, University of Tokyo,
Hongo 7-3-1, Bunkyo-ku, Tokyo 113, Japan

S. Watanabe

Aono Atomcraft Project, ERATO,
Research Development Corporation of Japan, Kaga 1-7-13,
Itabashi-ku, Tokyo 173, Japan

R. Wiesendanger

Institute of Applied Physics, University of Hamburg,
Jungiusstrasse 11, D-20355 Hamburg, Germany

1. Introduction

R. Wiesendanger and *H.-J. Güntherodt*

"Scanning is believing". This statement goes back to a time long before the invention of Scanning Tunneling Microscopy (STM) and related scanning probe methods and has been cited many times, for instance, in conjunction with Scanning Electron Microscopy (SEM). To facilitate our believing we certainly have to understand what we measure. Therefore a theoretical foundation has always been indispensable.

Since 1982, STM has allowed us to apparently 'see' atoms and even smaller features on surfaces. Of course, we do not actually 'see' atomic scale structures. In STM, an atomically sharp tip is brought in close proximity to a sample to be investigated, until a current of typically 1 nA is detected. Subsequently, the tip is raster scanned (usually) at a constant current over the sample surface and equi-current contours are measured being interpreted as real-space images. Similarly, contours of constant tip–sample interaction strength are obtained by using other types of scanning probe microscopes based on different types of interaction between a tip and a sample. By starting with well-defined and characterized test structures, the experimentalists initially have verified that the real-space images obtained, for instance by STM, correspond closely to the expected surface structures as deduced by using different experimental techniques. However, for the application of STM and related methods to previously unknown surface structures, a profound theoretical understanding is needed in order to relate the equi-current contours, or generally the contours of constant tip–sample interaction strength, to the spatial variation of physical properties of the samples under investigation.

A theory usually has to meet two important requirements:

i) It has to provide explanations for already existing experimental data.
ii) It has to make predictions for future experiments to be performed, thereby guiding the experimentalists in the design of appropriate experiments extracting the essential aspects of an unsolved issue.

STM and its related scanning probe techniques provide major challenges for a theoretical treatment, mainly because they require to join macroscopic and microscopic concepts in physics. Certainly, for an appropriate theoretical treatment of a microscopical technique such as STM, or contact force microscopy capable of providing atomic resolution, the atomic structure of the tip and that of the sample surface have to be taken into account and have to be modelled appropriately. To develop a microscopic theory, for instance for STM, it is

Springer Series in Surface Sciences, Vol. 29
Scanning Tunneling Microscopy III Eds.: R. Wiesendanger · H.-J. Güntherodt
© Springer-Verlag Berlin Heidelberg 1993

therefore natural to start with an atomic-level picture emphasizing the properties of individual atoms. On the other hand, the tip atom being closest to the sample surface and the surface atom being probed by the outermost tip atom at a particular time are not isolated, but are part of macroscopic solids which are usually described within a framework emphasizing the collective properties of many atoms (typically 10^{23}). Therefore, a major problem in the theoretical treatment of atomic-resolution microscopies such as STM seems to be the link between solid state and atomic physics or, in other words, to join the 'band' and the 'bond' picture, as well as the electronic density-of-states picture and the picture based on the electronic configuration of individual atoms. Corresponding links are required for the theoretical treatment of STM-related scanning probe methods.

1.1 Theoretical Concepts for Scanning Tunneling Microscopy

One of the most successful early theories of STM has been developed by *Tersoff* and *Hamann* [1.1, 2] who used a perturbative treatment of tunneling based on *Bardeen's* transfer Hamiltonian approach [1.3]. This method requires explicit expressions for the wavefunctions of the tip and the sample surface. Since the actual atomic structure of the tip is generally not known, Tersoff and Hamann started with the simplest possible model for the tip which was assumed to be spherically symmetric (s-wave tip wave function approximation). Within this model, Tersoff and Hamann showed that the equi-current contours measured by STM can be interpreted as contours of constant charge density from electronic states at the Fermi level, at the position of the tip. They also derived an expression for the spatial resolution achievable in STM [1.1, 2], which has been successfully applied to explaining measured corrugation amplitudes of large-period structures on reconstructed metal surfaces as a function of the effective radius of curvature of the tip [1.4]. On the other hand, the Tersoff–Hamann theory failed to explain the observed atomic resolution on close-packed metal surfaces [1.5, 6].

Within a microscopic view of STM, taking into account the atomic structure of the tip, *Chen* [1.7, 8] was able to explain the atomic resolution STM data on Al(111) [1.6] by assuming a d_{z^2}-tip state. First-principles calculations of the electronic states of several kinds of tungsten clusters used to model the microscopic structure of the outermost effective tip, indeed, revealed the existence of dangling-bond states near the Fermi level at the apex atom which can be ascribed to d_{z^2}-states [1.9, 10]. The important role of non-s-wave tip states for enhancing the spatial resolution in STM had been pointed out earlier by *Baratoff* [1.11] and *Demuth* et al. [1.12]. The microscopic view of STM introduced in detail by *Chen* [1.7, 8] is based on symmetry considerations of the electronic states of the tip and the sample surface and therefore emphasizes the atomic-level picture of STM.

A very elegant model which in several respects links the microscopic and macroscopic pictures of STM has been introduced by *Lang* [1.13, 14] who described both tip and sample by an adatom on jellium. The wave functions for the two electrodes are calculated self-consistently within density-functional theory. This model has been applied successfully for explaining the STM image contrast of single-atom adsorbates on metals [1.15]. However, it neglects the atomic structure of the bulk electrodes.

Non-perturbative treatments of tunneling based on the Green's function method have been important to assess the accuracy of the perturbative transfer Hamiltonian method [1.16, 17]. A perturbative treatment of tunneling certainly must break down if the tip–surface separation becomes extremely small, i.e. if the wavefunction overlap of tip and sample surface becomes significant. In this case, the independent-electrode approximation where tip and sample surface are described by unperturbed wave functions is no longer valid. The close proximity of tip and sample surface leads to a significant distortion of the local electronic structure accompanied by a significant charge rearrangement [1.18]. Site-specific Tip-Induced Localized States (TILS) appear which cause the tunneling current to deviate considerably from the proportionality to the local density of states of the unperturbed sample as deduced within the Tersoff–Hamann theory. The formation of TILS has been studied theoretically by ab initio Self-Consistent-Field (SCF) pseudopotential calculations within the Local Density Approximation (LDA) for periodically repeated supercells [1.19]. If the tip is further approached, a qualitative change in electron transport from tunneling to ballistic conductance sets in.

Though electron tunneling is the primary interaction mechanism probed by STM, there are usually several other types of interactions present between tip and sample. A force interaction is always present due to the long-range van der Waals forces and – at small tip–surface separations – also due to the short-range quantum-mechanical exchange and correlation forces. The force interaction between tip and sample can become important for the interpretation of STM data if soft elastic materials are to be investigated, as first noticed in STM experiments on graphite [1.20]. By mounting either tip or sample on a flexible cantilever beam, this force interaction can be studied in more detail, leading to the force microscopy method.

1.2 Theoretical Concepts for Force Microscopy

Force microscopy performed in the regime of contact between tip and sample can provide atomic resolution on conducting as well as insulating samples. The contours of constant force usually obtained have initially often been compared with corrugation profiles as deduced from helium scattering experiments.

Based on results of ab initio SCF pseudopotential calculations, *Ciraci* et al. [1.19] have theoretically investigated what is actually probed by contact force microscopy. They found that only at relatively large tip–surface separations does the tip probe the total charge density of the sample surface, whereas at small tip–surface separations corresponding to the strong repulsive force regime, where most force microscope studies are performed, the ion–ion repulsion determines the observed image contrast. Therefore, contact force microscope images can be interpreted in a direct way by attributing the observed maxima to atomic sites, whereas such an interpretation is generally not applicable for STM results. *Ciraci* et al. [1.19] also pointed out that the repulsive force on the outermost tip atom is always underestimated, because the long-range attractive forces felt by tip atoms far from the outermost one can contribute considerably to the total tip force. Therefore, even if the total tip force might be attractive, the outermost tip atom can still be in the strong repulsive force regime which probably causes local deformations of the sample surface. Such deformations are particularly likely to occur if soft elastic samples are to be investigated by contact force microscopy.

Tománek et al. [1.21] have developed a theory for contact force microscopy of deformable surfaces such as graphite, where the local distortions in the vicinity of a sharp tip were quantitatively determined as a function of the applied force. Limits of resolution in contact force microscopy of graphite were deduced as well [1.22]. This theory has again been based on a combination of microscopic and macroscopic concepts in solid state physics: while the bulk elastic constants of graphite were determined from first principles using ab initio calculations within LDA, the equilibrium deformations were determined using continuum elasticity theory.

More recently, the atomic scale variation of the lateral (frictional) forces [1.23] have become of increasing topical interest. Friction forces couple macroscopic mechanical degrees of freedom of two objects in contact to microscopic degrees of freedom which appear, for instance, as heat or plastic deformation. A theory of atomic scale friction [1.24] again has to provide a link between macroscopic concepts used so far to describe frictional phenomena and microscopic concepts. In particular, the mechanism which leads to the excitation of microscopic degrees of freedom, and hence to energy dissipation in the friction force microscope, has to be clarified [1.25].

Apart from theoretical treatments of STM/SXM, computer 'experiments' based on molecular-dynamics simulations employing a realistic interatomic interaction potential [1.26–28] have become more and more important, because the evolution of a physical system can be simulated with high temporal and spatial resolution via a direct numerical solution of the model equations of motion. This opens additional prospects for the investigation of the microscopic origins of material phenomena, a research field which has been pushed forward tremendously by the invention of the scanning probe methods.

References

1.1   J. Tersoff, D.R. Hamann: Phys. Rev. Lett. **50**, 1998 (1983)
1.2   J. Tersoff, D.R. Hamann: Phys. Rev. B **31**, 805 (1985)
1.3   J. Bardeen: Phys. Rev. Lett. **6**, 57 (1961)
1.4   Y. Kuk, P.J. Silverman, H.Q. Nguyen: J. Vac. Sci. Technol. A **6**, 524 (1988)
1.5   V.M. Hallmark, S. Chiang, J.F. Rabolt, J.D. Swalen, R.J. Wilson: Phys. Rev. Lett. **59**, 2879 (1987)
1.6   J. Wintterlin, J. Wiechers, H. Brune, T. Gritsch, H. Höfer, R.J. Behm: Phys. Rev. Lett. **62**, 59 (1989)
1.7   C.J. Chen: Phys. Rev. Lett. **65**, 448 (1990)
1.8   C.J. Chen: J. Vac. Sci. Technol. A **9**, 44 (1991)
1.9   S. Ohnishi, M. Tsukada: Solid State Commun. **71**, 391 (1989)
1.10 S. Ohnishi, M. Tsukada: J. Vac. Sci. Technol. A **8**, 174 (1990)
1.11 A. Baratoff: Physica **127** B, 143 (1984)
1.12 J.E. Demuth, U. Köhler, R.J. Hamers: J. Microsc. **152**, 299 (1988)
1.13 N.D. Lang: Phys. Rev. Lett. **55**, 230 (1985)
1.14 N.D. Lang: Phys. Rev. Lett. **56**, 1164 (1986)
1.15 D.M. Eigler, P.S. Weiss, E.K. Schweizer, N.D. Lang: Phys. Rev. Lett. **66**, 1189 (1991)
1.16 C. Noguera: J. Microsc. **152**, 3 (1988)
1.17 A.A. Lucas, H. Morawitz, G.R. Henry, J.-P. Vigneron, Ph. Lambin, P.H. Cutler, T.E. Feuchtwang: Phys. Rev. B **37**, 10708 (1988)
1.18 E. Tekman, S. Ciraci: Phys. Rev. B **40**, 10286 (1989)
1.19 S. Ciraci, A. Baratoff, I.P. Batra: Phys. Rev. B **41**, 2763 (1990)
1.20 J.M. Soler, A.M. Baro, N. Garcia, H. Rohrer: Phys. Rev. Lett. **57**, 444 (1986)
1.21 D. Tománek, G. Overney, H. Miyazaki, S.D. Mahanti, H.-J. Güntherodt: Phys. Rev. Lett. **63**, 876 (1989)
1.22 W. Zhong, G. Overney, D. Tománek: Europhys. Lett. **15**, 49 (1991)
1.23 C.M. Mate, G.M. McClelland, R. Erlandsson, S. Chiang: Phys. Rev. Lett. **59**, 1942 (1987)
1.24 W. Zhong, D. Tománek: Phys. Rev. Lett. **64**, 3054 (1990)
1.25 D. Tománek, W. Zhong, H. Thomas: Europhys. Lett. **15**, 887 (1991)
1.26 U. Landman, W.D. Luedtke, M.W. Ribarsky: J. Vac. Sci. Technol. A **7**, 2829 (1989)
1.27 U. Landman, W.D. Luedtke, A. Nitzan: Surf. Sci. **210**, L177 (1989)
1.28 U. Landman, W.D. Luedtke, N.A. Burnham, R.J. Colton: Science **248**, 454 (1990)

2. STM Imaging of Single-Atom Adsorbates on Metals

N.D. Lang

With 15 Figures

Consider the imaging of a single atom adsorbed on a metal surface in the STM. Ideally, the STM tip will also be one atom adsorbed on a group of other metal atoms. For theoretical purposes, we will model this system using two flat metallic electrodes, each of which has a single atom adsorbed on its surface, with one representing the tip and the other the sample. If we calculate the current that flows between these electrodes when a bias voltage is applied between them, then we can study theoretically many of the basic physical aspects of STM imaging.

In order to answer such questions as the degree to which the images of chemically different atoms differ in the STM, it is necessary to take full account of the electronic structure of the adsorbed atoms, but the so-called jellium model [2.1] can be used for the metal electrodes themselves, in which the ionic lattice of each metal is smeared out into a uniform positive charge background.

For tunneling between two metallic electrodes with a voltage difference V across the gap, only the states within $|e|V$ above or below the Fermi level can contribute to tunneling (at zero temperature), with electrons in states within $|e|V$ below the Fermi level on the negative side tunneling into empty states within $|e|V$ above the Fermi level on the positive side (e is the unit of electronic charge). As shown in Fig. 2.1, other states cannot contribute, either because there are no electrons to tunnel at higher energy, or because of the exclusion principle at lower energy.

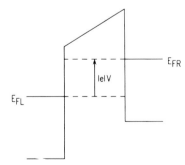

Fig. 2.1. Schematic of potential barrier between electrodes for vacuum tunneling. V is the applied bias, with the left electrode positive. The Fermi levels of left and right electrodes are respectively E_{FL} and E_{FR}, with $E_{FR} - E_{FL} = |e|V$

Springer Series in Surface Sciences, Vol. 29
Scanning Tunneling Microscopy III   Eds.: R. Wiesendanger · H.-J. Güntherodt
© Springer-Verlag Berlin Heidelberg 1993

2.1 Tunneling Hamiltonian Approach

In first-order perturbation theory, the current between the two electrodes is

$$I = \frac{2\pi|e|}{\hbar} \sum_{\mu,\nu} [f(E_\nu) - f(E_\mu)] |M_{\mu\nu}|^2 \delta(E_\nu + |e|V - E_\mu) , \tag{2.1}$$

where $f(E)$ is the Fermi function, V is the applied voltage, $M_{\mu\nu}$ is the tunneling matrix element between states ψ_μ and ψ_ν of the left and right electrodes respectively, calculated independently, E_μ is the energy of ψ_μ relative to the left-electrode Fermi level E_{FL}, and E_ν is the energy of ψ_ν relative to the right-electrode Fermi level E_{FR}. For most purposes, the Fermi functions can be replaced by their zero-temperature values, i.e. unit step functions. In the limit of small voltage, this expression then further simplifies to

$$I = \frac{2\pi}{\hbar} e^2 V \sum_{\mu,\nu} |M_{\mu\nu}|^2 \delta(E_\mu - E_F)\delta(E_\nu - E_F) . \tag{2.2}$$

Bardeen [2.2] showed that, under certain assumptions, the matrix elements in (2.2) can be expressed as

$$M_{\mu\nu} = \frac{\hbar^2}{2m} \int d\boldsymbol{S} \cdot (\psi_\mu^* \nabla \psi_\nu - \psi_\nu \nabla \psi_\mu^*) , \tag{2.3}$$

where the integral is over any surface lying entirely within the barrier region and separating the two half-spaces.

Now the ideal STM tip would consist of a mathematical point source of current, whose position we denote $\boldsymbol{r}_\mathrm{T}$. In that case, (2.2) for the current at small voltage would reduce to [2.3]

$$I \propto \sum_\mu |\psi_\mu(\boldsymbol{r}_\mathrm{T})|^2 \delta(E_\mu - E_F) \equiv \varrho(\boldsymbol{r}_\mathrm{T}, E_F) . \tag{2.4}$$

Thus the ideal STM would simply measure $\varrho(\boldsymbol{r}_\mathrm{T}, E_F)$, the local density of states (LDOS) at E_F. Note that the local density of states is evaluated for the sample surface in the absence of the tip, but at the position which the tip will occupy. Thus within this model, STM has a quite simple interpretation as measuring a property of the bare surface, without reference to the complex tip–sample system.

It is important to see how far this interpretation can be applied for more realistic models of the tip. *Tersoff* and *Hamann* [2.3] showed that (2.4) remains valid, regardless of tip size, so long as the tunneling matrix elements can be adequately approximated by those for an s-wave tip wave function [2.4]. The tip position $\boldsymbol{r}_\mathrm{T}$ must then be interpreted as the effective center of curvature of the tip, i.e. the origin of the s-wave which best approximates the tip wave functions.

2.2 Adsorbates on Metal Surfaces

When an atom interacts with a metal surface, the discrete levels of the atom which are degenerate with the metal conduction band broaden into resonances. The resonances formed in this way when Na, S, and Mo atoms interact with a high-electron-density metal (jellium model) are shown in Fig. 2.2, for the atoms at their calculated equilibrium distances [2.5]. Thus, this figure gives the total eigenstate density for the metal–adatom system minus that for the bare metal. Only the component with azimuthal quantum number $m = 0$ (s, p_z, . . .) is shown, because it is this that is most important to STM. For Na, the fact that the resonance, which corresponds to the 3s valence state of the free atom, is mostly above the Fermi level indicates a transfer of charge to the metal. (This resonance also includes $3p_z$, but we refer to it simply as 3s.) In the case of Mo, the large peak just below the Fermi level corresponds to 4d states of the free atom, and the smaller peak about 1 eV above the Fermi level corresponds to the 5s state. For S, the 3p peak is seen well below the Fermi level. Note that past about 1 eV above the Fermi level, the additional state density due to the presence of the S atom is negative; that is, some of the metal states that were present in this energy region are pushed out of it.

Before discussing STM images of these and other adsorbed atoms, we show first the current density distribution for an adsorbed Na atom, with the other electrode simply being a bare metal (jellium model) with no adsorbed atom. Now the tunneling Hamiltonian formalism as it stands cannot be used to obtain the current *density*, but it still proves possible to derive an expression for this quantity in terms of the separate wave functions of the two electrodes (this is done in Ref. [2.6]). The current density for the Na case [2.7] is shown in Fig. 2.3. The left and right edges of the box correspond to the positive background edges of the two electrodes. The presence of the atom is indicated schematically by two

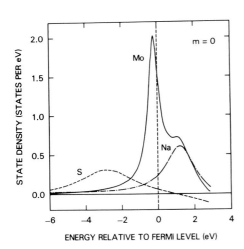

Fig. 2.2. Difference in eigenstate density between the metal-adatom system and the bare metal for adsorbed Na, Mo, and S. Only the $m = 0$ component is shown [2.5]

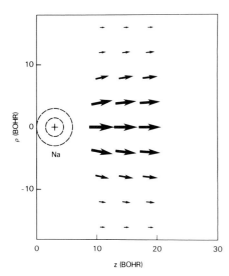

Fig. 2.3. Current density for the case in which a Na atom is adsorbed on the left electrode. The z and ϱ directions are, respectively, normal and lateral to the surface. The lengths and thicknesses of the arrows are proportional to $1 + \ln(j/j_0)$ evaluated at the spatial positions corresponding to the center of the arrows. The electrode separation shown in this figure is much larger than those typically used experimentally [2.7]

dashed circles with a cross which gives the computed equilibrium distance of the nucleus. Results are shown only in a strip in the center of the vacuum barrier; much closer to the surfaces, the equation used is not adequate. The current density j is represented by an arrow, whose length and thickness are made proportional to $1 + \ln(j/j_0)$, with j_0 the current density that would be present without the atom. This means that at large lateral distances, where $j = j_0$, a unit length arrow is shown. Note for example that in the right-hand column of arrows, the largest one represents roughly a factor of 25 in current density compared with the smallest (which corresponds approximately to j_0). The current distribution is quite sharp and shows a large enhancement due to the presence of the atom.

2.2.1 Topography

Now we return to the model described earlier in which both tip and sample are represented by an atom adsorbed on the jellium model of a surface. Calculations [2.8] using this model, together with (2.2), for the low-bias STM images of adsorbed sodium, sulfur, and helium atoms (with a sodium atom tip) are shown in Fig. 2.4.

We note first in this figure that the maximum change in tip distance Δs is much smaller for S than for Na. In part, this is due to the fact that S sits closer to the surface than Na, and in part to the fact that the Fermi-level state density for S is appreciably smaller than that for Na, as seen in Fig. 2.2. (For the low-bias case, only states near the Fermi level participate in the tunneling process.) In the case of He, the change in tip distance is seen to be slightly negative. The He atom sits rather far from the surface and so would cause a large enhancement of the

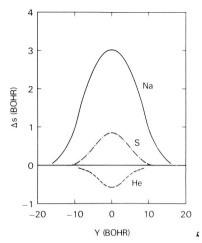

Fig. 2.4. Tip displacement Δs versus lateral separation Y for constant current. Tip atom is Na; sample adatoms are Na, S, and He. The component along the surface normal of the distance between nuclei of tip and sample atoms is 16 bohr at large Y (1 bohr = 0.529 Å.) [2.8]

tunneling current if there were to be even a small increase of the Fermi-level state density due to broadened levels of the atom. The closed valence shell is very far down in energy, however, and its only effect is to polarize metal states away from the Fermi energy, producing a decrease in Fermi-level state density, which leads to the negative tip displacement.

We now discuss the spatial distribution of the current in this two-atom problem, considering the particular case of a S atom on the left electrode and a Na atom on the right electrode. Figure 2.5 shows contour maps of j_z/j_0, with j_z the z component of the current density. Results are given only in a strip in the center of the vacuum barrier. At the largest lateral separation of the atoms shown (top left), the contour map is very similar to that which would be found for Na in the absence of the S atom. The part of the map for $\varrho < -5$ is essentially the same as it would be without the S; so it is clear from looking at the part of the map for $\varrho > -5$ that the presence of the S leads only to a small additional current. As the lateral distance between the atoms is decreased (other two maps), it is clear that the additional current that flows because of the lateral proximity of the atoms increases substantially. The fact that the z-axis displacement would have to be increased to maintain constant current as the lateral separation decreases is evident.

Returning to the tip displacement curves of Fig. 2.4, we note that in the case of Na, where everything is highly metallic, the theoretical image in Fig. 2.4 is found in the calculations of [2.8] to be almost indistinguishable from the contour of constant local density of states, confirming the s-wave tip model of *Tersoff* and *Hamann* [2.3]. In addition, both of these are found [2.8] to look like the contour of total charge density, which could be used as the definition of a surface topograph. For sulfur, as shown in Fig. 2.6, the image is again well reproduced by the local density of states contour, but these are noticeably different from the total charge density. So while the simple local density of states

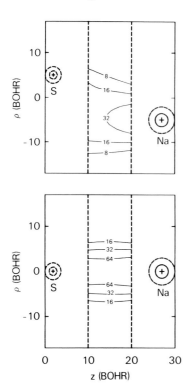

Fig. 2.5. Contour maps of j_z/j_0 for S adatom on left electrode and Na adatom on right electrode for three values of lateral separation Y (10 bohr, 6 bohr and 0), in the plane normal to the surfaces that includes both atomic nuclei. The electrode separation shown in this figure is much larger than those typically used experimentally [2.8]

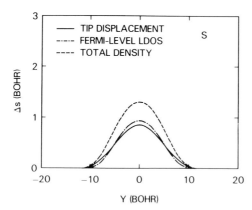

Fig. 2.6. Comparison of theoretical STM image for S adatom sample and Na adatom tip, with contours of constant Fermi-level local density of states and constant total charge density [2.8]

interpretation (2.4) remains valid, the image does not correspond as closely in this case to a topograph defined in terms of the total charge density.

It was found also in these studies that the local density of states at the Fermi level for certain adsorbed atoms in addition to helium, such as oxygen [2.9], could be lower than that for the bare metal; that is, the presence of these atoms

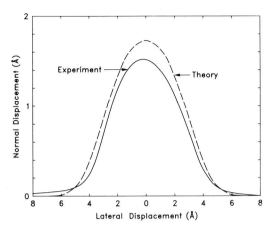

Fig. 2.7. A comparison of theoretical and experimental curves of normal tip displacement (Å) versus lateral tip displacement (Å) for Xe adsorbed on a metal surface [2.11]

can make a negative contribution to the local density of states at this energy. This of course would also cause these atoms to look like holes or depressions in the surface in a topographic image; this is seen in the studies of *Kopatzki* and *Behm* [2.10] for oxygen on Ni(100).

The heavier rare-gas atoms like Xe present an interesting test of the theory in that the low-bias images of adsorbed Xe are quite distinct [2.11], but yet Xe, like He, would seem to have no states near the Fermi level. Now if a contour of constant Fermi level local density of states is calculated [2.11] and compared with the measured tip trajectory across the center of a Xe atom, the two are seen to be very close (Fig. 2.7). The physical origin of the Fermi level state density which gives rise to the tip displacement is seen in Fig. 2.8. This plots the additional local density of states due to the presence of the Xe, at a point well outside the surface. A barrier penetration factor has been divided out. We see that the local state density at the Fermi level is due to the broadened 6s

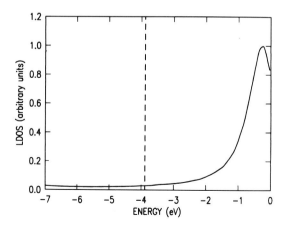

Fig. 2.8. Additional local density of states due to an adsorbed Xe atom, with a barrier penetration term divided out, versus energy measured from the vacuum level, calculated at a position 11 bohr directly above the center of the atom. Dashed line indicates Fermi level position [2.11]

resonance, the peak of which lies less than 0.5 eV below the vacuum level. (The adsorbed He atom shows no such resonance.)

Despite the fact that the 6s resonance is virtually unfilled, it dominates the local density of states on the vacuum side of the adsorbed Xe atom for two reasons. One is that Xe binds at a relatively large distance from the surface, and the other is that the 6s orbital is relatively large. Thus, even if the Fermi level is well out in the low-energy tail of the 6s resonance, the 6s *local* state density at this energy can be substantially greater than the bare-metal *local* state density on the vacuum side of the atom.

2.2.2 Spectroscopy

By analogy with the results of *Tersoff* and *Hamann* [2.3] for the low-bias limit, the approximation can be made for larger bias values that [2.12]

$$I \propto \int_{E_F}^{E_F + |e|V} dE \, \varrho_S(\mathbf{r}_T; E) \varrho_T(E - |e|V) , \tag{2.5}$$

where ϱ_T is the total state density of the tip, and ϱ_S is the local state density of the sample. Here E_F is the sample Fermi energy.

Now $\varrho_S(\mathbf{r}_T; E)$ will be roughly the product of the total sample density of states, $\varrho_S(E)$, and a barrier penetration factor, so that it might be something like

$$\varrho_S(\mathbf{r}_T; E) \sim \varrho_S(E) \exp\left[-\frac{2s}{\hbar} \sqrt{2m(W - E + \tfrac{1}{2}|e|V)} \right] ,$$

where W is the energy of the top of the sample surface potential barrier, if it is thought of as being square, and s is the tip-sample separation. If we consider dI/dV with a view to extracting out $\varrho_S(E)$, the quantity of primary interest, we find that in addition to having this sample state density multiplied by both the barrier-penetration factor and the tip state density, we also have terms involving the variation with V of the penetration factor and the variation with energy of the tip state density.

Stroscio et al. [2.13] considered the quantity $(dI/dV)/(I/V)$ in order to divide out a large part of the barrier penetration factor. There is still not a simple relationship between $(dI/dV)/(I/V) \equiv d\ln I/d\ln V$ and the state densities, except that sharp features in the sample (or tip) state densities will generally lead to sharp features in this derivative evaluated at the relevant bias.

This is illustrated in Fig. 2.9. The solid curve in the right panel gives $(dI/dV)/(I/V)$ for the case of a Ca sample adatom and a Na tip atom. The positions of the two Ca peaks and one Na peak in the spectrum correspond reasonably well to the features of the state-density curves shown at the left. Note also the negative values of $(dI/dV)/(I/V)$ for biases near $+2$ eV. Such "negative resistance" effects have been discussed by *Esaki and Stiles* [2.14], and have now been observed in the STM context by *Bedrossian* et al. [2.15] and *Lyo* and *Avouris* [2.16].

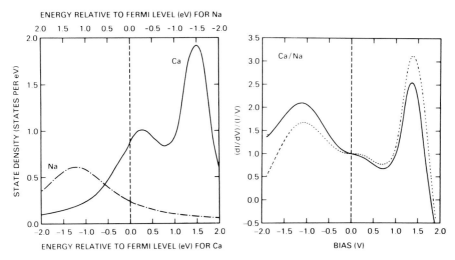

Fig. 2.9. *Left*: Curves of the difference in eigenstate density between the metal-adatom system and the bare metal for adsorbed Ca and Na. Note that the energy scale for Na (*top*) is reversed. The 3s resonance for Na is clearly evident. The lower-energy Ca peak corresponds to 4s, the upper to 3d (and some 4p). (Only $m = 0$ is shown.) *Right*: The solid line is the calculated curve of $(dI/dV)/(I/V)$ versus V for Ca/Na; the dotted line is the same quantity evaluated using a simple analytical model discussed in [2.12]. Center-to-center distance between the atoms is held fixed at 18 bohr (with zero lateral separation) [2.12]

In an experiment, it would be most convenient if the tip state density could be taken as relatively featureless, and thus be omitted from consideration. In some cases, tips are used which are purposely blunt (and probably disordered), and it is found that the tip state density appears to play no significant role in the results [2.13]. This seems quite reasonable, in view of the expectation that most of the sharp features in the density of states of such a tip would be washed out. Even if the tip were very sharp (a single atom), its state-density structure should be similar to the often broad resonances seen in the cases studied here; it would certainly not exhibit the complex surface-state structure that may be characteristic of an extended ordered surface made of these same atoms.

2.2.3 Voltage Dependence of Images – Apparent Size of an Adatom

The tip displacement in constant-current mode is studied in [2.5] for an isolated adatom for different bias voltages. We call the maximum vertical tip displacement due to an adatom the apparent vertical size of the adatom, Δs; thus this reference studies $\Delta s(V)$.

Curves of $\Delta s(V)$ for Na, Mo, and S sample adatoms are shown in Figs. 2.10–12 for V in the range -2 to $+2$ volts. Positive V again corresponds to the polarity in which electrons tunnel into *empty* states of the sample (i.e., sample positive). At the left in each figure, the corresponding sample state-density curve extracted from Fig. 2.2 for energies in the range -2 to $+2\,\text{eV}$ is given for comparison.

Figure 2.10 shows the state density and the $\Delta s(V)$ curve for the case of the Na sample adatom. The $\Delta s(V)$ curve clearly reflects the 3s resonance peak in the density of states. The curvature of $\Delta s(V)$ up towards the left in the negative bias region is simply an effect of the exponential barrier penetration factor, and would be present even if the state density of the sample, and tip as well, had no structure whatsoever.

The graph of $\Delta s(V)$ for Mo shown in Fig. 2.11 is very similar to that for Na, with the large peak at positive bias associated with the 5s resonance. It is striking how little evidence there is of the 4d resonance in the $\Delta s(V)$ curve [2.5]. The reason for this is that the valence d orbitals in the transition elements are in general quite localized relative to the valence s orbitals, and will thus have a much smaller amplitude at the tip. The relative unimportance of the contribution from d states in tunneling spectroscopy has been confirmed experimentally for Au and Fe by *Kuk* and *Silverman* [2.17].

Figure 2.12 gives $\Delta s(V)$ for S. Note that for a bias just above $+1\,\text{V}$, the S atom is essentially invisible, that is, it causes no displacement of the tip, and for biases much above this the tip displacement is negative. The negative displacement is a consequence of the fact that the additional state density due to the presence of the S is negative in the relevant energy region, just as in the cases of O and He mentioned earlier.

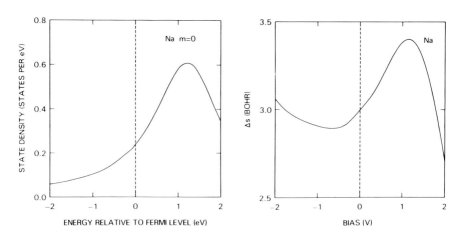

Fig. 2.10. *Left*: Difference in eigenstate density between metal-adatom system and bare metal for adsorbed Na ($m = 0$ component). The 3s resonance is clearly evident. *Right*: $\Delta s(V)$ for Na sample adatom [2.5]

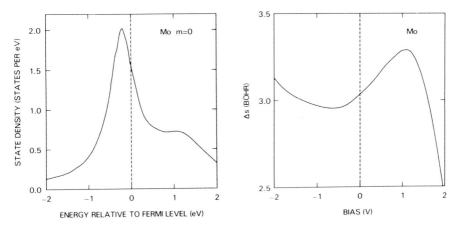

Fig. 2.11. *Left*: Mo eigenstate density difference ($m = 0$ component). The lower-energy Mo peak corresponds to 4d, the upper, smaller peak to 5s. *Right*: $\Delta s(V)$ for Mo sample adatom [2.5]

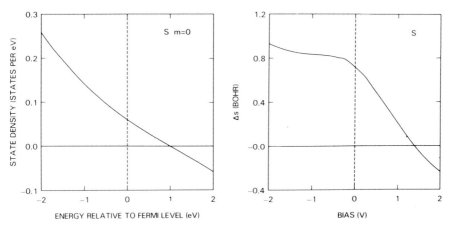

Fig. 2.12. *Left*: S eigenstate density difference ($m = 0$ component). *Right*: $\Delta s(V)$ for S sample adatom [2.5]

2.3 Close Approach of the Tip: The Strong-Coupling Regime

2.3.1 From Tunneling to Point Contact

Let us now consider what happens when we start to reduce the tip-sample distance, and pass into the region of transition from tunneling to point-contact. The initial contact takes place in the experiments we discuss between a single tip atom and the sample surface [2.17–19]. We consider the current that flows between the two flat metallic electrodes in the jellium model, one of which here has an adsorbed atom (the tip electrode), as a function of distance between them. For simplicity we consider the case of a small applied bias voltage.

Now in contrast to the studies described above, it is not possible here to use the tunneling-Hamiltonian formalism, since the overlap of the wave functions of the two electrodes is no longer small. It is necessary therefore to treat the tip and sample together as a single system in computing the wave functions [2.18].

The quantities of primary interest are the additional current density due to the presence of the atom, $\delta \boldsymbol{j}(\boldsymbol{r}) = \boldsymbol{j}(\boldsymbol{r}) - j_0$, where $\boldsymbol{j}(\boldsymbol{r})$ is the current density in the presence of the atom and j_0 that in its absence, and the total additional current δI which is obtained by integration of $\delta \boldsymbol{j}$ over an appropriate surface. We can define the additional conductance due to the presence of the atom as $\delta G = \delta I/V$, and it is convenient to define an associated resistance $R \equiv 1/\delta G$.

Consider as before the simple case of a Na tip atom. The tip-sample separation s is measured from the nucleus of the tip atom to the positive-background edge of the sample; the distance d between the center of the Na atom and the positive-background edge of the tip electrode is held fixed at 3 bohr, the equilibrium value for $s \to \infty$.

In Fig. 2.13, the resistance R defined above is shown as a function of separation s. At large separations, R changes exponentially with s. As s is decreased toward d, the resistance levels out at a value of $32\,000\,\Omega$. (For $s = d$, the atom is midway between the two metal surfaces, so this can in some sense be taken to define contact between the tip atom and the sample surface). We can understand this leveling out, including the order of magnitude of the resistance, from the discussions of *Imry* and *Landauer* [2.20]. These authors point out that there will be a "constriction" resistance $\pi h/e^2 = 12\,900\,\Omega$ associated with an ideal conduction channel, sufficiently narrow to be regarded as one-dimensional, which connects two large reservoirs. Our atom, in the instance in which it is midway between the two electrodes, contacting both, forms a rough approximation to this.

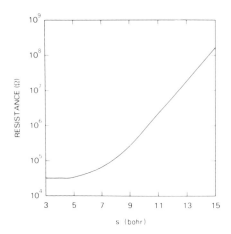

Fig. 2.13. Calculated resistance $R \equiv 1/\delta G$ as a function of tip–sample separation s for a Na tip atom [2.18]

Just such a plateau in the resistance was found in the experiments of *Gimzewski* and *Möller* [2.19] using an Ag sample surface and an Ir tip (though the identity of the tip atom itself was not determined). These authors fix the voltage on the tip at some small value, and measure the current as a function of distance as the tip is moved toward the sample. They find a plateau in the current at about the same distance from the surface as found in the calculation described here, with the resistance at the plateau $\sim 35\,\mathrm{k\Omega}$. Analogous results were found by *Kuk* and *Silverman* [2.17] with a minimum resistance of $\sim 24\,\mathrm{k\Omega}$.

2.3.2 Measuring the Tunneling Barrier

Now consider the height of the tunneling barrier as it is commonly measured in the STM [2.21]. For the simple case of a one-dimensional square barrier of height φ above the Fermi level, the tunneling current I at a small bias V is proportional to $V \exp(-2s\sqrt{2m\varphi}/\hbar)$, where s is the barrier thickness; thus for constant bias, $\varphi = (\hbar^2/8m)(\mathrm{d}\ln I/\mathrm{d}s)^2$. This relationship is often used to *define* an *apparent* barrier height (or local "effective work function")

$$\varphi_A \equiv \frac{\hbar^2}{8m}\left(\frac{\mathrm{d}\ln I}{\mathrm{d}s}\right)^2. \tag{2.6}$$

A number of authors, as noted above, present "local barrier-height" images of surfaces (that is, images of φ_A) as an alternative to the more commonly shown "topographic" images [2.22].

Reference [2.21] studies the apparent barrier height φ_A as a function of separation for a single-atom tip and a flat sample surface, using exactly the same model employed above to discuss the resistance. An experimental study of the dependence of apparent barrier height on separation has been reported by *Kuk* and *Silverman* [2.17]. Now it is clear that for tip-sample separation $s \to \infty$, $\varphi_A(s) \to \Phi$, the work function (more precisely, an appropriate average between sample and tip work functions). The question of interest is whether or not $\varphi_A(s)$ will be close to Φ for the range of tip–sample separations commonly used in scanning tunneling microscopy.

The apparent barrier height $\varphi_A(s)$ calculated in [2.21] using the above equation is shown in Fig. 2.14. Note that φ_A is well below the work function Φ except at the largest distances shown in the graph, where in fact it goes slightly above (it reaches a maximum of 4.3 eV at $s \sim 19$ bohr). That it is below Φ for intermediate distances results from the long range of both the exchange-correlation part of the surface potential (see the discussion in [2.21]), and the electrostatic potential due to the charge-transfer dipole that forms at the Na tip atom. That it is even further below at the shortest distances results from the complete absence at these separations of a potential barrier to electron tunneling, as we see below. Near the point of nominal contact ($s = 3$ bohrs), the tunneling current reaches a plateau, as seen above, and thus φ_A is zero. (It is strictly zero only when all tunneling to the substrate upon which the atom is adsorbed is

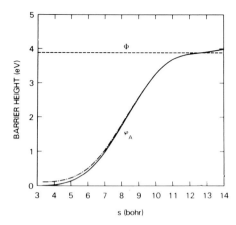

Fig. 2.14. Apparent barrier height φ_A as a function of tip–sample separation s for two electrodes, one representing the sample, the other, with an adsorbed Na atom, representing the tip. See text for distinction between solid and dot-dash curves. The work function Φ for the sample electrode by itself is shown for comparison [2.21]

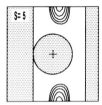 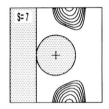 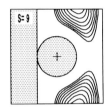 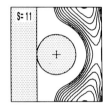

Fig. 2.15. Contour maps of the potential v_{eff} for two-electrode case with an adsorbed Na atom (tip). The presence of the atom is represented by a shaded circle with a cross at the position of the nucleus; the positive background regions of both tip and sample electrodes are shaded also. Maps are shown for four values of s (given in bohrs), which is the distance between the nucleus of the tip atom and the positive background edge of the sample electrode. The nucleus is at the center of each box, so that the sample electrode in fact lies entirely outside of the box for all but the smallest of the s values shown ($s = 5$ bohr). The contour closest to the atom in each case is that for $v_{eff} = E_F$; the other contours correspond to higher energy values [2.21]

neglected; including such tunneling in a physically reasonable manner gives the dashed curve in Fig. 2.14).

Figure 2.15 gives contour maps of the total potential seen by an electron, v_{eff}, for several distances s. It shows the contour $v_{eff}(\mathbf{r}) = E_F$ (the one closest to the atom), with E_F the Fermi energy, as well as a number of other contours for values above E_F; thus the contour-filled areas represent regions where a Fermi-level electron encounters a potential barrier. At $s = 5$ bohrs, for example, the electrons moving from the single-atom tip to the sample electrode encounter essentially no barrier whatsoever, while at $s = 11$ bohrs, there is a barrier for electrons tunneling along all directions.

For $s = 9$ bohrs, there is a barrier for tunneling in most directions, but not for tunneling nearly normal to the surface. However, even for tunneling in the normal direction there will be an *effective* barrier. The reason for this is that though there is a small opening in the barrier, whose transverse size we denote by a, electrons moving through this opening will, by the uncertainty principle, have a minimum transverse momentum of $O(\hbar/a)$. This in turn decreases the energy available for motion along the direction of the surface normal, and thus a Fermi-level electron will in fact have to tunnel through a barrier even in the region of this "opening" [2.23].

References

2.1   For a discussion of the jellium model, see e.g. N.D. Lang: Density functional approach to the electronic structure of metal surfaces and metal-adsorbate systems, in *Theory of the Inhomogeneous Electron Gas*, ed. by S. Lundqvist, N.H. March (Plenum, New York, 1983), pp. 309–389, or N.D. Lang: The density-functional formalism and the electronic structure of metal surfaces, in *Solid State Physics*, **28**, 225–300 (Academic, New York, 1973). The calculations shown in this article are all for a positive background density $\bar{n}$ such that $r_s = 2$ bohr, where $(4/3)\pi r_s^3 \equiv \bar{n}^{-1}$. This corresponds to a typical high-electron-density metal.

2.2   J. Bardeen: Phys. Rev. Lett. **6**, 57 (1961)

2.3   J. Tersoff, D.R. Hamann: Phys. Rev. B **31**, 805 (1985); Phys. Rev. Lett. **50**, 1998 (1983)

2.4   Discussions for tip wave functions of higher l values were given in [2.3] and by M.S. Chung, T.E. Feuchtwang, P.H. Cutler: Surf. Sci. **187**, 559 (1987); C.J. Chen: J. Vac. Sci. Technol. A **6**, 319 (1988); and W. Sacks, C. Noguera: Phys. Rev. B **43**, 11612 (1991)

2.5   N.D. Lang: Phys. Rev. Lett. **58**, 45 (1987)

2.6   N.D. Lang: Phys. Rev. Lett. **55**, 230 and 2925 (E) (1985)

2.7   N.D. Lang: IBM J. Res. Dev. **30**, 374 (1986)

2.8   N.D. Lang: Phys. Rev. Lett. **56**, 1164 (1986)

2.9   N.D. Lang: Comments Cond. Mat. Phys. **14**, 253 (1989)

2.10 E. Kopatzki, R.J. Behm: Surf. Sci. **245**, 255 (1991)

2.11 D.M. Eigler, P.S. Weiss, E.K. Schweizer, N.D. Lang: Phys. Rev. Lett. **66**, 1189 (1991)

2.12 N.D. Lang: Phys. Rev. B **34**, 5947 (1986)

2.13 J.A. Stroscio, R.M. Feenstra, A.P. Fein: Phys. Rev. Lett. **57**, 2579 (1986); but regarding the role of tip electronic structure in spectroscopy, see by contrast R.M. Tromp, E.J. van Loenen, J.E. Demuth, N.D. Lang: Phys. Rev. B **37**, 9042 (1988), where an important tip contribution was observed.

2.14 L. Esaki, P.J. Stiles: Phys. Rev. Lett. **16**, 1108 (1966)

2.15 P. Bedrossian, D.M. Chen, K. Mortensen, J.A. Golovchenko: Nature **342**, 258 (1989)

2.16 I.-W. Lyo, Ph. Avouris: Science **245**, 1369 (1989)

2.17 Y. Kuk, P.J. Silverman: J. Vac. Sci. Technol. A **8**, 289 (1990). Regarding the role of d states, see also discussion in J.E. Demuth, U. Koehler, R.J. Hamers: J. Microscopy **152**, 299 (1988).

2.18 N.D. Lang: Phys. Rev. B **36**, 8173 (1987)

2.19 J.K. Gimzewski, R. Möller: Phys. Rev. B **36**, 1284 (1987); and data of Gimzewski and Möller reproduced in [2.18]

2.20 Y. Imry: Physics of mesoscopic systems, in *Directions in Condensed Matter Physics: Memorial Volume in Honor of Shang-keng Ma*, ed. by G. Grinstein, G. Mazenko (World Scientific, Singapore, 1986), pp. 101–163; R. Landauer: Z. Phys. B **68**, 217 (1987)

2.21 N.D. Lang: Phys. Rev. B **37**, 10395 (1988)

2.22 G. Binnig, H. Rohrer: Surf. Sci. 126, 236 (1983); R. Wiesendanger, L. Eng, H.R. Hidber, P. Oelhafen, L. Rosenthaler, U. Staufer, H.-J. Güntherodt: Surf. Sci. **189/190**, 24 (1987); B. Marchon, P. Bernhardt, M.E. Bussell, G.A. Somorjai, M. Salmeron, W. Siekhaus: Phys. Rev. Lett. **60**, 1166 (1988)

2.23 Another instance of this occurs in the theory of point electron sources: N.D. Lang, A. Yacoby, Y. Imry: Phys. Rev. Lett. **63**, 1499 (1989)

3. The Scattering Theoretical Approach to the Scanning Tunneling Microscope

G. Doyen

With 16 Figures

In the interpretation of Scanning Tunneling Microscopy (STM) one often adopts the somewhat simplistic view that the tunnel current should be proportional to the local surface charge density at the position of the tip. More often than not this gives a reasonable interpretation of the STM-images. But in some cases it leads to an apparent contradiction with other available information on the sample surface. These contradictions are not always easy to resolve and some have been puzzling the STM-community for years.

If STM is considered as a probe of the local sample charge density at the Fermi level, this suggests a comparison to results obtained from He-scattering [3.1, 2] where the total sample charge density is probed at distances which are roughly equivalent to those accessible by STM. Table 3.1 contains this comparison, which demonstrates that for clean and adsorbate covered close-packed metal surfaces there are problems in interpretating the STM images. For the given examples a picture of the local charge density differs qualitatively from the images seen in STM. In these cases the vertical coordinate of the probe tip that maintains a constant tunneling current will not coincide with what is usually imagined as the topography of the surface. The local charge density picture can be theoretically justified in the case of large tip–sample separation where the interaction between the two electrodes is negligible [3.3]. It is hence concluded that tip–sample interaction plays an important role in some STM-experiments and that a more general theory is needed to achieve a general understanding of the STM imaging process.

A general theory has to reliably treat the quantum properties of electrons of a few eV kinetic energy. These electrons interact with structures of dimensions of the order of their wavelength. It has been recognized in the literature that realistic potentials and an exact solution of the three-dimensional scattering problem are required [3.4–8] in this situation. Various attempts have been made in this direction. *Tsukada* et al. [3.9] and *Ciraci* et al. [3.10] concentrated on realistic potentials but either did not calculate the current at all [3.10] or only in a rough approximation [3.11] or in perturbation theory [3.9]. *Noguera* and *Sacks* [3.12] stressed on the importance of multiple scattering effects, but used only simple model potentials.

In this article, theories of the Scanning Tunneling Microscope (STM) are reviewed which use a localized Green function approach to calculate the current beyond perturbation theory. The potential of the tip atom is expanded in a localized basis set. The scattering wave functions yielding the spatial current

Springer Series in Surface Sciences, Vol. 29
Scanning Tunneling Microscopy III Eds.: R. Wiesendanger · H.-J. Güntherodt
© Springer-Verlag Berlin Heidelberg 1993

Table 3.1. Comparison of local charge-density arguments with STM experiments

Sample	Expectation for STM from local charge density	STM experiment
Clean metal surface	Qualitative similar results as for He-scattering	Large corrugation for Al(111); Small corrugation for Ni(100)
Adsorbed atoms	Appear as protrusions	K-atoms not seen on Cu(110); Oxygen atoms appear often as indentations
Insulating organic layers	Not visible in STM	STM images reflecting the expected geometry

distribution near the tip region are calculated numerically exactly. The method can be applied to study the current to clean and adsorbate covered surfaces.

This chapter does not aim at giving a complete overview of the present state of the scattering theoretical approach to STM. Different aspects of this problem are reviewed in Chaps. 2 and 4. Also not all aspects of the tip–sample interaction are treated and the reader is referred to the other chapters for further information. Especially temperature and non-equilibrium effects and the interaction of the tunneling low-energy electrons with the solid state excitations (i.e., inelastic tunneling) are completely left out. Concerning the applications of the theory, the present report focusses on some problematic and unresolved questions which are related to the modification of the electronic structure by the tip.

3.1 The Theoretical Formalism

3.1.1 The Limits of Perturbation Theory

In the *Tersoff-Hamann* picture [3.3] the total tunnel current is proportional to the local charge density at the Fermi level evaluated at the center r_0 of the tip:

$$J \propto \sum_f |\psi_f(r_0)|^2 \delta(E_f - E_i) . \tag{3.1}$$

ψ_f indicates a wave function of the sample electrode unperturbed by the tip. The derivation of this relationship starts from *Bardeen's* perturbation theory [3.13] which evaluates the current inducing transition matrix element as an integral over a dividing surface ∂G within the barrier:

$$\langle f | V_{\text{tip}} | i \rangle = -i \int_{\partial G} J_{fi}(r) \cdot dA \tag{3.2}$$

where

$$J_{fi} = \frac{ie\hbar}{2m_e} [\psi_f(\boldsymbol{r})\nabla\psi_i^*(\boldsymbol{r}) - \psi_i^*(\boldsymbol{r})\nabla\psi_f(\boldsymbol{r})] \; .$$

ψ_i is a wave function of the isolated tip electrode. (e: electron charge, m_e: electron mass, dA: area element).

The validity of (3.2) depends on several approximations introduced by *Bardeen* [3.13]:

1) $\psi_i(\boldsymbol{r})$ is real
2) $H\psi_f = E_f\psi_f$ within the sample and within the barrier
3) $V_{tip}\psi_f \neq 0$

These approximating assumptions are found in many three-dimensional theories about STM. For this special choice of $\psi_f(\boldsymbol{r}) = \langle \boldsymbol{r}|f\rangle$ the volume integration on the l.h.s. of (3.2) can be restricted to the tip region which means that the boundary surface ∂G can lie anywhere in the barrier. Condition 2 implies that $|f>$ is a good solution of Schrödinger's equation inside the sample surface and within the barrier. This is, however, an approximation as the generalized Ehrenfest theorem (Sect. 3.1.2) defines the final state function $|f>$ uniquely and consequently $|f>$ cannot be a good solution within the barrier. Also the assumption of a real valued initial-state wave function cannot be made, if $|i+>$ is a proper scattering state with correct boundary conditions. Hence, these requirements are not fulfilled and lead to errors in the calculation, when the tip approaches the surface and the bare tip and the bare surface-wave functions overlap.

The important point to notice is that even if approximate wave functions $|i+>$ and $|f>$ are used, (3.2) is not in general valid with a boundary surface ∂G inside the barrier. Evaluating $\langle f|V_{tip-sample}|i+\rangle$ directly, or using the surface integral with ∂G inside the barrier, will in general lead to different results.

We conclude that the Tersoff-Hamann picture is based on three fundamental approximations: (i) an approximate wave function, (ii) an approximate evaluation of the transition matrix element and (iii) the assumption of a spherical tip.

To improve on the approximation (i) is the most difficult part of the problem and the rest of this chapter is devoted to it. The approximations (ii) and (iii) are easier to avoid. The approximate evaluation of the transition matrix element can be avoided by performing the volume integration directly. This has been done by *Doyen* and collaborators in their studies of STM at metal steps [3.14] and single adsorbates on metal surfaces [3.15, 16]. This work is not reviewed here as the theory of STM for adsorbates is treated in Chap. 2.

The shape of the tip is important in the vicinity of the sample. In fact, a tip represented by a d-orbital will even in the perturbation picture yield different information on the sample surface. This is illustrated in Fig. 3.1. If the sample wave function would have the form (applying to a simple potential step)

$$\psi(E_f, \boldsymbol{k}_{||}) \propto \sqrt{2E_f - k_{||}^2} \cdot e^{-\sqrt{2\phi + k_{||}^2}\, z} \cdot e^{i\boldsymbol{k}_{||}\cdot\boldsymbol{r}_{||}}$$

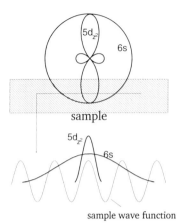

sample

sample wave function

Fig. 3.1. Schematic illustration of the overlap of d- and s-tip orbitals with sample wave functions. The lower half of the figure shows the amplitude of the wave functions versus the distance parallel to the surface along the cut indicated in the upper half of the figure

(ϕ: barrier height), we realize that tip d-orbitals will probe wave functions of larger momentum $k_\parallel$ parallel to the sample electrode compared to s-orbitals. Larger $k_\parallel$ means a shorter range of the sample wave function perpendicular to the surface. This implies then a more rapid decay of the tunnel current perpendicular to the surface and therefore a larger apparent barrier height [3.17]:

$$\phi_{\mathrm{app}} = \frac{\hbar^2}{8m_{\mathrm{e}}} \left(\frac{\mathrm{d}\ln J(z)}{\mathrm{d}z} \right)^2 . \tag{3.3}$$

Chen has generalized the perturbation approach by relaxing the restriction to a spherical tip [3.18]. He discussed the implications of using p- and d-states as tip orbitals. In particular he stressed the importance of these tip states for explaining the anomalous corrugation observed on close packed metal surfaces. More details can be found in his chapter in this monograph.

3.1.2 Tunneling as a Scattering Process

If the perturbation theoretical approach breaks down, one has to concentrate on a proper description of the interacting tip–sample system which is schematically illustrated in Fig. 3.2. The theoretical formalism we describe applies to STM-models consisting of two flat electrodes and an atom (referred to as *tip atom*) or a group of atoms (referred to as *tip cluster*) adsorbed on one electrode. This cluster of atoms models the apex of the tip. In the following we assume an applied bias voltage such that the current flows from the tip to the sample. A quite general point of view is to visualize the tunnel event as a scattering process, where the electron is indicent from the interior of the tip base material, then scatters from the tunnel junction and has a certain probability of penetrating into the sample surface. The scattering process is described by a wave function $|i+ >$ which is an eigenfunction of the total hamiltonian H including the tip potential V_{tip}:

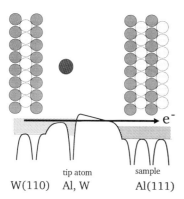

Fig. 3.2. Schematic illustration of a theoretical STM setup with a one atom tip with atomic structure. Tip and sample composition are chosen according to the model hamiltonian calculations described in Sect. 2.13

$$H = H_0 + V_{\text{tip}} \ . \tag{3.4}$$

H_0 is the hamiltonian of a two-electrode system (capacitor) without tip atom. V_{tip} is the potential induced by the tip atom or the tip cluster, respectively, which would break any symmetry that might be given for the two-electrode system. When measuring the current, the scattered electron with final momentum f is detected in the wave function $|f>$, which does not experience the potential V_{tip} in the barrier region. This means that $|f>$ has to be an eigenstate of H_0 satisfying *outgoing* scattering boundary conditions, i.e., $|f>$ describes a wave with plane-wave components running away from the tunnel junction in the direction towards the interior of the sample and scattered-wave components collapsing towards the scattering centers. It is essential that $|f>$ is calculated with these boundary conditions in order to obtain correct results for the tunnel current. The probability for tunneling through the barrier is:

$$P = |\langle f|i+\rangle|^2 \ .$$

i indicates the incoming momentum of the electron. The $+$ sign indicates here *incoming* scattering boundary conditions, i.e., $|i+>$ describes a wave with plane-wave components propagating from the interior of the tip electrode towards the tunnel junction and with scattered-wave components emerging from the scattering centers. The tunnel current is then:

$$J = e\frac{\mathrm{d}}{\mathrm{d}t}|\langle f|i+\rangle|^2 \ .$$

The current itself is the number of electrons per unit time being detected in the state $|f>$ times the electron charge e. According to Lippmann's generalization [3.19] of Ehrenfest's theorem [3.20] the tunnel current can be written in the form (temperature effects are neglected):

$$J = 4\frac{\pi e}{\hbar}\sum_{f,i}|\langle f|V_{\text{tip}}|i+\rangle|^2\,\delta(E_f - E_i) \ . \tag{3.5}$$

The sum runs over all states between the two Fermi levels having the correct scattering boundary conditions as described above. A factor 2 for spin degeneracy is included. This expression is exact. It reduces to Fermi's golden rule, if the exact scattering state $|i+>$ is replaced by a state $|i_p>$ which does not include the tip–sample interaction.

V_{tip} is restricted to the region around the tip atom which is spanned by the localized basis set $\{|A>\}$. Using the *Lippman–Schwinger* equation [3.21] one obtains:

$$|i+\ > \ = |i_p> \ + \sum_{A,B} G^{cap}|B> \ \langle B|V_{tip}|A\rangle\langle A|i+\rangle$$

$$= |i_p> \ + \sum_{A,B} G|B> \ \langle B|V_{tip}|A\rangle\langle A|i_p\rangle \ . \qquad (3.6)$$

If not explicitly indicated, the Green operators G^{cap} and G correspond to incoming boundary conditions (usually denoted by $G^{cap,+}$ and G^+). G^{cap} is the Green operator of the two-electrode system without tip atom or tip cluster obeying translational invariance parallel to the surfaces. G is the Green operator of the total hamiltonian and is referred to as the *exact* Green operator. In order to evaluate the amplitude of the scattering-wave function in the tip region

$$\langle A|i+\rangle = \langle A|i_p\rangle + \sum_B G_{AB} V_{Bi}$$

the Green operator G for the hamiltonian including the tip is needed. The Green operators G and G^{cap} for the two systems (without and including the tip) are connected by Dyson's equation. This equation is especially useful when the perturbing potential is limited to a small area in space, as is the case for the tip. Solving explicitly for the wave function we need to perform a matrix inversion. The size of the matrix depends on the basis set. We have the advantage that matrix inversion is only necessary in the subspace $\{|A>\}$, where the potential V_{tip} induced by the tip cluster is non-zero:

$$\langle A|i+\rangle = \sum_B (1 - G^{cap} V_{tip})^{-1}_{AB} \langle B|i_p\rangle \ . \qquad (3.7)$$

The transition matrix element needed in the generalized Ehrenfest theorem is

$$\langle f|V_{tip}|i+\rangle = V_{fi} + \sum_{A,B} V_{fA} G_{AB} V_{Bi} \ .$$

The important task is now to evaluate the *exact Green functions* G_{AB} in the tip region. The matrix elements of the Green operators between basis functions of the local subspace $\{|A\}$ are referred to as *tip Green functions*.

It has to be stressed that the local basis set (consisting, e.g., of Gaussian orbitals) is only used to describe the tip potential. The decay of the wave functions at large distances is important in STM calculations. The basis set used to describe the wave functions must be adopted for this purpose.

Once we have calculated the exact Green functions we can evaluate the tunnel current exactly. This strategy can be used independently of the theoretical model we use to describe the interacting tip–sample system.

3.1.3 Current Density and Generalized Ehrenfest Theorem

The current density for a single scattering-wave function is defined by:

$$j(r) = \frac{2e\hbar}{m_e} \operatorname{Im} \left\{ \psi_i^*(r) \nabla \psi_i(r) \right\} \tag{3.8}$$

with $\psi_i(r) = \langle r | i + \rangle$.

The current density in position space will be a superposition of the waves emanating from the tip and the scattering waves originating from the atoms in the sample as schematically illustrated in Fig. 3.3. During their way through the barrier the electrons are focussed, which is important for the resolution and the imaging characteristics. *Gottlieb* et al. discussed electron emission from tips ending at a single atom [3.22]. If the sample consists of only few atomic layers, the current density can experimentally be visualized on a channel plate positioned at a macroscopic distance behind the sample [3.23].

In order to calculate the current density we make use of the fact that once the exact Green operator is known in the local subspace defining the tip region, the wave function outside this area can be obtained by a simple matrix multiplication. Expanding into plane waves $|k>$ one obtains:

$$\langle k | i + \rangle = \langle k | i_p \rangle + \sum_{A,B} \langle k | G^{\text{cap}} | B \rangle \langle B | V_{\text{tip}} | A \rangle \langle A | i + \rangle . \tag{3.9}$$

$\langle A | i + \rangle$ is calculated according to (3.7). From this expression the current density is evaluated by summing over all incident and final waves on the energy

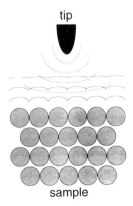

tip

sample

Fig. 3.3. Illustration of the wave description of the tunneling electron. The current density in the region between tip and sample results from interfering illuminating and reflected waves

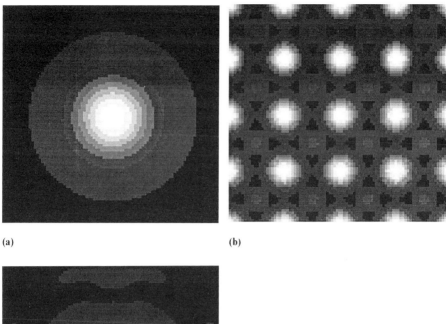

(a)   (b)

(c)

Fig. 3.4. Grey scale representation of the current densities for (**a**) tip with a structureless sample, (**b**) two electrode system (capacitor without tip) with Pd(100) surface, (**c**) the tip over a sample atom of a Pd(100) surface. The barrier width is the same in all cases (4.1 Å)

shell for both spin orientations:

$$j(r) = \frac{2eh}{m_e} \sum_{i,k,l} k\, \mathrm{e}^{i(k-l)r} \langle i + |l\rangle \langle k|i + \rangle \ .$$

Figure 3.4 displays the variation of the perpendicular component of the current density in a plane parallel to the electrode surfaces for three different situations calculated by the method described in Sect. 3.3.1 [3.24]. The plane is situated where the potential step of the barrier occurs near the sample surface. Figure 3.4a shows the results for a tip with structureless sample and Fig. 3.4b for a two-electrode system (capacitor without tip) with a jellium and a Pd(100) surface. In both calculations the same width of the square barrier (4.1 Å) was used.

For the tip with the structureless sample, the current density is cylindrically symmetric with respect to an axis passing through the tip atom. The current densities for the capacitor with the Pd(100) sample electrode clearly exhibit maxima at the positions of the Pd atoms. On Pd(100) even at the positions of the second layer of Pd atoms some enhancement of the current density perpendicular to the sample surface can be observed. The current density for the tip plus the Pd(100) sample surface (displayed in Fig. 3.4c, again with the same barrier width) arises from a holographic-like superposition of the wavefields of the two situations of Fig. 3.4a, b. The position of the Pd atoms is still discernible in the current density. All multiple-scattering effects are included for these calculations.

The intuitive approach to calculating the current in STM-theory is to first evaluate the current density in position space and then to integrate it over an interface parallel to the electrodes. It has been recognized in the literature that the connection to the transfer hamiltonian formulation [3.13] is not well established. *Pendry* et al. [3.8] addressed this problem by starting from the current density point of view and deriving *Bardeen's* transfer hamiltonian expression [3.13] in the perturbation limit.

The current density is not an observable as it implies a simultaneous sharp measurement of position and momentum. It is usually maintained that integration over an interface yields the observable "total current". It has been demonstrated recently that the integral over the current density can be reduced to the corresponding basic expression of scattering theory [3.26], i.e.,

$$J = \int_{\partial G} \boldsymbol{j}(\boldsymbol{r}) \cdot \mathrm{d}\boldsymbol{A} = \frac{4\pi e}{\hbar} \sum_{i,f} |\langle f| V_{\mathrm{tip}} |i+\rangle|^2 \, \delta(E_f - E_i) \; .$$

This offers two possibilities of evaluating the total tunnel current. The integration in position space is for example used in the method developed by *Derycke* et al. [3.7], which constructs a transfer matrix in a mixed real- and momentum-space representation. Often the generalized Ehrenfest theorem allows a more efficient numerical evaluation.

3.1.4 Local Charge Density at the Fermi Level and Tunnel Current

The discussion in this subsection follows that in [3.24]. The generalized Ehrenfest theorem (3.5) yields the *exact* tunnel current for a given one-electron potential. Introducing the expansion of V^{tip} discussed in Sect. 3.1.2 into (3.5) yields an expression for the tunnel current which contains quantities that are easily accessible to physical interpretation:

$$J = \frac{4\pi e}{\hbar} \frac{1}{\Delta E} \sum_{A,B,C,D} S^{\mathrm{tip}}_{AB} S^{\mathrm{cap}}_{CD} \tilde{V}_{BC} \tilde{V}_{DA} \tag{3.10}$$

with

$$S_{AB}^{tip} = \sum_i \langle A|i+\rangle\langle i+|B\rangle \qquad (3.11)$$

$$S_{CD}^{cap} = \sum_f^{capacitor} \langle C|f\rangle\langle f|D\rangle . \qquad (3.12)$$

ΔE is the energy difference between the two Fermi levels. The calculations to be described later are usually performed in the limit $\Delta E \to 0$. In this limit the tunnel current becomes zero, but one obtains a finite value for the *tunnel conductivity* $eJ/\Delta E$. The summation in (3.11) is carried out over eigenstates of the system without the tip atom present, which resembles a capacitor of two plane electrodes. $\tilde{V}_{BC}$ is a matrix element of the operator

$$S^{-1}V^{tip}S^{-1} = \sum_{A,D} |B > S_{BA}^{-1}\langle A|V^{tip}|D\rangle S_{DC}^{-1} < C| .$$

S_{AB}^{tip} is a matrix element of the projection operator $\Sigma_i|i+> <i+|$ in the localized basis set. The projection is on the eigenfunctions of the total hamiltonian with boundary conditions corresponding to incident waves from the tip side. S_{AB}^{cap} is the corresponding projection on the wave functions without tip, i.e. on eigenfunctions of H_0 with boundary conditions corresponding to outgoing waves $|f>$ inside the sample. The diagonal matrix elements might be interpreted as partial charge densities in the tip region integrated over the energy range between the two Fermi levels. S_{AA}^{tip} is the projection of the local function $|A>$ in the barrier on those exact scattering eigenstates that are generated by incident waves running towards the sample surface and will be termed *Tip Projected Local Density* (TIP-LOD). S_{AA}^{cap} is the projection of the local state on eigenfunctions of the two electrode system without tip atom and is called *Capacitor Projected Local Density* (CAP-LOD). It may be interpreted as a partial local charge density at the Fermi level of the unperturbed two-electrode system (sample plus free-electron electrode).

The set of functions $\{|A>\}$ localized in the barrier is chosen to define the tip region and shape, i.e. the region in position space, where the tip potential acts. Hence, the meaning of the capacitor and tip projected local electron densities is just the electron density at the Fermi level projected onto the region, where the tip potential is effective for the system with or without tip. The CAP-LODs thus include information about the electronic structure of the sample, whereas the additional information gained from the TIP-LODs can be attributed to the tip-sample interaction. This means that the comparison of the tunnel current contrast with the variations of CAP-LOD and TIP-LOD yields information about the influence of tip–sample interactions. An example is discussed in Sect. 3.3.3.

It might be tempting to substitute the quantities CAP-LOD and TIP-LOD by the negative imaginary parts of the Green functions of the capacitor (matrix elements of G^{cap}) and of the exact Green functions (matrix elements of G), respectively:

$$-\frac{1}{\pi} \, \text{Im} \, \{G^{cap}_{AA}(E_F)\} = \sum_f \langle A|f\rangle\langle f|A\rangle\delta(E_F - E_f) \tag{3.13}$$

$$-\frac{1}{\pi} \, \text{Im} \, \{G_{AA}(E_F)\} = \sum_i \langle A|i+\rangle\langle i+|A\rangle\delta(E_F - E_i) \;. \tag{3.14}$$

The important difference is that the summations in TIP-LOD and CAP-LOD, as defined by (3.13, 14), are not over a *complete* set of eigenfunctions on the energy shell as they are for the imaginary parts of the respective Green functions.

In *Bardeen's* perturbation theory [3.13], where the interaction of the tip and the tip electrode with the sample surface is neglected and $|i+ >$ is replaced in the generalized Ehrenfest theorem by an eigenfunction of standing-wave character of the unperturbed tip electrode, S^{tip}_{AA} does not vary with lateral tip position. In this limit S^{cap}_{AA} is reduced to S^{sample}_{AA} and the current would be proportional to

$$S^{sample}_{AA} = \sum_k^{sample} \langle A|k\rangle\langle k|A\rangle \;. \tag{3.15}$$

S^{sample}_{AA} is essentially the local-charge density at E_F of the unperturbed sample, described by eigenfunctions $|k >$, averaged over the tip orbital $|A >$. It is termed *Sample Projected Local Density* (SAP-LOD). For an s-type tip orbital $|A >$, at a sufficiently large distance from the sample surface this represents essentially the *Tersoff-Hamann* approximation [3.3].

3.1.5 Resonance Tunneling

Resonance tunneling might experimentally be observed as a strongly bias-dependent structure in the tunneling conductivity. It has been suggested as the imaging mechanism for the voltage-dependent images of liquid crystals on graphite [3.28]. It has also been proposed as a possible explanation [3.29] of the anomalously large corrugation observed in STM-imaging of Al(111) [3.30]. Resonance tunneling is usually discussed in a one-dimensional model, where a potential well is inserted in the barrier [3.28, 31]. If the well has bound levels, then the transmission probability will be near unity for tunneling energies close to these bound levels.

A three-dimensional discussion can be given in the framework of the outlined scattering theory [3.25]. For this purpose we assume that the tip potential is represented by a single local orbital $|A >$ and write the TIP-LOD in terms of the exact Green functions including the interaction with the sample surface:

$$S^{tip}_{AA} = \langle A|G^-|A\rangle\langle A|G^+|A\rangle \, W^2(A) \sum_k^{tip} \langle A|k\rangle\langle k|A\rangle \;.$$

Here $W(A) = \langle A|V^{tip}|A\rangle$. Introducing $\text{Im} \, q_A$ as the imaginary part of the self-energy of the tip orbital, which determines the width of the resonance, and ε_A as the center of the resonance, one obtains:

$$\langle A|G^-|A\rangle\langle A|G^+|A\rangle = [(E_{\text{tunnel}} - \varepsilon_A)^2 + (\text{Im}\, q_A)^2]^{-1} \ .$$

A small variation of ε_A might obviously lead to a large variation of the TIP-LOD, if the width of the resonance is sufficiently small.

The behaviour of the tunnel conductivity in the resonance situation can be derived from the generalized Ehrenfest theorem. If the tip potential is represented by a single orbital, $V^{\text{tip}} = |A > V_A < A|$, the tunnel conductivity reads:

$$\frac{dJ}{dV} = \frac{4\pi e^2}{\hbar} |\langle f|A\rangle|^2 |\langle A|i+\rangle|^2 |V_A|^2 \varrho_f \varrho_i \ .$$

The summation over initial and final states is here replaced by the densities ϱ_i and ϱ_f of the initial and final states. We now assume the symmetric situation that the tip atom represented by the single tip orbital $|A>$ is situated midway between two identical electrodes. In this case we have $\varrho_f = \varrho_i = \varrho/2$, where ϱ is the total density of states at the Fermi level. Using (3.7) with the resonance condition $\text{Re}\{1 - G_{AA}^{\text{cap}} V_A\} = 0$, i.e., $\text{Re}\{G_{AA}^{\text{cap}}\} = 1/V_A$ one obtains

$$\langle A|i+\rangle = -\frac{\langle A|i_p\rangle}{V_A \pi |\langle A|i_p\rangle|^2 \varrho} \ .$$

For the assumed symmetrical situation one has $|\langle f|A\rangle|^2 = |\langle i_p|A\rangle|^2$ and therefore

$$\frac{dJ}{dV} = \frac{e^2}{\pi\hbar} \ .$$

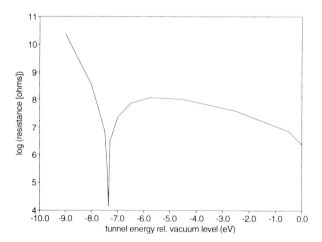

Fig. 3.5. Tunnel resistance versus tunnel energy for a tip potential of gaussian shape midway between two jellium electrodes separated by a square barrier of 10 eV height and 5.3 Å width

This is the resonance conductivity, which corresponds to $(12\,906\,\Omega)^{-1}$, the so-called Sharvin contact conductivity [3.32, 33]. A similar discussion of the contact resistance has been given by *Martin-Rodero* et al. [3.34].

One should observe that the maximum in the tunnel current does not occur exactly at $\varepsilon_A = E_{tunnel}$ though very near to it. The resonance behaviour is displayed in Fig. 3.5, which shows the tunnel resistance as a function of the tunnel energy for a tip atom midway between two jellium surfaces [3.35].

3.2 Tunneling Through Thick Organic Layers

3.2.1 The Experimental Situation

STM [3.36, 37] and AFM [3.40] have been shown to be promising techniques for the elucidation of the structure of organic substances in general and of Langmuir–Blodgett (LB) films in particular at the molecular level. STM pictures obained on bilayers transferred on Highly Oriented Pyrolithic Graphite (HOPG) and on gold substrates have allowed to visualize LB films in direct space.

The observations in STM-imaging are that lattice spacings as well as the size of the embedded protein molecules are in almost quantitative agreement with published electron and X-Ray diffraction work. Artifacts were excluded, such as imaging of the underlying substrate with altered and distorted magnification. However, the mechanism for electron tunneling through a 50 Å-thick organic layer has not yet been clearly identified. An important observation is that the LB film only becomes conducting upon very close approach of the tip, when pressure is exerted on the organic layer [3.38]. The situation is not completely clear as new experiments [3.39] indicate that paraffins lie flat on an HOPG substrate.

We investigate the related problem of tunneling through insulating paraffin layers [3.41]. Here the experimental system is complicated, but we try to explain a qualitative effect: a change of orders of magnitude in the tunnel current. We are not interested here in the details of the imaging process and we can therefore use a simple model hamiltonian which illustrates the main principles.

One would imagine that with strong bulk insulators STM cannot be used because electrons remain trapped at the surface. Paraffin molecules, also called *n*-alkanes, consist of long saturated nonpolar chains, $(CH_3)-(CH_2)_n-(CH_3)$, with $n > 10$. They solidify into orthorhombic or monoclinic layered crystals, which are highly insulating in the bulk.

The resistance found in STM experiments is of the order 10^7 to $10^8\,\Omega$, the resistivity for 50 Å chains is $10^3\,\Omega\,cm$. Typically the applied voltage is between 40 mV and a few 100 mV. The current is more than ten orders of magnitude larger than expected from the measured electrical conduction of planar metal/fatty acid/metal junctions, where the resistivity is 10^9 to $10^{15}\,\Omega\,cm$. The paraffin film becomes totally invisible to STM (very probably meaning nonconducting) when it melts, but it promptly reappears at recrystallization although bulk paraffins do not change their bulk conductivity significantly at melting.

3.2.2 A Simple Soluble Model

It should be possible to understand these enormous variations in the conductivity in qualitative terms if these are a real consequence of the STM imaging process. A simple model hamiltonian, which is illustrated in Fig. 3.6, is used to study the tunneling through insulating alkane layers [3.42]:

$$
\begin{aligned}
H = & \sum_i \varepsilon_i n_i + \sum_i (V_{i\alpha} c_i^+ c_\alpha + \text{herm. conj.}) + E_\alpha n_\alpha \\
& + (V_{\alpha C_1} c_\alpha^+ c_{C_1} + \text{herm. conj.}) \\
& + \sum_j E_{C_j} n_{C_j} + \sum_{jk} (V_{C_j C_k} c_{C_j}^+ c_{C_k} + \text{herm. conj.}) \\
& + \sum_f (V_{C_N f} c_{C_N}^+ c_f + \text{herm. conj.}) + \sum_f \varepsilon_f n_f \; .
\end{aligned}
\tag{3.16}
$$

Here the notation of second quantization has been used. The model describes two three-dimensional gold electrodes. An atom adsorbed on one electrode represents the tip (first line of (3.16)). The alkane chain is described·in a tight-binding aproximation (third line). The alkane chain is coupled to the tip atom [second line of (3.16)] and the gold substrate (fourth line) by one-electron hopping matrix elements. The tip atom is represented by a single orbital which is coupled to a continuum of metal states as in the Anderson-Newns picture [3.43]. For this model an exact analytic solution can be obtained within the developed formalism [3.42]. In fact, an increase in conductivity by ten orders of magnitude is found, if a strong compression of the alkane layer by the tip is assumed.

Figure 3.7 shows the logarithm of the tunnel resistance as a function of the tunneling energy, i.e., we vary the Fermi energy of the model system. We assume that we have a small applied bias voltage and that therefore only electrons near the Fermi level contribute to the current.

Whenever the tunnel energy agrees with a molecular level of the alkane chain, the transmitivity is one and we have the quantum limit of resistance of

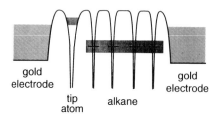

Fig. 3.6. Schematic illustration of a simple model which can be solved analytically in order to investigate the problem of tunneling through insulating alkane layers

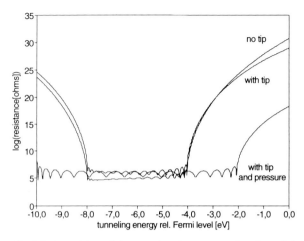

Fig. 3.7. Logarithm of the tunnel resistance as a function of the tunnel energy for the model of Fig. 3.6 [3.42]

$12\,906\,\Omega$ (cf. Sect. 3.1.5). Hence the minima in the resistance agree with the alkane molecular levels.

An interacting chain of CH_2-groups gives rise to a set of discrete closely spaced alkane molecular levels in the energy range between 9 and 4 eV below the Fermi level. Photoemission spectroscopy indicates that the first ionization potential of polyethylene is 9 eV with respect to the vacuum level. For the gold surface we assume a work function of 5 eV. The width of the C–H-band of polyethylene (which is the highest lying band) is 6 eV. This band has a fine structure consisting of four bands at 9, 7.8, 6.4, and 4.9 eV below the Fermi level and ranges from 10 to 3 eV below Fermi [3.44].

Interaction of the alkane with the gold electrode broadens the molecular levels and smears them out into a continuous band. Therefore the transmittivity does not decrease to zero, if the Fermi level is in between two alkane molecular levels.

If we now introduce a tip atom on one of the gold electrodes, the resistance at the bottom of the gold conduction band and near the Fermi level decreases by about two orders of magnitude, i.e., the STM current increases. This arises from the chemical interaction of the tip resonance state near the Fermi level with the alkane molecular levels and the latter gain some spectral weight at the Fermi level. Because chemical interaction means forming of bonding and anti-bonding states, the spectral weight of the alkane molecular levels at the bottom of the conduction band increases as well. The increase of conductivity connected with this tip interaction is, however, not large enough to explain the experimental findings.

A similar calculation performed by *Joachim* and *Sautet* [3.46] yielded a much smaller tunnel resistance. They considered an alkane chain between two gold wires in contrast to the calculation described above where a three-dimensional gold electrode and a three-dimensional tip was modelled. Therefore their gold conduction band is too shallow (only 3 eV). Also the HOMO of their alkane chain is too high compared to experimental data. It is situated only 1 eV below the Fermi level.

Attempts in the literature to explain the high conductivity are based on the observation that the tunneling transmittance via a resonant state is exactly one for any barrier thickness. *Lindsay* et al. [3.45] suggest that some kind of empty state shifts by pressure to an energetic position close to the Fermi level. This fits with the disappearance of current at melting, since the liquid will not support tip pressure. This explanation can only be valid, if the resonance level does not have to be located right at the Fermi level. A significant effect must already be obtained, if the HOMO moves towards the Fermi level by a few eV.

If we increase the hopping matrix elements in the model hamiltonian (3.16) by 1 eV simulating a strong pressure exerted by the tip, we find in fact an increase in the tunnel current by 10 orders of magnitude. Whether this reflects the experimental situation, is not clear at present. Anyhow, this model study clearly demonstrates the importance of the tip–sample interaction.

3.3 Scanning Tunneling Microscopy at Metal Surface

The motivation of studying tip–sample interaction for clean metal surfaces becomes obvious, if we come back to the comparison to He-scattering experiments, which supplied the following results:

- Similar corrugation for close packed Ag, Cu, Ni, Pd surfaces (of order 0.1 to 0.2 Å)
  no corrugation for Al(111) (less than 0.02 Å)
- Corrugation results from variation of occupied d-electron charge density [3.2]
- He-scattering probes the *total* occupied local charge density [3.1, 2]

If STM probes the local charge density at the Fermi level, then we expect small corrugation for Al, Cu, Au and large corrugation for Ni, Pd. There is, however, clear evidence that on close packed Al, Cu, and Au surfaces the lateral STM resolution is definitely better than purely topographic resolution [3.30, 49–51].

In order to approach a theoretical understanding of these qualitatively different images we now need a theoretical STM-model which explicitly accounts for the atomic structure of the sample electrode.

Two complementary theoretical methods have been developed at the Fritz-Haber-Institute in Berlin, both of which use a localized Green function approach to calculate the current beyond perturbation theory. The first one concentrates on a realistic description of the sample surface [3.24, 25], whereas the second one attempts to model a more realistic transition metal tip [3.17].

3.3.1 A Method Based on the Korringa–Kohn–Rostocker (KKR) Band Theory

In the first approach the sample surface is built of muffin-tin potentials accounting for the atomic structure and a realistic electron density. The other electrode carrying the tip is a planar free-electron metal surface. The tip potential is expanded in a gaussian basis set as described in Sect. 3.1.2. The method consists of three steps [3.24, 25]:

(i) Construct the Green function G^0 for the sample surface without barrier and without tip. Here standard methods based on the layer-KKR formalism, which is commonly used in LEED theory, are applied [3.27, 52, 53].
(ii) Include the barrier potential V_{bar} between two plane electrodes:

$$G^{cap} = G^0 + G^0 V_{bar} G^{cap}$$

This can also be handled in the framework of the layer-KKR method by means of the layer-doubling technique [3.27].
(iii) Calculate the wave function for the system including the tip:

$$\psi(i) = \psi_p(i) + G^{cap} V_{tip} \psi(i)$$

V_{tip} is restricted to the region α within the hemisphere on the jellium. The outer region is denoted β. In the region α a matrix inversion has to be performed, see (3.7). In a short symbolic notation this might be written as

$$\psi_\alpha(i) = (1 - G_{\alpha\alpha}^{cap} V_{\alpha\alpha}^{tip})^{-1} \psi_\alpha^{(2)}(i) \ .$$

In the region β, where the tunnel current has to be calculated, only a matrix multiplication is needed:

$$\psi_\beta(i) = \psi_\beta^{(p)}(i) + G_{\beta\alpha}^{cap} V_{\alpha\alpha}^{tip} \psi_\alpha(i)$$

This method permits the calculation of the current density in position space. Examples are given in Fig. 3.4.

3.3.2 Including the Atomic Structure of the Tip: Model Hamiltonian Approach

An important unresolved question in understanding STM concerns the role played by the nature of the tip. By manipulating the tip in a usually not very well defined way, the experimentalists achieve atomic resolution, change the distance dependence of the tunnel current and even produce image inversion, where protrusions turn into indentations and vice versa.

The lack of information about the shape and the atomic structure of the tip is also responsible for the fact that the exact tip sample separation for typical operation conditions is still uncertain. A way out of this dilemma would be to measure simultaneously with the tunnel current, additional characteristic phys-

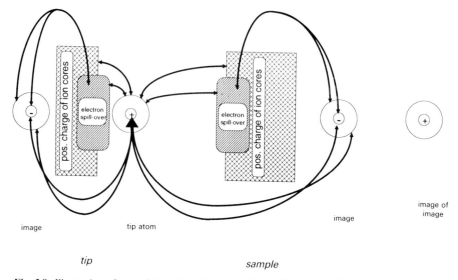

Fig. 3.8. Illustration of some interactions in the model hamiltonian used to investigate tips with atomic structure

ical quantities such as the interaction force, dissipative effects, etc. A comparison to a theory which calculates these quantities could then yield additional insight into these open questions.

This poses the challenge to the theory of STM to evaluate within the same model the electronic structure, the interaction forces and the tunnel current for a realistic tunnel junction. First principle methods are not yet in a shape to tackle this difficult task, but careful model studies can already yield valuable information and add to the understanding of this important subject.

A model hamiltonian approach has been developed which studies a metal tip consisting of a single atom (W, Al) adsorbed on a metal surface [3.54]. A chemisorption model [3.47], which is solved in a self-consistent spin-unrestricted mean field approximation, is used to evaluate the electronic structure, the interaction forces, and the tunnel current. The p- and d-electrons on the tip atom and in the tip base are taken into account. The elastic deformation of the tip when approaching the sample can be included in this theory.

Figure 3.8 summarizes some essential characteristics of the model hamiltonian, which treats the local region near the tip atom in a quite explicit way. A full set of one- and two-electron integrals is included which allows to account for all important physical interactions like electron–core attraction, electron–electron repulsion and core–core repulsion. The electronic structure of the separated system (tip base and sample surface on its own and isolated tip atom) is not recalculated, but is taken from other ab-initio calculations and reliable experimental information is used to compensate for errors like neglected correlation effects. The model then calculates the interaction of the tip atom with the two electrodes in a self-consistent way.

The polarisation of the more distant metal electrons leads to image effects which can be handled by coupling the local charge density to the plasmons. Multiple-image effects arising from the polarisation of one electrode by the induced charge density in the other electrode are included as well.

The tunnel current is evaluated exactly within the single-particle picture as described above.

3.3.3 Close Packed Metal Surface

A study of the close packed surface of Al and Pd has been performed in the layer-KKR based scheme described above [3.24]. The following model system is used which is schematically illustrated in Fig. 3.9: The semi-infinite crystal is built from muffintin potentials taken from self-consistent bulk calculations [3.48]. Although individual wave functions are modified significantly at the surface, the potentials are only little affected. This holds at least for close packed, unreconstructed metal surfaces. Therefore the same potentials have been used at the surface layers as for the bulk.

The following results have been found: For the clean metal surfaces atomic resolution is obtained with a magnitude of the lateral tunnel current variation as often found in STM operations without special tip preparation. At typical tip–sample separations ($\geq 3\,\text{\AA}$) the substrate atoms appear as protrusions. Quite remarkably, the contrast is found to be larger in the images of close packed Al-surfaces than in those of close packed Pd-surfaces. This is a consequence of the tip–sample interaction. At *close* distances the *interstitial regions* appear as maxima in the variation of the tunnel current due to an effect of the tip–sample interaction. This is demonstrated in Fig. 3.10 which shows the lateral variation of the tunnel current for Pd(100) at a tip–sample separation of 1.8 Å. In order to analyze the situation we plot in Fig. 3.11 the CAP-LODs and the TIP-LODs (Sect. 3.1.3) for small tip–sample separation on Pd(100). The CAP-LOD is similar to the unperturbed charge density of the sample surface. The behaviour of the TIP-LOD is different and reflects the influence of the tip–sample interaction. If the tip atom is close enough to the sample surface, covalent interactions play some role. A chemical bond formation between tip

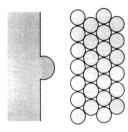

Fig. 3.9. Schematic representation of the potential for the model STM consisting of a half-infinite muffin-tin sample and a free-electron electrode with a model tip

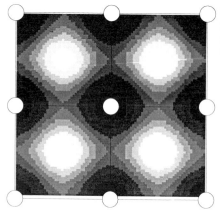

Fig. 3.10. Variation of the total tunnel current parallel to a Pd(100) surface for a tip-sample separation of 1.8 Å showing image inversion. The empty circles indicate Pd atoms in the first layer

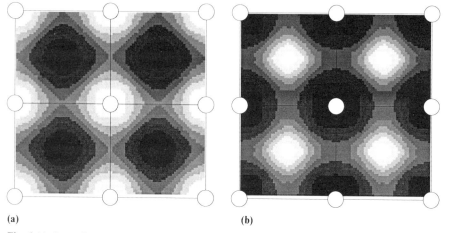

(a)   (b)

Fig. 3.11. Lateral variation of CAP-LOD (**a**) and TIP-LOD (**b**) for Pd(100) for a tip-sample separation of 1.8 Å

and sample increases the density of states at the Fermi level, because the anti-bonding state has some spectral weight here, and therefore leads to a larger tunnel current. The bond is somewhat stronger in the hollow position compared to the top position.

The effects of chemical bond formation between the tip atom and the sample surface are present for the Al(111) surface as well, but they are smaller, because no d-electrons are present.

Table 3.2 summarizes the results for the variation of the total tunnel current compared to the variation of the local charge density at the Fermi level. The latter quantity has been evaluated as the imaginary party of the Green function

Table 3.2. Variation of total tunnel current and charge density at the Fermi level at the tip position (in brackets) between top and fcc-hollow sites. Positive values indicate maximum on top sites.

tip–sample separation [Å]	Al(111)	Pd(111)
1.8	5.6% (38%)	− 2.5% (42%)
4.5	4.5% (3%)	2.2% (4%)
6	1.2% (0.8%)	0.7% (0.7%)
8	0.2% (0.2%)	0.1% (0.1%)

matrix element $-\operatorname{Im}\{G_{AA}^{cap}\}$ for a contracted gaussian at the tip center at different lateral positions relative to the sample.

At large separations the variation of the total tunnel current is equal to the variation of the local charge density. At somewhat smaller separations (4.5 Å) the tip–sample interaction acts on Pd to diminish the variations in the tunnel current relative to the charge density variations, whereas on Al(111) the variations of the tunnel current are enhanced compared to the charge density. At very close separations (1.8 Å) the tip–sample interaction diminishes the corrugation on both metal surfaces.

As a consequence the contrast in the total current variation is larger on Al(111) than on Pd surfaces. This results from the fact that the chemical bond formation reflected in the TIP-LODs has an anti-corrugating effect and tends to enhance the tunnel current at tip positions of higher coordination.

An investigation of the Al(111) sample surface has also been performed in the model hamiltonian approach described in Sect. 3.1.2. This theory allows to explore the influence of the tip on the tunnel current [3.54]. The sample surface is Al(111) and the tip is modelled as a metal atom adsorbed on a W(110) or Al(111) surface. Figure 3.12 shows the calculated tunnel conductivity as a function of the tip–sample separation, which is the distance between the center of the tip atom and the first layer of Al-atoms in the sample surface. For the tips with the Al-atom at the apex the conductivity increases at large distances exponentially with an apparent barrier height, see (3.3), of about 4.5 eV and then flattens off at closer distances. Near 2 Å the conductivity increases by one order of magnitude indicating ballistic transport due to the collapse of the barrier. A qualitative change of the electronic configuration is obtained in the theory, because a new potential energy curve crosses. Such a sudden increase in conductivity is often found in experiment near the point of contact [3.57].

Just before contact, a decrease of the tunnel current is found arising from interference effects due to strong multiple scattering between tip and sample. A maximum in the current versus distance curve has also been obtained in a theory based on a tight-binding hamiltonian [3.55]. This phenomenon might be connected with the fine structure sometimes observed in experimental current-versus-distance curves just before the instability [3.56].

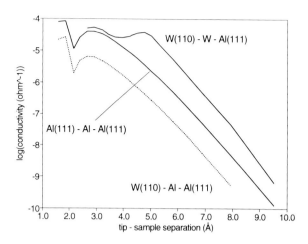

Fig. 3.12. Tunnel conductivity versus tip-sample separation for the model hamiltonian approach describing various tip models with an Al(111)-sample surface: Al and W tip atom on a W(110) and an Al tip atom on an Al(111) surface

The theoretical curves can be compared to experimental data by *Gimzewski* and *Möller* [3.57] who used an Ir tip on an Ag film and to data by *Dürig* et al. [3.59–62] who investigated an Al film with an Ir tip. Material transport from the sample to the tip during tip preparation is a possibility which is suggested by the experimentalists. Due to the changing tip characteristics, different conductivity curves may be obtained within the same experimental setup.

Lang [3.58] has been able to reproduce one of the experimental curves of *Gimzewski* and *Möller* by calculating the conductivity for a Na-atom between two jellium ($r_s = 2$) surfaces. The high conductivity in his theoretical model is explained by the high 3s spectral weight at the Fermi level.

The experimental data are displayed in Fig. 3.13 (dotted curves) in comparison to the calculated conductivities. As the tip–sample separation is unknown in the experiments, the experimental curves of *Gimzewski* and *Möller* have been adjusted so that the instabilities in the conductivities coincide with those of the theoretical curves. These jumps in the conductivities are often interpreted as *mechanical* instabilities, whereas the theory discussed above explains it as an electronic instability. In the experiments by *Dürig* et al. the point of contact was not reached and we need a different method for calibrating this experimental curve against the theoretical one. In these latter experiments the force gradient between tip and sample was measured simultaneously and calibration can be done by comparison with the theoretical force gradient. The measured force gradient is slightly attractive at further distances and turns repulsive several Ångstroms before the plateau in the conductivity is reached. The same behaviour is found in the theoretical force gradient [3.54]. Only at smaller distances (where the measurements became unstable) does the force become strongly repulsive.

The absolute values of the tunnel current are considerably larger for the W/W(110) tip compared to the Al/W(110) tip at all distances. The differences arise from the different electronic structure induced by the different tip atoms. Figure 3.14 displays the spectral resolution of a major current providing tip

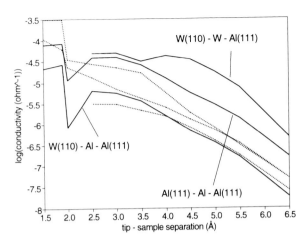

Fig. 3.13. Comparison of the model hamiltonian results (*solid curves*) with experimental data taken from [3.57] (*the two upper dotted curves*) and from [3.61] (*lower dotted curve*)

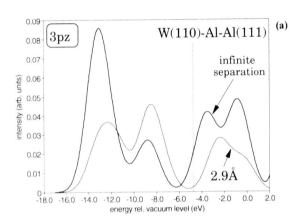

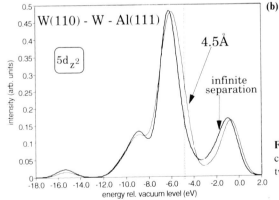

Fig. 3.14a, b. Spectral resolution of current providing tip orbitals for the two model tips

orbital, namely the $3p_z$-orbital for Al/W(110) and the $5d_{z^2}$-orbital for W/W(110). The spectral weight of this tip orbital right at the Fermi level is important for the magnitude of the tunnel current. The larger this spectral weight the larger the tunnel current.

For the Al tip, the spectral distribution shifts to higher energies (towards the Fermi level) due to multiple-image effects and electron–electron repulsion. This lowers the spectral weight at the Fermi level gradually as the tip approaches and accounts for the soft flattening of the conductivity curve.

For the W tip atom the situation is qualitatively different. The major peak of the $5d_{z^2}$-spectral distribution is centered approximately 1 eV below the Fermi level and therefore the spectral weight at the Fermi level of the $5d_{z^2}$-orbital is considerably larger than that of the $3p_z$ Al-orbital even at infinite separation which explains the larger tunnel current. This is a resonance like situation (Sect. 3.1.5). Upon approach of the tip towards the sample, the main peak of the $5d_{z^2}$-spectral distribution shifts to higher energies for the reason mentioned above which increases the spectral weight at the Fermi level further.

At smaller tip–sample separations the W-atom develops a chemical bond with the sample surface which increases the splitting between the bonding (below the Fermi level in Fig. 3.14b) and the anti-bonding peak (above the Fermi level). This bond formation starts around 4.5 Å from the sample and leads to a quite abrupt decrease of the spectral weight at the Fermi level and to the levelling off of the conductivity versus distance curve in Fig. 3.12.

When the tip approaches the sample surface, forces act on the tip atom and this will lead to a new equilibrium position of the tip atom between the two electrodes [3.17]. This induces variations in the interaction potential as a function of lateral position and also to variations in the force gradient which are more pronounced than the variations in the tunnel current. Hence the force gradient might well be capable of detecting topographic details as has already been demonstrated by *Dürig* [3.63].

3.3.4 Open Metal Surfaces

Large corrugation is expected for open metal surfaces as e.g. the (1×2) reconstructed Au(110) surface. Already the early investigations by *Binnig* et al. [3.64] clearly resolved the atomic structure of the densely packed [110] rows of this surface. A quantitative interpretation was given in terms of the local charge density at the Fermi level [3.3].

Later detailed investigations showed, however, that the corrugation amplitude does not increase monotonically with decreasing tip–sample separation, but it rather shows a maximum and then decreases again upon further approach of the tip [3.65]. This behaviour has been demonstrated to be a consequence of the tip–sample interaction [3.65, 68]. A calculation has been performed within the KKR-layer based scattering theory for a missing-row reconstruction of Cu(110). The missing-row reconstruction of Cu(110) does not occur spontaneously as for example for Au(110), but is induced by K-adsorption [3.66, 67].

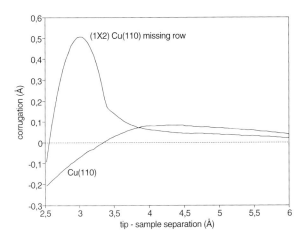

Fig. 3.15. Calculated corrugations for the Cu(110) and the (1×2) missing-row reconstructed Cu(110) surfaces

Calculating first the missing-row structure allows to later directly trace the influence of adsorbed potassium.

Figure 3.15 displays the calculated corrugations (i.e., the amplitude of the tip excursion when a constant tunnel current is maintained) for the Cu(110) and the (1×2) missing-row reconstructed Cu(110) surfaces. The calculations for the latter surface show the same qualitative behaviour as *Gritsch* [3.65] measured for the reconstructed Au(110) surface. The theory explains the maximum in the corrugation amplitude as a consequence of the tip–sample interaction, which at shorter distances increases the tunnel current in the hollow position due to the stronger chemical interaction at this site. The explanation is quite similar to the image inversion for Pd(100) discussed in Sect. 3.3.3.

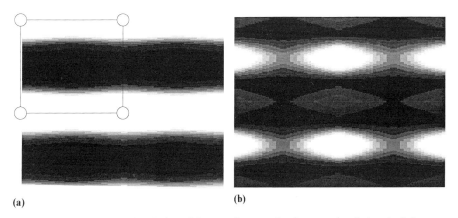

(a)   **(b)**

Fig. 3.16. Calculated lateral variation of the tunnel current for the potassium induced missing-row structure on Cu(1 1 0) at **(a)** 4.4 Å and **(b)** 2.6 Å. The circles indicate the positions of the Cu atoms in the first layer. At the larger distance the Cu rows are seen whereas at the smaller distance the rows of K atoms appear as protrusions

3.3.5 Imaging Adsorbed Alkali Atoms: K/Cu(110)

This method has also been applied to study the potassium induced missing row structure on Cu(110) [3.65, 68]. As K is buried relatively deeply inside the first Cu layer, the 4s-resonance is energetically centered well below the Fermi level and therefore the modification of the local charge density at the Fermi level is quite small. Therefore the K atoms are not discernible in the STM image at larger distances. Figure 3.16a displays the calculated lateral variation of the tunnel current. Similar STM images have been obtained experimentally [3.69]. At small tip–sample separations the situation changes because the tip interacts with the K atoms. This leads to an inversion of the STM picture where now the rows of K atoms appear as protrusions (Fig. 3.16b). This image inversion has been observed in experiment as well [3.69].

3.4 Summary and Conclusions

This article should have highlighted how in the field of scanning tunneling microscopy theory and experiment are complementary. A correct interpretation of STM images requires a close collaboration of both disciplines.

Close surface–tip interactions require a non-perturbative approach to STM theory. Modelling the tip as a single atom electron source appears to be a good starting point and gives a qualitatively reasonable description of STM properties. The following points have been stressed in this review:

- A scattering formalism and Green function technique can be used to calculate exactly (within the one-electron picture) the tunnel current from a tip to a surface when both electrodes have an atomic structure.
- The presence of the tip can induce qualitative changes in the Fermi level charge density distribution of the sample. This might lead to image inversion where at small tip–sample separations the sample atoms are imaged as indentations. The tip can exert pressure which changes the electronic structure of organic adsorbates and increases the tunnel current by several orders of magnitude.
- Transitions from electron tunneling to point contact is indicated by a sudden jump in the conductivity which can be reproduced by theory if a realistic tip model is treated in the scattering approach.

An accurate prediction of the tunneling current might be possible for realistic tip models. This is a delicate point as the current can change by orders of magnitude over 1 Å distance due to the exponential dependence on the barrier height. A correct description of the decay of the wave functions in the barrier is important in order to obtain the correct magnitude of the tunnel current for a given tip–sample separation. The mechanical interactions between tip and surface are one of the most important and least understood outstanding issues in STM.

The purpose of this article was to identify and discuss problems in the present understanding of STM and to propose scattering theory as a sound approach for a deeper understanding.

Acknowledgement. Financial support by the "Sonderforschungsbereich 6" is acknowledged. The author is grateful to D. Drakova, E. Koetter, and M. Scheffler for useful discussions.

References

3.1   K.H. Rieder: In Springer Ser. in Surf. Sci. *Dynamical Phenomena at Surfaces, Interfaces and Superlattices*, ed. by F. Nizzoli, K.-H. Rieder, R.F. Willis Vol. 3 (Springer, Berlin, Heidelberg 1985) Chap. 1

3.2   D.Drakova, G. Doyen, F. von Trentini: Phys. Rev. B **32**, 6399 (1985)

3.3   J. Tersoff, D.R. Hamann: Phys. Rev. Lett. **50**, 1998 (1983); Phys. Rev. B **31**, 805 (1985)

3.4   A.A. Lucas, H. Morawitz, G.R. Henry, J.-P. Vigneron, Ph. Lambin, P.H. Cutler, T.W. Feucht-wang: Solid State Commun. **65**, 1291 (1988)

3.5   A.A. Lucas, H. Morawitz, G.R. Henry, J.-P. Vigneron, Ph. Lambin, P.H. Cutler, T.E. Feucht-wang: Phys. Rev. B **37**, 10708 (1988)

3.6   J.P. Vigneron, M. Scheffler, Th. Laloyaux, I. Derycke, A.A. Lucas: Vacuum, **41**, 745 (1990)

3.7   I. Derycke, J.P. Vigneron, Ph. Lambin, Th. Laloyaux, A.A. Lucas: Intern. J. Quantum Chem. **25**, 687–702 (1991)

3.8   J.B. Pendry, A.B. Prêtre, B.C.H. Krutzen: J. Phys. Cond. Matt. **3**, 4313 (1991)

3.9   M. Tsukada, K. Kobayashi, N. Isshiki, H. Kageshima: Surf. Sci. Rep. **13**, 265 (1991)

3.10 S. Ciraci, A. Baratoff, Inder P. Batra: Phys. Rev. B **41**, 2763 (1990); ibid. B **42**, 7618 (1990)

3.11 S. Ciraci: Phys. Rev. B **40**, 11969 (1989)

3.12 C. Noguera: Phys. Rev. B **42**, 1629 (1990); C. Noguera, J. Microsc. 152, Part 1, 3 (1988); W.S. Sacks, C. Noguera, ibid. **152**, Part 1, 23 (1988); C. Noguera: J. Phys. (Paris) **50**, 2587 (1989); W.S. Sacks, C. Noguera: Phys. Rev. B **43**, 11612 (1991); J. Vac. Sci. Technol. B **9**, 488 (1991)

3.13 J. Bardeen: Phys. Rev. Lett. **6**, 57 (1961)

3.14 G. Doyen, D. Drakova: Surf. Sci. **178**, 375 (1986)

3.15 G. Doyen, D. Drakova, E. Kopatzki, R.J. Behm: J. Vac. Sci. Technol. A **6**, 327 (1988)

3.16 E. Kopatzki, G. Doyen, D. Drakova, R.J. Behm: J. Microsc. **151**, 687 (1988)

3.17 G. Doyen, E. Koetter, J. Barth, D. Drakova: In: *Basic Concepts and Applications of Scanning Tunneling Microscopy (STM) and Related Techniques*, Ed. by R.J. Behm, N. Garcia, H. Rohrer (Kluwer, Dordrecht 1990) p. 97

3.18 C.J. Chen: Phys. Rev. B **42**, 8841 (1990); Phys. Rev. Lett. **65**, 448 (1990); J. Phys. Cond. Matt. **3**, 1227 (1991)

3.19 B.A. Lippmann: Phys. Rev. Lett. **15**, 11 (1965); ibid. **16**, 135 (1966)

3.20 P. Ehrenfest: Z. Physik, **45**, 455 (1927)

3.21 B.A. Lippmann, J. Schwinger: Phys. Rev. **79**, 469 (1950)

3.22 B. Gottlieb, M. Kleber, J. Krause: Z. Phys. A **339**, 201 (1991)

3.23 H.-W. Fink, H. Schmid, H.J. Kreuzer, A. Wierzbicki: Phys. Rev. Lett. **67**, 1543 (1991)

3.24 G. Doyen, D. Drakova, M. Scheffler: Phys. Rev. B **47**, 15 April 1993

3.25 G. Doyen, E. Koetter, J.P. Vigneron, M. Scheffler: Appl. Phys. A **51**, 281 (1990)

3.26 G. Doyen: submitted to J. Phys. Cond. Matt.

3.27 J.B. Pendry: *Low Energy Electron Diffraction* (Academic, London 1974)

3.28 W. Mizutani, M. Shigeno, M. Ono, K. Kajimura: Appl. Phys. Lett. **56**, 1974 (1990)

3.29 N.J. Zheng, I.S.T. Tsong: Phys. Rev. **41**, 2671 (1990)

3.30 J. Wintterlin, J. Wiechers, H. Brune, T. Gritsch, H. Höfer, R.J. Behm: Phys. Rev. Lett. **62**, 59 (1989)

3.31 C.B. Duke, M.E. Alferieff: J. Chem. Phys. **46**, 923 (1967)
3.32 V. Kalmeyer, R.B. Laughlin: Phys. Rev. B **35**, 9805 (1987)
3.33 R. Landauer: Z. Phys. B **68**, 217 (1987)
3.34 A. Martin-Rodero, J. Ferrer, F. Flores: J. Microsc. **152**, Pt. 2, 317 (1988); E. Louis, F. Flores, P.M. Echenique: Rad. Eff. Defects in Solids, **109**, 309 (1989)
3.35 I. Derycke, J.P. Vigneron, D. Drakova, G. Doyen, M. Scheffler: in preparation
3.36 D.P.E. Smith, A. Bryant, C.F. Quate, J.P. Rabe, Ch. Gerber, L.D. Swalen:, Proc. Natl. Acad. Sci. U.S.A. **84**, 969 (1987)
3.37 J.K.H. Hörber, C.A. Lang, T.W. Hänsch, W.M. Heckl, H. Möhwald: Chem. Phys. Lett. **145**, 151 (1988)
3.38 N.S. Maslova, Yu.N. Moiseev, V.I. Panov, S.V. Savinov, S.I. Vasilev, I.V. Yaminski: physica status solidi (a) **131**, 35 (1992)
3.39 J.P. Rabe, S. Buchholz: Science **253**, 424 (1991)
3.40 L. Bourdieu, P. Silberzan, D. Chatenay: Phys. Rev. Lett. **67**, 2029 (1991)
3.41 B. Michel, G. Travaglini, H. Rohrer, C. Joachim, M. Amrein: Z. Physik B **76**, 99 (1989)
3.42 G. Doyen, D. Drakova, V. Mujica: Phys. Stat. Sol. (a) **131**, 107 (1992)
3.43 D.M. Newns: Phys. Rev. **178**, 1123 (1969)
3.44 J. Delhalle, J.M. Andre, S. Delhalle, J.J. Pireaux, R. Caudano, J.J. Verbist: J. Chem. Phys. **60**, 595–600 (1974)
3.45 S.M. Lindsay, O.F. Sankey, Y. Li, C. Herbst, A. Rupprecht: J. Phys. Chem. **94**, 4655 (1990)
3.46 C. Joachim, P. Sautet: In: *Basic Concepts and Applications of Scanning Tunneling Microscopy (STM) and Related Techniques*, Ed. by R.J. Behm, N. Garcia, H. Rohrer (Kluwer 1990) p. 377
3.47 D. Drakova, G. Doyen, R. Hübner: J. Chem. Phys. **89**, 1725 (1988)
3.48 V.L. Moruzzi, J.F. Janak, A.R. Williams: *Calculated Electronic Properties of Metals* (Pergamon, New York 1978)
3.49 V.M. Hallmark, S. Chiang, J.F. Rabolt, J.D. Swalen, R.J. Wilson, Phys. Rev. Lett. **59**, 2879 (1987)
3.50 Ch. Wöll, S. Chiang, R.J. Wilson, P.H. Lippel: Phys. Rev. B **39**, 7988 (1989)
3.51 P.H. Lippel, R.J. Wilson, M.D. Miller, Ch. Wöll, S. Chiang: Phys. Rev. Lett. **62**, 171 (1989)
3.52 K. Kambe, M. Scheffler: Surf. Sci. **89**, 262 (1979)
3.53 F. Maca, M. Scheffler: Computer Phys. Commun. **38**, 403 (1985); Computer Phys. Commun. **52**, 381 (1988)
3.54 E. Koetter: Theoretische Untersuchung zum Einfluß der elektronischen und geometrischen Struktur der Spitze in der Raster-Tunnel-Mikroskopie. Dissertation Tu-Berlin, 1993
3.55 J. Ferrer, A. Martin-Rodero, F. Flores: Phys. Rev. B **34**, 10113 (1988)
3.56 G. Binnig, H. Rohrer, Ch. Gerber, E. Weibel: Appl. Phys. Lett. **40**, 178 (1982)
3.57 J.K. Gimzewski, R. Möller: Phys. Rev. B **36**, 1284 (1987)
3.58 N. Lang: Phys. Rev. B **36**, 8173 (1987)
3.59 U. Dürig, J.K. Gimzewski, D.W. Pohl: Phys. Rev. Lett. **57**, 2403 (1986)
3.60 U. Dürig, O. Züger, D.W. Pohl: J. Microscopy **152**, Pt. 1, 259 (1988)
3.61 U. Dürig, O. Züger: Vacuum **41**, 382 (1990)
3.62 U. Dürig, O. Züger, D.W. Pohl: Phys. Rev. Lett. **65**, 349 (1990)
3.63 U. Dürig: private communication
3.64 G. Binnig, H. Rohrer, Ch. Gerber, E. Weibel: Bull. Am. Phys. Soc. **28**, 461 (1983)
3.65 G. Doyen, D. Drakova, J.V. Barth, R. Schuster, T. Gritsch, R.J. Behm, G. Ertl: Submitted to Phys. Rev. B
3.66 Z.P. Hu, B.C. Pan, W.C. Fan, A. Ignatiev: Phys. Rev. B **41**, 9692 (1990)
3.67 R. Schuster, J.V. Barth, G. Ertl, R.J. Behm: Phys. Rev. B **44**, 13 689 (1991)
3.68 G. Doyen, D. Drakova: unpublished results
3.69 R. Schuster, J. Barth, G. Ertl, R.J. Behm: Surf. Sci. **247**, L229 (1991)

4. Spectroscopic Information in Scanning Tunneling Microscopy

C. Noguera

With 8 Figures

During the last ten years, the Scanning Tunneling Microscope (STM) was proved to be a very powerful tool to obtain local structural information on metallic and semiconducting surfaces [4.1]. Many studies were devoted to the imaging of steps, roughness, reconstructions, or adsorbates of larger and larger complexity on these surfaces. Moreover, the spectroscopic mode of the STM, in which the tunneling current I is recorded as a function of the applied bias V, was shown to give information on the local electronic structure of the surfaces. In some cases, it allowed to discriminate the contributions of inequivalent atoms at the surface [4.2]. The $I-V$ curves, or $d \log I / d \log V$ depending upon the authors, are usually interpreted in terms of the local density of states at the observed point of the surface. A rough one to one correspondence with ultra-violet photoemission spectra gave a qualitative experimental support to such a conjecture [4.3]. It is of course highly desirable to assess whether this correspondence may be put on quantitative grounds, so that finer and finer information could be extracted from spectroscopic data.

The most widely used theory of STM is that of *Tersoff* and *Hamann* [4.4], who adapted *Bardeen*'s calculation [4.5] of the tunneling current between weakly-coupled electrodes, to the particular geometry of a tip close to a surface. Their result states that the tunneling conductance at low bias and low temperature is indeed proportional to the Local Density Of States (LDOS) of the sample: $\sigma \propto \varrho(r_0, E_F)$ evaluated at the Fermi level E_F and at the tip curvature center r_0. This theory has allowed one to clarify many important questions such as the resolution of the microscope [4.6], the interpretation of STM images [4.7–10], and, to some degree, the role of the tip [4.11, 12].

While, for weakly coupled electrodes, both spectroscopic and topographic STM data can be interpreted in terms of the LDOS of the free surface, it is often necessary to go beyond this perturbation limit. One straightforward reason has to do with the tip-surface distance d, which, in practice, may become of the order of the penetration length of the electrons in the vacuum barrier ($1/\kappa \sim 1$ Å). In that case, the electrodes become strongly coupled, through multiple-scattering events taking place in the vacuum barrier, with non-negligible weight $\exp(-n\kappa d)$. Another possible reason for the breakdown of the perturbation approximation lies in the strength of the reflection coefficients at the electrode surfaces [4.13].

Theories going beyond the perturbation approach have been developed in the past, for the study of metal–insulator–metal junctions [4.14–16]. As con-

Springer Series in Surface Sciences, Vol. 29
Scanning Tunneling Microscopy III Eds.: R. Wiesendanger · H.-J. Güntherodt
© Springer-Verlag Berlin Heidelberg 1993

cerns the STM, the recent interest falls mainly in the following areas: (1) study of the surface states on metal surfaces [4.17], which has progressed less quickly than on semiconductor surfaces [4.18, 19] (2) understanding of the occurrence of coupled tip–surface states [4.20, 21] (3) explanation of the giant corrugations observed on graphite [4.20], or on metal surfaces [4.22, 23], which are much larger than the LDOS corrugations, (4) transition to point contact [4.24–26], and (5) image simulation of complex systems, such as adsorbed atoms [4.27] or molecules [4.28].

We present here a Green's function method that we have developed for calculating the tunneling current in STM [4.29]. It goes beyond the perturbation approximation, in the sense that it fully treats the coupling between the tip and the surface electronic states. From this point of view, it helps the understanding of the STM spectroscopic mode, although some effects, such as the precise description of the vacuum barrier with or without an electric field, are clearly beyond its scope. The method, which relies upon a matching procedure of the free surface electron states, is developed in Sect. 4.1, either when a single or when two matching surfaces are considered. From the expression of the tunneling current that we obtain, we are able to reproduce the transfer Hamiltonian result, in the perturbative limit, and thus assess more precisely its limits of validity [4.13]. This is done in Sect. 4.2. Then, in Sect. 4.3, we apply the formalism to various one-dimensional models [4.30, 31] which help understanding the relevant microscopic quantities. Finally, in Sect. 4.4, we present three-dimensional situations, in which the tip is treated in a simplified way [4.32–34].

4.1 Green's Function Method

The method is a generalization of the treatment used by *Caroli* et al. [4.14], *Feuchtwang* [4.15] and *Combescot* [4.16] for planar junctions, which expresses the tunneling current as a function of quantities characteristic of the electrodes in contact with a vacuum. Two ways of partitioning space are possible: the first one uses a single surface of separation located in the vacuum barrier, while, in the second one, the total system is divided into three parts: left electrode, vacuum barrier and right electrode, with two surfaces of separation. In our opinion, the second method is best suited to a discussion of the contributions of the electronic states of both electrodes and of the multiple-reflection events in the barrier, while the first one allows an easy link with the transfer Hamiltonian approach. We will first make a derivation of the tunneling current using a single matching surface, then we will consider two surfaces of separation.

4.1.1 Matching at a Single Surface

Let us consider the system described in Fig. 4.1. We call I and II the two half-spaces separated by a surface Σ located in the vacuum barrier, and we

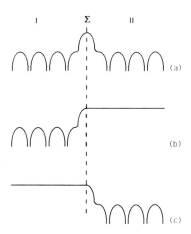

Fig. 4.1. Geometry of the tunneling system with the two electrodes and the vacuum barrier; in (a) the system is shared into two parts called I and II, by a boundary Σ entirely located in the barrier; in (b) left free electrode, in which the inner potential is equal to that of (a) on the left of Σ, and with vacuum extending to infinity on the right of Σ; in (c) right free electrode, in which the inner potential is equal to that of (a) on the right of Σ, and with vacuum extending to infinity on the left of Σ

assume that we know the electronic properties of I and II when the vacuum potential extends to infinity (Fig. 4.1b, c): in other words, we know the characteristics of the free electrodes. We define two kinds of Green's functions: first, there are the advanced or retarded ones, which in terms of the stationary states Ψ_n of energies E_n of the system read

$$g^{a,r}(x, x') = \sum_n \frac{\Psi_n(x)\Psi_n^*(x')}{\omega - E_n \pm i\eta} \tag{4.1}$$

and the g^+ Green's function introduced by *Keldysh* [4.35] for out-of-equilibrium situations, which, for our concern, contains all the information upon occupied states

$$g^+(x, x') = 2i\pi \sum_{n\ occ} \Psi_n(x)\Psi_n^*(x')\delta(\omega - E_n) \ . \tag{4.2}$$

We work in Fourier space as concerns the time variables, although we have not introduced the frequency ω explicitly in the notations of $g^{a,r}(x, x')$ and $g^+(x, x')$, for the sake of simplicity. The gradients of the advanced and retarded Green's functions are discontinuous at $x = x'$, because of the 'delta function' in the right hand side of the differential equation that governs their behaviour. As a consequence, when necessary, we will add a $+$ or $-$ superscript to indicate that the limit $x' \to x$ is taken from the right or from the left. These superscripts are unnecessary for the gradients of g^+ functions which present no discontinuity at $x = x'$.

We wish to calculate the G^+ Green's function of the whole system {tip + surface} (capital letters refer to the whole system Fig. 4.1a), from which the current density J may be derived

$$J(x) = \int d\omega \ \text{Re}\{\nabla_x G^+(x, x' \to x)\} \ . \tag{4.3}$$

A matching procedure at the surface Σ allows to relate G^+ to the Green's functions $g_{\mathrm{I,II}}(x, x')$ of the free electrodes I and II, displayed in Fig. 4.1b and c. This procedure, usually developed for the advanced and retarded Green's functions [4.36], is extended without difficulty to G^+, taking into account its specific differential equation [4.35], and yields the following result (unless otherwise stated, the ∇ sign refers to the x coordinates and integration is performed on Σ; Rydberg atomic units are used throughout $\hbar = 2m = e = 1$)

$$G^+(x_1, x_2) = g_{\mathrm{I}}^+(x_1, x_2)$$

$$- \int d\boldsymbol{S}[\nabla g_{\mathrm{I}}^{\mathrm{a}}(x_1^-, x)G^+(x, x_2) - g_{\mathrm{I}}^{\mathrm{a}}(x_1, x)\nabla G^+(x, x_2)]$$

$$+ \int d\boldsymbol{S}[g_{\mathrm{I}}^+(x_1, x)\nabla G^{\mathrm{r}}(x, x_2^-) - \nabla g_{\mathrm{I}}^+(x_1, x)G^{\mathrm{r}}(x, x_2)]$$

if $x_2 \in \mathrm{I}$ and $x_1 \in \mathrm{I}$ (4.4)

$$G^+(x_1, x_2) = \int d\boldsymbol{S}[\nabla g_{\mathrm{II}}^{\mathrm{a}}(x_1^+, x)G^+(x, x_2) - g_{\mathrm{II}}^{\mathrm{a}}(x_1, x)\nabla G^+(x, x_2)]$$

$$- \int d\boldsymbol{S}[g_{\mathrm{II}}^+(x_1, x)\nabla G^{\mathrm{r}}(x, x_2^-) - \nabla g_{\mathrm{II}}^+(x_1, x)G^{\mathrm{r}}(x, x_2)]$$

if $x_2 \in \mathrm{I}$ and $x_1 \in \mathrm{II}$.   (4.5)

The knowledge of the advanced and retarded Green's functions $G^{\mathrm{a,r}}$ of the whole system is required in order to solve (4.4, 5). The $G^{\mathrm{a,r}}$ are self-consistently determined by ($u = \mathrm{a, r}$)

$$G^u(x_1, x_2) = g_{\mathrm{I}}^u(x_1, x_2)$$

$$- \int d\boldsymbol{S}[\nabla g_{\mathrm{I}}^u(x_1^-, x)G^u(x, x_2) - g_{\mathrm{I}}^u(x_1, x)\nabla G^u(x, x_2^-)]$$

if $x_2 \in \mathrm{I}$ and $x_1 \in \mathrm{I}$ (4.6)

$$G^u(x_1, x_2) = \int d\boldsymbol{S}[\nabla g_{\mathrm{II}}^u(x_1^+, x)G^u(x, x_2) - g_{\mathrm{II}}^u(x_1, x)\nabla G^u(x, x_2^-)]$$

if $x_2 \in \mathrm{I}$ and $x_1 \in \mathrm{II}$.   (4.7)

This system of coupled equations may be put in a more tractable form by eliminating $g_{\mathrm{I,II}}^+$ thanks to the relationship

$$g^+(x_1, x_2) = [g^{\mathrm{r}}(x_1, x_2) - g^{\mathrm{a}}(x_1, x_2)]\Theta(\mu - \omega)$$ (4.8)

in which μ_{I} (or μ_{II}) is the Fermi energy of the electrode I (or II) in the presence of the bias V and the Heaviside function $\Theta(\mu_{\mathrm{I,II}} - \omega)$ tells that only occupied states of energy $\omega < \mu_{\mathrm{I,II}}$ contribute to g^+, at zero temperature ($\Theta(x) = 0$ if $x < 0$ and

1 if $x > 0$). Equations (4.6, 7) are thus rewritten in the following way

$$G^+(x_1, x_2) + \int dS [\nabla g_I^a(x_1^-, x) G^+(x, x_2) - g_I^a(x_1, x)\nabla G^+(x, x_2)]$$

$$= -2i\Theta(\mu_I - \omega)(\mathrm{Im}\{g_I^a(x_1, x_2)\}$$

$$+ \int dS[\mathrm{Im}\{g_I^a(x, x_1^-)\}\nabla G^r(x, x_2)$$

$$- \mathrm{Im}\{\nabla g_I^a(x_1, x)\}G^r(x, x_2^-)])$$

if $x_1 \in I$ and $x_2 \in I$ (4.9)

$$G^+(x_1, x_2) - \int dS[\nabla g_{II}^a(x_1^+, x) G^+(x, x_2) - g_{II}^a(x_1, x)\nabla G^+(x, x_2)]$$

$$= -2i\,\Theta(\mu_{II} - \omega)$$

$$\times (-\int dS[\mathrm{Im}\{g_{II}^a(x_1^+, x)\}\nabla G^r(x, x_2)$$

$$- \mathrm{Im}\{\nabla g_{II}^a(x_1, x)\}G^r(x, x_2^-)])$$

if $x_1 \in I$ and $x_2 \in II$.   (4.10)

It is convenient to introduce an operator form for the Green's function

$$\langle x|G|x'\rangle = G(x, x') \tag{4.11}$$

together with a symbolic notation for the integration on the surface Σ

$$\langle x|G|x_1\rangle\langle x_1|G'|x'\rangle = \int_\Sigma dS_1\, G(x, x_1)G'(x_1, x') . \tag{4.12}$$

The resolution of (4.6, 7, 9, 10) yields the following expression for the total tunneling current across Σ [4.13]

$$J = \frac{4}{\pi} \int_{\mu_I}^{\mu_{II}} d\omega \langle x_1|\mathrm{Im}\{X_I^a\}|x_1'\rangle\langle x_1'|G^a|x_1''\rangle$$

$$\times \langle x_1''|\mathrm{Im}\{X_{II}^a\}|x_1'''\rangle\langle x_1'''|G^r|x_1\rangle . \tag{4.13}$$

J is symmetric with respect to the two electrodes, as it should be. It involves an integration on energies, bound by the Fermi energies of the two electrodes, and four integrations on the boundary Σ. It is expressed as a function of the advanced and retarded Green's functions $G^{a,r}$ of the whole system, and as a function of two operators X_I and X_{II} which characterize the two free electrodes. We call them: logarithmic derivatives of the Green's function because they are defined as

$$X_I^a = \tilde{g}_I^a|x_1\rangle \quad \langle x_1|(I + \nabla_2 g_I^a) \tag{4.14}$$

with $\tilde{g}$ a pseudo-inverse of g

$$\langle x|g|x_1\rangle\langle x_1|\tilde{g}|x'\rangle = \langle x|I|x'\rangle = \delta(x - x') \qquad (4.15)$$

defined for $x \in \Sigma$ and $x' \in \Sigma$, and ∇_2 is the gradient operator relative to the second coordinate of g. The operator X_1 (and X_{11}, respectively) has an interesting property, which comes out of the matching equations. Its matrix elements do not depend on what is on the right (and left, respectively) of Σ, be it vacuum or a second electrode. For example, they are equal to the matrix elements of the logarithmic derivative of the Green's function of the whole system

$$\langle x|\tilde{g}_1|x_1\rangle\langle x_1|(I + \nabla_2 g_1)|x'\rangle = \langle x|\tilde{G}|x_1\rangle\langle x_1|(I + \nabla_2 G)|x'\rangle . \qquad (4.16)$$

To complete the resolution of the matching equations, the matrix elements of G on Σ are found equal to

$$\langle x_1|G|x_1'\rangle = \langle x_1|\overbrace{X_1 + X_{11}}|x_1'\rangle \qquad (4.17)$$

so that the total current can be calculated provided one knows the characteristics $X_{1,11}$ of the free electrodes.

4.1.2 Matching at Two Surfaces

It is sometimes convenient to proceed in a different way and consider the two electrodes and the vacuum barrier as three distinct media, as displayed in Fig. 4.2. The matching procedure has to be applied at two surfaces S_1 and S_2, which limit the L (left), M (middle) and R (right) media.

The choice of S_1 and S_2 is not straightforward. Actually, it is necessary that, on the left (right) of S_1 (S_2) the electrons experience the same potential as in the free electrode. This presents no difficulty if an abrupt potential step exists between the inner electrode potential and the outside. But it is known that exchange and correlation effects (image forces) soften the discontinuity. In addition, when the distance between the electrodes decreases significantly, the vacuum barrier may ultimately vanish. We will not try to deal here with such a limiting case in which only a self-consistent numerical method may yield the tunneling current, and in which the electrodes are so perturbed by each other that there is no hope of deducing information on them from an STM measurement.

One may start from the procedure developed in Sect. 4.1.1 and conceptually identify S_1 with Σ. L is thus associated with X_L identical to the X_1 of the preceding section. X_{11} has to be reexpressed as a function of the Green's function of the whole system and of the right electrode. This can be done by a matching procedure at the surface S_2. The final result reads

$$J = \frac{4}{\pi}\int_{\mu_1}^{\mu_{11}} d\omega \langle x_1|\mathrm{Im}\{X_L^a\}|x_1'\rangle\langle x_1'|G^a|x_2\rangle\langle x_2|\mathrm{Im}\{X_R^a\}|x_2'\rangle\langle x_2'|G^r|x_1\rangle .$$

$$(4.18)$$

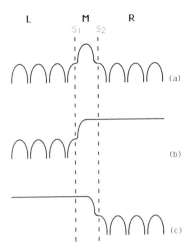

Fig. 4.2. Same system as in Fig. 4.1: the system is now shared into three parts called left (L) medium (M) and right (R) and S_1 and S_2 are the surfaces of contact between the two electrodes and the vacuum (or insulator). In (**b**) and (**c**) the left and right free electrodes are defined in a way similar to Fig. 4.1

This expression of the tunneling current is formally very similar to that obtained with a single matching surface, except that now it involves two integrations on the boundary S_1 and two on S_2. Aside from the logarithmic derivatives of the Green's functions for the two electrodes, evaluated at the electrode surfaces, J depends upon the propagator $\langle x_1|G^a|x_2\rangle$ through the vacuum barrier, in the presence of the electrodes, i.e. including all multiple scattering events. The equivalence between (4.13) and (4.18) may be checked by using the property of conservation of the logarithmic derivatives of the Green's function stressed in the preceding section, and using the fact that there is no propagating state in the barrier at the energy under consideration. The interesting output of this approach lies in the fact that J is now expressed as a function of the surface characteristics of the free electrodes.

In conclusion, the tunneling current J has been calculated in a one-electron picture, without relying on any perturbation approximation. J depends on quantities X which characterize the electronic structure of the free electrodes and on the Green's function G of the whole system {tip + surface}. We will present in Sect. 4.3 and Sect. 4.4 specific examples which help to understand which information is contained in these quantities.

4.2 Derivation of the Transfer Hamiltonian Approach

Since the tunneling current has been obtained without the help of a perturbation approximation, we now compare its expression to that given by the transfer

Hamiltonian approach, in order to demonstrate the limits to which this latter may be safely applied.

4.2.1 The Transfer Hamiltonian Approach

The transfer Hamiltonian method was first introduced by *Bardeen* [4.5] to calculate the current across a planar tunnel junction in the presence of a small bias. It considers the partitioning of space with a single matching surface (Fig. 4.1) and makes use of a time-dependent perturbation approximation for the wave functions of the whole system, assuming that the set of wave functions Ψ_n and Φ_m characterizing the two free electrodes I and II is known. The wave function of an electron, initially in the left electrode ($\Psi(t = 0) = \Psi_n$), gets progressively mixed with states Φ_m of the right electrode

$$\Psi(t) = \Psi_n(t) + \sum_m a_m(t)\Phi_m(t) \ . \tag{4.19}$$

The coefficients $a_m(t)$, calculated by time-dependent perturbation theory, are shown to be proportional to an effective tunneling matrix element M_{nm} equal to

$$M_{nm} = - \int_\Sigma dS(\Phi_m^* \nabla \Psi_n - \Psi_n \nabla \Phi_m^*) \tag{4.20}$$

so that the tunneling current is given by the golden rule

$$J = 2\pi \sum_{m,n} |M_{nm}|^2 f(E_m)[1 - f(E_n - eV)]\delta(E_m - E_n) \tag{4.21}$$

which makes use of the Fermi Dirac function $f(E)$. J is independent of the choice of the matching surface Σ.

Lang [4.37] generalized this result to account for the current density in the junction

$$J(r) = 2\pi \sum_{m,n} M_{nm}[\Phi_m(r)\nabla\Psi_n^*(r) - \Psi_n^*(r)\nabla\Phi_m(r)]$$
$$\times f(E_m)[1 - f(E_n - eV)]\delta(E_m - E_n) \ . \tag{4.22}$$

4.2.2 Tersoff and Hamann's Theory

After the invention of the scanning tunneling microscope, *Tersoff* and *Hamann* [4.4] have applied the transfer Hamiltonian approach to the geometry of the STM. They represent the tip (right electrode) as a sphere with s-wave functions and they assume that the surface (left electrode) presents a perfect periodicity in the plane parallel to the surface [actually, the result (4.23) does not require this assumption; we recover it in the following, whatever the geometry of the left electrode]. Making a two-dimensional Fourier expansion of the wave functions, and taking advantage of their exponential character in the vacuum barrier, they prove that the tunneling current is proportional to the Fermi level density of states of the left electrode, calculated at the center of curvature r_0 of the tip

$$J \propto V \sum_n |\Psi_n(r_0)|^2 \delta(E_n - E_F) . \tag{4.23}$$

Later, *Chen* [4.38, 39] extended this approach to tip wave functions of higher angular momenta. He showed that the current involves the first (and second respectively) partial derivatives of the surface wave functions in the case of p (and d, respectively) tip states.

4.2.3 New Derivation of the Transfer Hamiltonian Approach

We now rederive the transfer Hamiltonian expression for the tunneling current, starting from the formalism developed in Sect. 4.1.1. For this purpose, it is necessary to make the two following assumptions:

a) at all energies under consideration ($\mu_I < \omega < \mu_{II}$), the vacuum advanced and retarded Green's functions $g_0^{a,r}$ are assumed to be real, and thus equal to each other

$$g_0(x_1, x_2) = \frac{e^{-\kappa|x_1 - x_2|}}{|x_1 - x_2|} . \tag{4.24}$$

The penetration length $1/\kappa$ is a function of the difference in energy between the vacuum level V_0 and the electron energy ω via: $\kappa^2 = V_0 - \omega$. This first hypothesis restricts us to low bias below the Fowler-Nordheim regime.

b) The advanced and retarded Green's functions for the free electrodes I and II may be rewritten without loss of generality under the following form ($u = a, r$)

$$g_I^u(x_1, x_2) = g_0(x_1, x_2) + f_I^u(x_1, x_2)$$

$$g_{II}^u(x_1, x_2) = g_0(x_1, x_2) + f_{II}^u(x_1, x_2) \tag{4.25}$$

where the gradients of f_I and f_{II} no longer present discontinuities at $x_1 = x_2$. f_I and f_{II} are now assumed to have smaller values than g_0, whatever the coordinates x_1 or x_2 on Σ. We will come back to the validity of this statement in Sect. 4.2.4 but, roughly speaking, it requires that the barrier width be large and that the matching surface Σ be far enough from both electrodes.

With the help of these two asumptions we are now able to solve the system of coupled equations (4.6, 7, 9, 10) to successive orders of approximations, and find $G^{a,r}(x_1, x_2)$ and $G^+(x_1, x_2)$ for any position of x_1 and x_2 on Σ

zeroth order:

$$G^{a,r}(x_1, x_2) = g_0(x_1, x_2)$$

$$G^+(x_1, x_2) = 0 \tag{4.26}$$

1st order:

$$G^{a,r}(x_1, x_2) = g_0(x_1, x_2) + f_1^{a,r}(x_1, x_2) + f_{11}^{a,r}(x_1, x_2)$$

$$G^+(x_1, x_2) = -2i[\Theta(\mu_1 - \omega)\text{Im}\{f_1(x_1, x_2)\}$$

$$+ \Theta(\mu_{11} - \omega)\text{Im}\{f_{11}(x_1, x_2)\}] . \qquad (4.27)$$

At this order of approximation $G^+(x_1, x_2)$ is imaginary, so that the tunneling current is equal to zero; it is necessary to go to the next order for G^+ which makes use of the first-order approximation of $G^{a,r}$.

2nd order:

$$\text{Re}\{G^+(x_1, x_2)\} = 2[\Theta(\mu_1 - \omega) - \Theta(\mu_{11} - \omega)]$$

$$\times \int d\mathbf{S}[\text{Im}\{\nabla f_1(x_1, x)\}\text{Im}\{f_{11}(x, x_2)\}$$

$$- \text{Im}\{f_1(x_1, x)\}\text{Im}\{\nabla f_{11}(x, x_2)\}$$

$$+ \text{Im}\{\nabla f_{11}(x_1, x)\}\text{Im}\{f_1(x, x_2)\}$$

$$- \text{Im}\{f_{11}(x_1, x)\}\text{Im}\{\nabla f_1(x, x_2)\}] . \qquad (4.28)$$

The current density at any point y of Σ (the integration is performed with respect to the x variables and the scalar product $d\mathbf{S}$. ∇_x is considered for each of the four terms) thus reads

$$J(y) \propto \int_{\mu_1}^{\mu_{11}} d\omega \int d\mathbf{S}[\text{Im}\{f_{11}(x, y)\}\nabla_x\nabla_y\text{Im}\{f_1(x, y)\}$$

$$- \nabla_y\text{Im}\{f_1(x, y)\}\nabla_x\text{Im}\{f_{11}(x, y)\} + \text{Im}\{f_1(x, y)\}\nabla_x\nabla_y\text{Im}\{f_{11}(x, y)\}$$

$$- \nabla_y\text{Im}\{f_{11}(x, y)\}\nabla_x\text{Im}\{f_1(x, y)\}] . \qquad (4.29)$$

Since g_0 is real, the imaginary parts of $f_{1,11}$ and their gradients are equal to the imaginary parts of $g_{1,11}$ and their gradients. As a consequence, they may be developed on the basis of the eigenstates of the left or right free electrodes, as follows

$$\text{Im}\{f_1(x, x')\} = \pi \sum_n \Psi_n^*(x)\Psi_n(x')\delta(E_n - \omega)$$

$$\text{Im}\{f_{11}(x, x')\} = \pi \sum_m \Phi_m^*(x)\Phi_m(x')\delta(E_m - \omega) . \qquad (4.30)$$

Finally, the tunneling current density reads

$$J(y) \propto \int_{\mu_1}^{\mu_{11}} d\omega \sum_{m,n} \delta(E_n - \omega)\delta(E_m - \omega)$$

$$\times [\Psi_n(y)\nabla\Phi_m^*(y) - \nabla\Psi_n(y)\Phi_m^*(y)]$$

$$\times \int d\mathbf{S}[\Psi_n^*(x)\nabla\Phi_m(x) - \Phi_m(x)\nabla\Psi_n^*(x)] . \qquad (4.31)$$

This expression is exactly the transfer Hamiltonian result (4.22), which proves that (a) and (b) are sufficient conditions to assess the validity of the perturbation development.

4.2.4 Validity of the Transfer Hamiltonian Approach

The whole validity of the transfer Hamiltonian method thus relies on the comparison between $g_0(x_1, x_2)$ and both $f_I(x_1, x_2)$ and $f_{II}(x_1, x_2)$ defined in (4.25), when x_1 and x_2 lie on the matching surface Σ. To understand under which circumstances f_I and f_{II} are small compared to g_0, we first note that they come from the electronic waves reflected on the electrode surfaces (S_1 or S_2 in Fig. 4.2)

$$f_I(x_1, x_2) = \int_{S_1} dS \int_{S_1} dS' g_0(x_1, x) A(x, x') g_0(x', x_2) \ . \tag{4.32}$$

The non-local function A contains information on the reflectivity of the electrode surface for waves traveling in the vacuum. Since for $\omega < V_0$ (assumption a) of Sect. 4.2.3) the electronic waves are exponentially damped, f_I is weighted by a factor of the order of $\exp(-2\kappa d_1)$ where d_1 is the distance of closest approach between Σ and S_1. As a consequence, the perturbative development with respect to f_I and f_{II} requires that: (i) the barrier width be large compared to $1/\kappa$, (ii) Σ be not chosen too close to either electrode; and (iii) the reflection matrix be non-singular.

The first requirement is not very restrictive as long as large biases are not applied to the electrodes: in usual materials, for electrons at the Fermi level, $V_0 - \omega$ is of the order of a few eV, so that $1/\kappa$ is close to 1 Å, while the microscope is generally operated at tip-to-surface distances of the order of ten Ångstroems. Yet, this is no longer true at smaller distances or close to the Fowler–Nordheim regime, where $1/\kappa$ becomes large. Indeed, in the one-dimensional models [4.32] illustrated in Sect. 4.3.4, we find large discrepancies between the transfer Hamiltonian result and the current (4.13), in these two cases.

The second requirement tempers the general belief that the choice of the matching surface Σ is completely free in the transfer Hamiltonian approach: actually, it is incorrect to calculate the transfer matrix element for example on one of the surfaces of the electrodes (S_1 or S_2), since, there, g_0 and f_I are of similar magnitude.

The third condition involves the reflection matrix A at the surface of the electrodes, and more precisely its imaginary part. This latter is non-zero for energies ω associated with propagating states in the bulk band structure of the electrodes, but also for discrete energies in the gaps of the band structure, when a special matching between evanescent waves inside and outside the crystal may be performed. This is the general criterium which reveals the existence of surface states, well recognized and observed e.g. in photoemission. This point is discussed in Sect. 4.3.3.

In conclusion we have been able to rederive the transfer Hamiltonian result, starting from the Green's function approach, provided that three conditions are fulfilled: (i) the barrier width is large compared to the penetration length $1/\kappa$ of the electrons in the barrier; (ii) the surface Σ on which the tunneling current is evaluated is not chosen too close to either electrode; and (iii) the reflection operator of the free electrodes is non-singular at the energies being considerated.

4.3 One-Dimensional Models

Although the STM is actually a three-dimensional system, it is instructive to first apply the tunneling current formalism to simple one-dimensional models, in order to learn what spectroscopic information may be derived.

4.3.1 Free Electron Model with a Square Barrier

The calculation of the transmission coefficient for a square barrier of width d and height U with respect to the bottom of the two free electron conduction bands, is known to yield the exact result

$$T(E) = \frac{1}{1 + (1 + \varepsilon)^2 \sinh^2(\kappa d)/4\varepsilon^2} \tag{4.33}$$

where $\varepsilon = k/\kappa$, $k^2 = E$ and $\kappa^2 = U - E$. To demonstrate it, we start from the expression of the tunneling current (4.13, 17), using a single matching surface

$$J = \frac{4}{\pi} \int_{E_F}^{E_F + eV} dE \, \frac{\text{Im}\{X_I\} \, \text{Im}\{X_{II}\}}{|X_I + X_{II}|^2} \tag{4.34}$$

in which the integrand is proportional to the transmission coefficient

$$T(E) = 4 \frac{\text{Im}\{X_I\} \text{Im}\{X_{II}\}}{|X_I + X_{II}|^2} . \tag{4.35}$$

We chose Σ at mid-distance from each electrode, so that the logarithmic derivatives of the Green's function, $X_{I,II}$ are equal to each other

$$X_{I,II}(d/2) = \kappa \frac{i\varepsilon - \tanh(\kappa d/2)}{1 - i\varepsilon \tanh(\kappa d/2)} \tag{4.36}$$

and the exact result (4.33) follows.

At large barrier thicknesses $\kappa d \gg 1$, a perturbative development of $T(E)$ gives

$$T(E) \approx \frac{(\text{Im}\{X_I\})^2}{\kappa^2} \approx 16 \frac{k^2 \kappa^2}{(k^2 + \kappa^2)^2} \exp(-2\kappa d) . \tag{4.37}$$

Within this limit, $\text{Im}\{X_1\}$ is as expected proportional to the local density of states of the left electrode, evaluated at $d/2$.

This one-dimensional square barrier model between two free electron media, usefully stresses the requirement $\kappa d \gg 1$ necessary to recover the transfer Hamiltonian result. But, in order to understand which spectroscopic information is contained in the functions X, we are obliged to go to systems of more elaborate band structure.

4.3.2 One-Dimensional Array of Square Well Potentials

In this section, we consider a single free electrode, and evaluate the logarithmic derivative of its Green's function at its surface. The electrode is described as a one-dimensional semi-infinite array of square-well potentials, in abrupt contact with the vacuum at a position x_1 equidistant from two wells. This is one of the simplest models which displays bands of energy separated by forbidden gaps, in each of which exists a surface state. This appears in Fig. 4.3a, b in which the bulk local density of states (equal to the imaginary part of the infinite array Green's function, evaluated at x_1), and the surface local density of states (for the semi-infinite array) are represented correspondingly.

The low dimensionality of the system is revealed by the existence of strong Van Hove singularities in $1/\sqrt{|E - E_n|}$ at the band edges E_n, in the bulk density of states. These singularities give a zero contribution to the logarithmic derivative of the Green's function $\text{Im}\{X(x_1, x_1)\}$ represented in Fig. 4.3c, although they are associated with an infinite density of states in the bulk, because of the absence of group velocity for the electrons at these energies. This emphasizes the fact that only states carrying flux contribute to the tunneling current. Similarly, states associated with a small normal group velocity or high effective mass, as d states, would contribute to the current less than more propagating s or p states, despite their higher densities of states. This point was already stressed by *Tersoff* and *Hamann* [4.4] from the transfer Hamiltonian approach, and

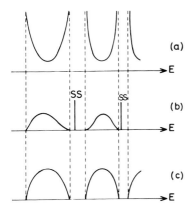

(a)

(b)

(c)

Fig. 4.3. Local density of states of a one-dimensional array of square well potentials evaluated at x_1, midpoint between two wells, in the case (a) of an infinite array and (b) of a semi-infinite array with an abrupt contact with vacuum at x_1; (c) imaginary part of the logarithmic derivative of the Green's function at x_1 valid for both situations (a) and (b)

more quantitatively established by *Lang* [4.40] in a simple model of an atom on a jellium surface. In our formalism, this information is hidden is $\text{Im}\{X\}$.

$\text{Im}\{X(x_1, x_1)\}$ vanishes for all energies in the gap, even at the energies of the surface states. This could have been inferred from the fact that X is independent on the value of the potential on the right of the matching surface S_1 (here x_1), while the existence and the energy of the surface states strongly depend upon this latter. In other words, surface states do not contribute to the tunneling current because they have a zero-group velocity, with wave functions decreasing exponentially on both sides of the surface. We will develop this point in the next section.

In summary, the study of this simple one-dimensional model of a free electrode has helped understanding that the energy dependence of $\text{Im}\{X\}$ is not similar to that of the local density of states, because $\text{Im}\{X\}$, involving gradients of the wave functions, strongly relies upon the value of the group velocity at the energy being considerated.

4.3.3 The Question of the Surface States

We now emphasize that, at the energy of the surface states, not only does the transfer Hamiltonian approach give *quantitatively* wrong results, but that it predicts a contribution to the tunneling current which does not exist. For this purpose, we consider the general problem of a free electrode (left electrode) in one dimension, with a surface S_1, located at x_1, and we express both the Green's function $g_1(x_0, x_0)$ and $X_1(x_0, x_0)$ at a point x_0 in vacuum ($x_0 > x_1$), thanks to the matching procedure. Denoting X_0 and X_L the logarithmic derivatives of the vacuum and left electrode Green's functions evaluated at x_1, in agreement with the general expression (4.32), we find that

$$g_1(x_0, x_0) = g_0(x_0, x_0) + g_0(x_0, x_1)A(x_1, x_1)g_0(x_1, x_0) \tag{4.38a}$$

with the reflection matrix A equal to

$$A(x_1, x_1) = \frac{1}{g_0(x_1, x_1)} \frac{(X_0 - X_L)}{(X_0 + X_L)}. \tag{4.38b}$$

The local density of states at the surface of the left electrode is obtained from the imaginary part of g_1 when $x_0 = x_1$ which, as expected, reads $g_1(x_1, x_1) = 1/(X_0 + X_L)$. Below the vacuum level, $\text{Im}\{g_1(x_1, x_1)\}$ is non-vanishing (i) at the energies of the left electrode propagating states, when $\text{Im}\{X_L\}$ is non-zero, and (ii) at discrete energies such that $X_0 + X_L = 0$. Let us call such surface state energy E_{ss} and let us develop $g_1(x_1, x_1)$ at energies ω close to E_{ss}, in the following way

$$g_1(x_1, x_1) = \frac{W_{ss}}{(\omega - E_{ss} - i\eta)}. \tag{4.39}$$

The local density of states at the surface of the electrode presents a divergence at $\omega = E_{ss}$ of weight W_{ss}, due to the divergence of the reflection coefficient. A similar divergence characterizes the density of states at x_0, and thus the imaginary part of the function f_1 introduced in (4.25)

$$\text{Im}\{f_1(x_0, x_0)\} = -2X_0 \frac{g_0(x_1, x_0)^2}{g_0(x_1, x_1)} \pi W_{ss} \delta(\omega - E_{ss}) . \tag{4.40}$$

This clearly invalidates the use of perturbation theory to calculate the tunneling current.

On the other hand, $X_1(x_0, x_0)$ reads ($X_0 = \kappa$)

$$X_1(x_0, x_0) = \kappa \frac{1 - \dfrac{(X_0 - X_L)}{(X_0 + X_L)} e^{-2\kappa(x_0 - x_1)}}{1 + \dfrac{(X_0 - X_L)}{(X_0 + X_L)} e^{-2\kappa(x_0 - x_1)}} . \tag{4.41}$$

At energies close to E_{ss}, X_1 is real, because of a cancellation between the numerator and the denominator in (4.41). This cancellation effect is destroyed and the divergence is restored, when one makes a perturbative development of X_1, in the spirit of the transfer Hamiltonian approach

$$X_1(x_0, x_0) = \kappa \left[1 - 2\frac{(X_0 - X_L)}{(X_0 + X_L)} e^{-2\kappa(x_0 - x_1)} \right] . \tag{4.42}$$

This is reminiscent of the cancellation effect stressed by *Harrison* [4.41] which forbids to express the tunneling current uniquely in terms of density of states.

To sum it up, in one dimension, surface states appear at energies such that the reflection coefficients diverge. Consequently, the perturbation development, with respect to this quantity, made in the transfer Hamiltonian approach is unauthorized. Surface states do not contribute to the tunneling current in one dimension because the group velocity of the electrons in these states is equal to zero.

4.3.4 Resonant States in the Barrier

Actually, the above conclusion is valid, not only for surface states, but also for any state (due to defects . . .) which is associated with a discrete pole in the Green's functions $g_{1,\text{II}}$. We can physically assign this result to the fact that the tunneling current does not simply depend upon the density of the states available for tunneling, but also upon the normal group velocity of the electrons in these states. Although we have not written down explicitly the expression of the group velocity in our derivation, one should remember that in one dimension, or in three dimensions at a given $k_\parallel$ value, it is inversely proportional to the surface density of states. A discrete pole in the Green's function is thus necessarily associated with a state localized in real space, whose normal group velocity is

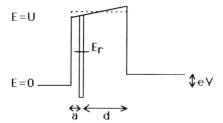

Fig. 4.4. Model for a square barrier with a resonant state E_r in the barrier at a distance a from the left electrode; the energy reference is at the bottom of the band; the barrier height is U and an average barrier approximation is used to treat the effect of the electric field in the barrier (*dotted line*)

equal to zero. For such discrete poles of the Green's function, the perturbative development is never valid.

On the contrary, a resonant state located in the vacuum barrier, does contribute to the tunneling current, because the electrons are not trapped in it during an infinite time. We have explored this situation in one dimension as will now be developed, and also in three dimensions (Sect. 4.4.2). This question is relevant when individual atoms are adsorbed on a surface [4.44], or in the presence of defects in the metal–insulator–metal junctions [4.42, 43].

We consider the model depicted in Fig. 4.4 in which a localized state is present in the barrier, associated with a δ potential $-W_0 \delta(x-a)$, with respect to the vacuum level, at distances a and d from the free-electron counterelectrodes. In the following, we focus on the manifestation of a resonance in the transmission factor $T(E)$ and in the tunnel conductance dJ/dV as a function of the various parameters involved, and we make a comparison with the transfer Hamiltonian results, using one matching surface Σ just on the right of the resonant state. The local density of states $\mathrm{Im}\{g_1(a,a)\}$, as a function of the left electrode Green's function g_L (g_L is defined in the absence of the localized state) reads

$$\mathrm{Im}\{g_1(a,a)\} = \frac{\mathrm{Im}\{g_L(a,a)\}}{[1 + W_0\,\mathrm{Re}\{g_L(a,a)\}]^2 + [W_0\,\mathrm{Im}\{g_L(a,a)\}]^2}. \qquad (4.43)$$

It displays a resonance of width proportional to $\mathrm{Im}\{g_L(a,a)\}$, approximately located at an energy $E_r = U - W_0^2/4$, with respect to the bottom of the free-electron bands. The logarithmic derivative of the Green's function $g_1(a,a)$, on the other hand, is equal to

$$X_1(a,a) = X_L(a,a) + W_0. \qquad (4.44)$$

Consequently the transmission coefficient of the barrier reads

$$T(E) =$$
$$\frac{4\,\mathrm{Im}\{X_L(a,a)\}\,\mathrm{Im}\{X_{II}(a,a)\}}{[W_0 + \mathrm{Re}\{X_L(a,a)\} + \mathrm{Re}\{X_{II}(a,a)\}]^2 + [\mathrm{Im}\{X_L(a,a)\} + \mathrm{Im}\{X_{II}(a,a)\}]^2}. \qquad (4.45)$$

This can be compared with the transfer Hamiltonian result

$$T_B(E) = \frac{4}{g_0(a, a)^2} \frac{\text{Im}\{g_L(a, a)\} \, \text{Im}\{g_{II}(a, a)\}}{[1 + W_0 \text{Re}\{g_L(a, a)\}]^2 + W_0^2 [\text{Im}\{g_L(a, a)\}]^2} \quad . \tag{4.46}$$

The position and the width of the resonance in $T(E)$ depend on both $g_L(a, a)$ and $g_{II}(a, a)$ contrary to those in $T_B(E)$ which are only functions of $g_L(a, a)$. Yet both results converge when the two following conditions are fulfilled

$$\frac{|g_{II}(a, a) - g_0(a, a)|}{|g_0(a, a)|} \ll 1 . \quad \left| \frac{X_{II}(a, a) + \kappa}{X_I(a, a) - \kappa} \right| \ll 1 . \tag{4.47}$$

This resembles the condition of validity of the transfer Hamiltonian approach, because both the barrier thickness and the electronic structure of the electrodes are involved. We have examined numerically the following three cases [4.31]:

- close to the resonance energy, where we show that the discrepancy between $T(E)$ and $T_B(E)$ increases as the resonance width gets smaller (e.g. when the parameter a increases). This is in an analogy with the problem of surface states.
- near the bottom of the band, where both conditions are easily satisfied at large barrier thickness.
- near the top of the barrier, where the first condition is nearly never satisfied, because the penetration length $1/\kappa$ of the electrons in vacuum gets larger and larger, increasing the probability of multiple scattering events.

This discussion is relevant for understanding the tunneling current value at finite bias provided that the average barrier approximation be used, as shown in dotted line in Fig. 4.4. Figure 4.5 demonstrates that the perturbation approximation dJ_B/dV systematically overestimates the tunneling conductance in a more and more dramatic way as the tip–surface distance d decreases. In addition, the resonance in dJ/dV is much broadened and shifted compared to that in $\text{Im}\{g_I(a, a)\}$ or dJ_B/dV, because of the coupling between the tip and the sample. This illustrates the fact that the tunneling conductance not only depends upon $\text{Im}\{X_{I,II}\}$, but also contains spectroscopic information on the Green's function of the whole system {tip + surface} $G = 1/(X_I + X_{II})$, see (4.13). We will come back to this point in Sect. 4.5.

To sum up, thanks to these one-dimensional models, we have been able to understand the physical meaning of the two main ingredients of the tunneling current (4.13), namely the imaginary part of the Green's function $\text{Im}\{X_{I,II}\}$ of the two free electrodes, evaluated on the matching surface, and the Green's function G of the whole system {tip + surface}. The former were shown to differ significantly from the local density of states of the free electrodes, because they also involve the value of the normal group velocity of the electrons. The states associated with a large (d states) or infinite (Van Hove singularities, surface

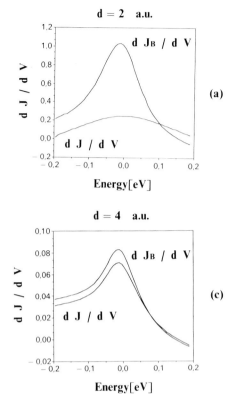

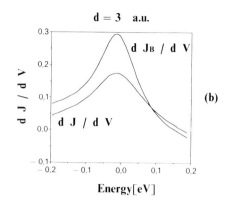

Fig. 4.5a–c. Spectroscopy of the resonant state in the barrier for three distances d = 2, 3, 4 a.u: comparison between the tunneling conductance dJ/dV and the one using the transfer Hamiltonian approximation dJ_B/dV (in units of $4/\pi$); for the smaller tip-surface distance, the resonant peak has mostly disappeared through the large coupling between the electrode; the Fermi energy is taken at the position of the resonant state E_r

states) effective mass, contribute little, or nothing, to the tunneling current. The Green's function of the whole system contains information on the electronic states of the coupled electrodes: this coupling gets larger as the barrier thickness decreases or when the reflection coefficients at the electrode surfaces are large. It induces a shift and a broadening of the structures present in the local densities of states of the free electrodes, and eventually the occurrence of bonding and antibonding states, as will appear more clearly in the following section.

4.4 Three-Dimensional Models

In three dimensions, it is not so easy to calculate the quantity $X(x, x')$ as in the one-dimensional case, for several reasons. First it depends upon the actual shape of the surfaces S_1 or S_2. Second it involves an "inversion" of the Green's operators on the boundary. To our knowledge, there exist various numerical codes to calculate a Green's operator, but none for a two-dimensional inverse Green's operator.

There is a possible path to evaluate this latter, which makes use of the property of conservation of X. As we have stressed previously, X_1 does not depend upon the potential value on the right of the surface S_1. It could thus be calculated, for example, taking this latter infinite [4.14, 16]. This offers some advantage, because, then, X is proportional to the Green's function of this new system, evaluated at an infinitely small distance from S_1 [4.29]. As a consequence, if one applied the existing numerical codes to this unphysical system, one could get the X value of the actual free electrode.

We have not yet tried this procedure. We have rather explored a simpler situation, in which it is assumed that the whole tunneling current passes through a sphere at the extremity of the tip. This may be achieved when the tip is very sharp, the sphere being a single atom, or when a structureless tip is assumed, as done by *Tersoff* and *Hamann* [4.4]. Of course, within such approximation for the tip, 'interference effects' such as those displayed by a cluster of three or four atoms at the end of the tip [4.21, 45] are never present, but it still yields a generalization of the most widely used formula of the tunneling current [4.4]. In [4.32–34], we have used a wave-function approach to study this situation. We will now present the main steps leading to the results, applying the Green's function formalism with a single boundary.

4.4.1 Formalism for a Spherical Tip

Let us consider the system represented on Fig. 4.6. The extremity of the tip is a spherical potential well (radius R, and constant depth W_0 with respect to the vacuum level) centered at r_0, in the vicinity of the surface of the left electrode, and the surface of the well is taken as the matching surface Σ. The simplification inherent in this model comes from the simple shape of the boundary, which allows to explicitly perform the "inversion" of the Green's function, thanks to a development in spherical harmonics. Let us calculate successively the operators X_1 and X_{11} associated respectively with the left free electrode and with the tip.

In order to evaluate X_1 at Σ, we recall that the Green's function reads

$$g_1(r, r') = g_0(r, r') + \int_{S_1} g_0(r, r_1) \int_{S_1} A(r_1, r_2) g_0(r_2, r') \qquad (4.48)$$

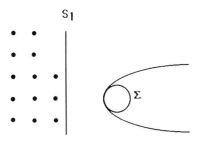

Fig. 4.6. Geometry of the three-dimensional model system, in which the whole tunneling current passes through the sphere at the extremity of the tip

when r and r' are in the vacuum. A is the reflection matrix at the surface S_1 of the left electrode. g_1 is developed in spherical harmonics with respect to r_0, in the following way

$$g_1(r, r') = i\kappa \sum_{l,m} \sum_{l',m'} Y_l^m(\hat{r}_<) Y_{l'}^{m'*}(\hat{r}_>) j_l(i\kappa r_<)$$

$$\times [h_{l'}^+(i\kappa r_>)\delta_{ll'}\delta_{mm'} + A_{ll'}^{mm'} j_{l'}(i\kappa r_>)] \tag{4.49}$$

with $\kappa^2 = -E$ ($E < 0$ for evanescent waves in the vacuum) and $r_<$ and $r_>$ being, respectively, defined as the minimum and maximum values of $|r - r_0|$ and $|r' - r_0|$. $\delta_{ll'}$ is the Kronecker symbol.

The diagonal term, with respect to the angular momenta, comes from the vacuum Green's function $g_0(r, r')$, while the second is due to the reflection at the electrode surface. As usual in such a development j_l and h_l^+ are the spherical Bessel functions [4.46] with convergent character at the origin for j_l and outgoing character for h_l^+. The $Y_l^m(\Omega)$ are the spherical harmonics. An interesting point to notice is that the value of the reflection part δg_1 of g_1 ($g_1 = g_0 + \delta g_1$) at the center of the well reads

$$\delta g_1(r_0, r_0) = i\kappa A_{00}^{00} \tag{4.50}$$

so that the local density of state of the left electrode evaluated at the center of the tip, is non-zero only when $\mathrm{Re}\{A_{00}^{00}\}$ is non-zero. We will restrict ourselves in the following to the case in which only s waves are relevant. A generalization to higher angular momenta may be found in [4.33]. When only s waves are considered, a single coefficient A_{00}^{00} plays a role (that we will hencefore call A_0), and the inversion of g_1 on the matching surface yields the following value for X_1 on Σ

$$X_1(r, r') = -\frac{1}{4\pi} \frac{\dfrac{dh_0^+(i\kappa R)}{dR} + A_0 \dfrac{dj_0(i\kappa R)}{dR}}{h_0^+(i\kappa R) + A_0 j_0(i\kappa R)} \tag{4.51}$$

and for its imaginary part

$$\mathrm{Im}\{X_1(r, r')\} = \frac{1}{4\pi\kappa R^2} \frac{\mathrm{Re}\{A_0\}}{|h_0^+(i\kappa R) + A_0 j_0(i\kappa R)|^2} \tag{4.52}$$

or

$$\mathrm{Im}\{X_1(r, r')\} = \frac{1}{4\pi\kappa^2 R^2} \frac{\mathrm{Im}\{\delta g_1(r_0, r_0)\}}{|h_0^+(i\kappa R) + A_0 j_0(i\kappa R)|^2} . \tag{4.53}$$

The Green's function of the isolated well at the end of the tip reads

$$g_{II}(r, r') = k \sum_{l,m} Y_l^m(\hat{r}_<) Y_l^{m*}(\hat{r}_>) h_l^+(k r_>) j_l(k r_<) \tag{4.54}$$

with $k^2 = W_0 + E (E < 0)$. Keeping only s waves, the logarithmic derivative of g_{II} on the surface Σ is equal to

$$X_{II} = \frac{1}{4\pi} \frac{\dfrac{dj_0(kR)}{dR}}{j_0(kR)} \; . \tag{4.55}$$

The matching on Σ requires the evaluation of $X_1 + X_{II}$

$$X_1 + X_{II} = \frac{1}{4\pi} \frac{W[\,j_0(kR), h_0^+(i\kappa R)\,]}{j_0(kR)[\,h_0^+(i\kappa R) + A_0 j_0(i\kappa R)\,]} [1 - \varrho A_0] \tag{4.56}$$

in which ϱ is the well reflection coefficient for waves coming from outside

$$\varrho = - \frac{W[\,j_0(kR), j_0(i\kappa R)\,]}{W[\,j_0(kR), h_0^+(i\kappa R)\,]} \tag{4.57}$$

and $W(f, g) = g\, df/dx - f\, dg/dx$ is the Wronskian of f and g. Equation (4.56) shows that $X_1 + X_{II}$ is proportional to $1 + \lambda\, \delta g_1(r_0, r_0)$ in which $\delta g_1(r_0, r_0)$ characterizes the left electrode and $\lambda = i\varrho/\kappa$ characterizes the tip reflection properties.

As expected, there is no current when the well is isolated because λ is real. When the well is coupled to a continuum of states, we have shown [4.32–34] that λ is complex and that the current reads

$$J = 4V \frac{\text{Im}\{\lambda\}\, \text{Im}\{\delta g_1(r_0, r_0)\}}{|1 + \lambda\, \delta g_1(r_0, r_0)|^2} \; . \tag{4.58}$$

The numerator is essentially the result of *Tersoff* and *Hamann* [4.4], with $\text{Im}\{\delta g_1(r_0, r_0)\}$ the Local Density Of States (LDOS), while the denominator $|1 + \lambda\, \delta g_1(r_0, r_0)|^2$ departs from unity because of multiple-reflection events in the barrier between the tip and the sample. As noticed previously, the multiple reflections couple the electronic states of the free electrodes. It proves convenient to define a tip-Modified Local Density (MLDOS) $\varrho_{\text{mod}}(r_0)$ such that the current reads

$$J = 4V \text{Im}\{\lambda\}\, \varrho_{\text{mod}}(r_0) \tag{4.59}$$

in analogy with the result of Tersoff and Hamann. It is then clear that the STM images should be interpreted in terms of

$$\varrho_{\text{mod}}(r_0) = \frac{\text{Im}\{\delta g_1(r_0, r_0)\}}{|1 + \lambda\, \delta g_1(r_0, r_0)|^2}$$

a quantity which depends not only on the surface under study, but also on the electronic characteristics of the tip.

The MLDOS is proportional to the product of the Local Density Of States (LDOS) of the surface times the square of the Green's function of the whole system {tip + surface} $G = 1/(X_I + X_{II})$ on Σ. Through the poles or resonances of this latter, the MLDOS contains information on the Tip-Induced Localized States, or TILS, introduced by *Tekmann* and *Ciraci* [4.20]. The tip-induced modification first shifts and broadens the structures in the LDOS at large separation. Then, as the tip–surface distance decreases, it leads to the onset of a covalent interaction between the tip and the sample with a possible existence of bonding and anti-bonding states, as we will see in the following section. The splitting of the bonding and antibonding states has a contribution which depends exponentially on the tip–surface separation, so that the MLDOS presents a strong distance dependence for small separations. In this distance range, the spectroscopic mode of the STM does not yield information on LDOS of the sample (such as one would obtain in photoemission spectra). Rather, it is the electronic states of the interacting tip and surface system which are being probed.

4.4.2 Application to an Adsorbate on a Surface

In order to illustrate these concepts, we have considered a flat metallic surface with a single atomic adsorbate, constituting the left electrode. This has been thoroughly examined by Lang, using the transfer Hamiltonian formalism. He has calculated the tunneling current density [4.37], the imaging of single atoms [4.40], the effects of the bias voltage [4.47, 48], and the transition to point contact [4.40]. An important theme of his work is the influence of the chemical identity of the atom, and of the nature of its interaction with the substrate, on STM measurements. Our approach yields a discussion along similar lines, but at smaller tip-to-surface separations where the transfer Hamiltonian approach is no longer valid.

We have considered two kinds of tips: an electropositive atom, such as Na, adsorbed on a flat jellium surface (tip A), and an electronegative one (tip B), possibly arising from a tip contaminant. In each case we have assumed that the whole current passes through the tip adsorbate (but not necessarily through the left electrode adsorbate). In Fig. 4.7a, b, we display the LDOS and the MLDOS of a left electrode with a Na adsorbate, for these two kinds of tips located directly above the adsorbate at a height of 3.6 Å (as measured from the jellium edge). The state density for the metallic flat surface is also given, and serves as a reference. The LDOS displays a peak above the Fermi level, corresponding to the adsorption level, in resonance with the jellium continuum of states. At E_F the density of states is approximately 2 orders of magnitude larger than the bare substrate value, so that a bump of approximately 2 Å in the constant current contour is expected in the framework of the transfer Hamiltonian approach.

Figure 4.7a, b indicate that the MLDOS is strongly dependent upon the chemical nature of the tip. In the case of tip A, the MLDOS displays two distinct peaks because the tip and adsorbate states, located at the same energy (tip A and

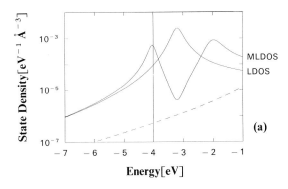

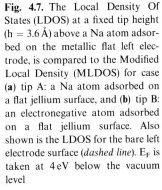

Fig. 4.7. The Local Density Of States (LDOS) at a fixed tip height ($h = 3.6$ Å) above a Na atom adsorbed on the metallic flat left electrode, is compared to the Modified Local Density (MLDOS) for case (a) tip A: a Na atom adsorbed on a flat jellium surface, and (b) tip B: an electronegative atom adsorbed on a flat jellium surface. Also shown is the LDOS for the bare left electrode surface (*dashed line*). E_F is taken at 4 eV below the vacuum level

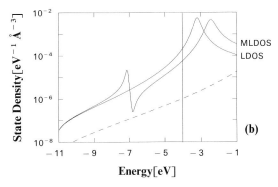

the left electrode are identical), have sufficient overlap to result in a bonding anti-bonding pair. At E_F, there is an *enhancement* of the MLDOS by almost a factor of 6, with respect to the substrate DOS. One concludes that the tunneling current must also be proportionally larger than the value using perturbation theory. As concerns the electro-negative tip atom (tip B), the overlap is similar, but the tip and the surface adsorbate states are no longer at the same energy: the MLDOS thus retains the two individual contributions, slightly shifted and broadened. The state density at the Fermi level is *reduced*, contrary to the case of tip A.

Figure 4.8 shows constant height ($h = 3.6$ Å) scans for the two kinds of tips together with the LDOS variation (dashed line). One sees clearly the effect of the enhancement of the MLDOS for tip A and the reduction for tip B.

When an electronegative atom is adsorbed on the left electrode surface, associated with an affinity level below the Fermi level, the preceding discussion remains valid, simply interchanging the roles of the surface and tip adsorbed atoms.

In summary, for the geometry of the tip assumed in this section, the tunneling current has been expressed under a form which represents a generalization of the formula by Tersoff and Hamann. J is interpreted in terms of a density of states modified by the presence of the tip. The tip interaction with

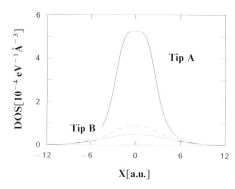

Fig. 4.8. Constant height scans (h = 3.6 Å) across the adsorbate for the examples of Fig. 4.7. The dashed line, is the LDOS variation, which shows the result of the transfer Hamiltonian approach. The LDOS bump is enhanced or reduced for tip A or B respectively

the surface causes a distinct modification of the bias dependence of J. For small distances, this modification is different depending on the identity of the tip.

4.5 Conclusion

The formalism developed in this chapter, in which the Green's functions of the free electrodes are matched on a surface located in the vacuum barrier, allows to obtain a compact expression for the tunneling current J. Although at this stage, it does not treat many body effects, nor includes a proper description of the electric field in the barrier, it presents the advantage of not requiring a weak coupling between the electrode electronic states, as the transfer Hamiltonian approach does.

The tunneling current J depends upon two operators $X_{I,II}$ which characterize the electronic structure and geometry of the free electrodes (at an infinite distance from each other), and upon the Green's function G of the whole system {electrode 1 + electrode 2}, with arguments r and r' in the vacuum barrier, which may eventually be expressed as a function of $X_{I,II}$ and the barrier width and shape.

Up to now, it has not been possible to achieve a calculation of J in a three-dimensional case of a realistic tip in front of a realistic surface. Yet, several one-dimensional models, as well as a simplified three-dimensional one, teach us the following points concerning the information that one may expect from a spectroscopic STM experiment:

i) the operators $X_{I,II}$ and, more specifically, their imaginary parts, contain information not only on the density of available states in the electrodes at the energy upon consideration, but also on the group velocity of the electrons in these states. The strength of s versus d states, of defect states, of Van Hove singularities, of resonant states, of surface states, is thus different in $Im\{X_{I,II}\}$, and in the local density of states of the free electrodes (LDOS).

ii) both electronic quantities (the LDOS and the group velocity) are combined in $X_{I, II}$ in the same way as they are combined in the reflection operators of the free electrodes for waves coming from outside.

iii) the role of the Green's function of the whole system is to wash out the resonances found in $X_{I, II}$ and transform them into new states of the coupled electrode system. The transformation depends upon the coupling strength, which is related to the probability of an electron travelling in the barrier and reflection at the surfaces of the electrodes. The coupling is thus said to be weak when (1) the penetration length $1/\kappa$ of the electrons in the barrier is small compared to the barrier width, and (2) when the reflection coefficients at the electrode surfaces are non-singular. Then, the structures present in the LDOS of the free electrodes are a little perturbed. When the coupling increases, i.e. when one or the other of the above assumptions breaks down (the barrier width decreases, the energy of the electrons comes closer to the vacuum level, a resonance in the reflection coefficient appears), the structures present in the LDOS of the free electrodes, progressively become broadened and shifted, and eventually transform into structures characteristic of a covalent (or ionocovalent) bond between the electrodes, which were called TILS in the literature.

iv) when the coupling strength between the electrodes increases, the perturbation treatment used in the transfer Hamiltonian approach, becomes unauthorized: its application in this limit yields not only quantitatively but sometimes qualitatively wrong results.

We believe that this formalism, although difficult to apply in realistic cases, provides a unifying frame to understand the various effects obtained from numerical calculations in the literature.

References

4.1   R.J. Behm, N. Garcia, H Rohrer (eds.): *Scanning Tunneling Microscopy and Related Methods* Nato ASI series, (Kluwer Dortrecht 1990)

4.2   R.J. Hamers, R.M. Tromp, J.E. Demuth: Phys. Rev. Lett. **56**, 1972–1975 (1986)

4.3   B. Reihl, J.K. Gimzewski, J.M. Nicholls, E. Tosatti: Phys. Rev. B **33**, 5770–5773 (1986)

4.4   J. Tersoff, D.R. Hamann: Phys. Rev. B **31**, 805–813 (1985)

4.5   J. Bardeen: Phys. Rev. Lett. **6**, 57–59 (1961)

4.6   J. Tersoff: Phys. Rev. B **39**, 1052–1057 (1989)

4.7   J. Tersoff: Phys. Rev. B **40**, 11990–11993 (1989)

4.8   G. Binnig, H. Rohrer, F. Salvan, C. Gerber, A. Baro: Surf. Sci. **157**, L373–378 (1985)

4.9   R.M. Feenstra, J.A. Stroscio, J. Tersoff, A.P. Fein: Phys. Rev. Lett. **58**, 1192–1195 (1987)

4.10 Y. Kuk, P. Silverman: Applied Phys. Lett. **48**, 1597–1599 (1986)

4.11 J. Tersoff: Phys. Rev. B **41**, 1235–1238 (1990)

4.12 R.M. Tromp, E.J. van Loenen, J.E. Demuth, N.D. Lang: Phys. Rev. B **37**, 9042–9045 (1988)

4.13 C. Noguera: J. de Phys. **50**, 2587–2599 (1989)

4.14 C. Caroli, R. Combescot, P. Nozières, D. Saint-James: J. Phys. C **4**, 916–929; C. Caroli, R. Combescot, D. Lederer, P. Nozières, D. Saint-James: J. Phys. C **4**, 2598–2610 (1971)

4.15 T.E. Feuchtwang: Phys. Rev. B **10**, 4121–4134 (1974)

4.16 R. Combescot: J. Phys. C **4**, 2611–2622 (1971)
4.17 A. Brodde, S. Tosch, H. Neddermeyer: J. Microsc. **152**, 441–448 (1988)
4.18 J.A. Stroscio, R.M. Feenstra, A.P. Fein: Phys. Rev. Lett. **57**, 2579–2582 (1986)
4.19 R.M. Feenstra, J.A. Stroscio, A.P. Fein: Surf. Sci. **181**, 295–306 (1987); J.A. Stroscio, R.M. Feenstra, D.M. Newns, A.P. Fein: J. Vac. Sci. Technol. A **6**, 499–507 (1988)
4.20 E. Tekman, S. Ciraci: Phys. Rev. B **40**, 10286–10293 (1989); and Ref A.1
4.21 F. Gautier, H. Ness, D. Stoeffler: Ultramicroscopy **42–44**, 91–96 (1992)
4.22 G. Doyen, E. Koetter, J.P. Vigneron, M. Scheffler: Appl. Phys. A **51**, 281 (1990)
4.23 N.J. Zheng, I.S.T. Tsong: Phys. Rev. B **41**, 2671–2677 (1990)
4.24 S. Ciraci, I. Batra: Phys. Rev. B **36**, 6194–6197 (1987)
4.25 S. Ciraci, A. Baratoff, I. Batra: Phys. Rev. B **41**, 2763–2775 (1990)
4.26 J. Ferrer, A. Martin-Rodero, F. Flores: Phys. Rev. B **38**, 10113–10115 (1988)
4.27 E. Kopatzki, G. Doyen, D. Drakova, R. Behm: J. Microsc. **152**, 687–695 (1988)
4.28 P. Sautet, C. Joachim: Ultramicroscopy **42–44**, 115–121 (1992)
4.29 C. Noguera: Phys. Rev. B **42**, 1629–1637 (1988)
4.30 C. Noguera: J. of Microsc. **152**, 3–9 (1988)
4.31 W. Sacks, C. Noguera: J. of Microsc. **152**, 23–33 (1988)
4.32 W. Sacks, C. Noguera: J. Vac. Sci. Technol. B **9**, (2) 488–491 (1991)
4.33 W. Sacks, C. Noguera: Phys. Rev. B **43**, 11612–11622 (1991)
4.34 W. Sacks, C. Noguera: Ultramicroscopy **42–44**, 140–145 (1992)
4.35 L.V. Keldysh: Sov. Phys. JETP **20**, 1018–1026 (1965)
4.36 J.E. Inglesfield: J. of Phys. C **4**, L14–17 (1971)
4.37 N.D. Lang: Phys. Rev. Lett. **55**, 230–233 (1985)
4.38 J. Chen: Phys. Rev. B **42**, 8841–8857 (1990)
4.39 J. Chen: Phys. Rev. Lett. **65**, 448–451 (1990)
4.40 N.D. Lang: Phys. Rev. Lett. **56**, 1164–1167 (1986); Phys. Rev. B **36**, 8173–8176 (1987)
4.41 W.A. Harrison: Phys. Rev. **123** n°1, 85–89 (1961)
4.42 J.P. Hurault: J. de Phys. **32**, 421–426 (1971)
4.43 J. Halbritter: Surf. Sci. **122**, 80–98 (1982)
4.44 B. Persson, A. Baratoff: Phys. Rev. Lett. **59**, 339–342 (987)
4.45 N. Isshiki, K. Kobayashi, M. Tsukada: Surf. Sci. L **238**, L 439–445 (1990)
4.46 M. Abramowicz, I. Stegun: *Handbook of Mathematical Functions* (Dover, New York 1972)
4.47 N.D. Lang: Phys. Rev. Lett. **58**, 45–48 (1987)
4.48 N.D. Lang: Phys. Rev. B **34**, 5947–5950 (1986)

5. The Role of Tip Atomic and Electronic Structure in Scanning Tunneling Microscopy and Spectroscopy

M. Tsukada, K. Kobayashi, N. Isshiki, S. Watanabe, H. Kageshima,
and *T. Schimizu*

With 20 Figures

Based on the first-principles Local Density Functional (LDA) calculation of the electronic states both for the tip and the sample surface, theoretical simulation of scanning tunneling microscopy and spectroscopy has been performed for various surface systems. For the tip, cluster models made of 10–20 atoms are utilized and for the sample surface slab models with several atomic layers are adopted. It is found that most of the tunnel current is concentrated on a single apex atom, if the other atoms on the top of the tip are not located on the same level. In such a case the STM image is normal with an atomic resolution. However, if the apex of the tip is formed by more than one atom, abnormal images tend to be formed. We can verify this feature by the numerical results for graphite, Si(100), and Si(111)/Ag surfaces. Due to the interplay between the tip and surface electronic states, some exotic behavior of electron tunneling can be observed in STM/STS. As examples we discuss the negative tunneling conductance observed in a nano-scale region, and the light emission from STM. These phenomena are explained based on realistic calculations of electronic states of the tip/sample system.

5.1 Background

The invention of Scanning Tunneling Microscopy (STM) has had an enormous impact on the basic research in surface science [5.1]. At the same time STM demonstrated a remarkable potentiality for a wide variety of application fields [5.2]. There are so many kind of reconstructed structures and chemisorption structures on ideal clean solid surfaces. The clarification of their atomistic structure as well as their electronic mechanism has been a central target of basic research in surface science. Introduction of STM opened a new evolutional stage of this sort of research field. A typical example of this is found in the clarification of the long-standing issue of the structure of the Si(111)7 × 7 surface [5.3]. Traditional experimental methods for the surface structure, except Field Ion Microscopy (FIM) mostly provided information of the wave number space, or information not specific to individual sites. In contrast, STM provides very detailed information about the direct local atomic structure of the surface. This is a tremendous advantage of STM compared with the other methods, since we can observe the surface structures without any periodicity. For example, the atomistic structure around a step, or the structure of various kind of defects can

Springer Series in Surface Sciences, Vol. 29
Scanning Tunneling Microscopy III Eds.: R. Wiesendanger · H.-J. Güntherodt
© Springer-Verlag Berlin Heidelberg 1993

be observed at the atomic level for the first time. Furthermore, STM provided the means for the study of the intermediate stage of a chemical reaction [5.4], or the crystal growth process at an atomic level.

Scanning Tunneling Spectroscopy (STS) [5.5] measures the tunnel spectrum at any desired position of the STM tip, and therefore provides information about the surface electronic structure resolved in real space on the atomic scale. So STS can be said to be a complementary method to the Angle Resolved Ultraviolet Photoemission Spectroscopy (ARUPS) which provides information on a surface electronic structure resolved in the wave number space [5.6]. This unique feature of STS is quite powerful for resolving the properties of the surface local structure.

In spite of this enormous potentiality, it is by no means trivial to reveal all the rich and profound information which STM/STS can, in principle, provide. The major reason comes from the fact that the tip and the sample surface interact directly. But the microscopic nature of the tip in the working condition cannot be known in usual cases. Probe-source problems cannot be ignored for the analysis of experimental data, in contrast to other methods using an electron beam source, such as LEED, or the conventional type electron microscopes. In fact, the STM image is a sort of convolution between the tip and the sample surface structure, and the STS spectrum is a convolution between the electronic structure of the sample surface and the tip. For this reason, one can easily understand the important role played by a reliable theoretical analysis based on electronic-state calculation in the understanding of the experimental data. Moreover a lot of exotic phenomena have been observed in the STM experiments. To mention some examples, light emission from STM [5.7, 8], local negative differential resistance [5.9, 10], standing-wave excitation [5.11], second-harmonic generation [5.12] and various others. These new phenomena are supposed to be significantly influenced by the microscopic state of the tip, and their clarification cannot be performed without the theory based on the microscopic electronic states.

Over the past few years we have developed a method of theoretical simulation of STM image and STS spectrum based on the first-principles Local Density Functional (LDA) theory [5.13, 14]. In our simulation we use cluster models for the tip, since its atomistic structure really governs the results. Results of simulation for many kinds of surfaces with various models for the tip revealed interesting effects of the tip on the experimental data, as well as the real microscopic mechanism of STM/STS.

In this chapter, we first point out some important features of STM/STS based on analytic arguments by the perturbation theory. In Sect. 5.2, a formalism of the first-principles theoretical simulation is provided. From Sect. 5.3 through 5.6, we discuss important features of the tip effect on STM/STS taking examples of various surface systems of current interest. In Sect. 5.7, we discuss the possibility of a microscope using light emission by STM.

5.2 Formalism of Theoretical Simulation of STM/STS

If one wants to calculate the tunnel current taking into account the real atomistic structure of both the tip and the sample surface, the perturbation theory of *Bardeen* [5.15] provides a reliable and still feasible method. The tunnel current can be expressed in this method as

$$I = \frac{2\pi e}{\hbar} \int dE [f(E) - f(E - eV)] A(\boldsymbol{R}, E, E - eV) , \tag{5.1}$$

$$A(\boldsymbol{R}, E, E') = \int_{\Omega_{\mathrm{T}}} d\boldsymbol{r} \int_{\Omega_{\mathrm{T}}} d\boldsymbol{r}' \, V_{\mathrm{T}}(\boldsymbol{r}) V_{\mathrm{T}}(\boldsymbol{r}') \mathscr{G}^{\mathrm{S}}(\boldsymbol{r} + \boldsymbol{R}, \boldsymbol{r}' + \boldsymbol{R}; E) \mathscr{G}^{\mathrm{T}}(\boldsymbol{r}', \boldsymbol{r}; E') , \tag{5.2}$$

where $\mathscr{G}^{\mathrm{S}}(\boldsymbol{r}, \boldsymbol{r}'; E)$ and $\mathscr{G}^{\mathrm{T}}(\boldsymbol{r}, \boldsymbol{r}'; E)$ are the imaginary part of the Green's function of the sample surface and tip, respectively. The energy density of the tunnel current is given in the framework of the Linear Combination of Atomic Orbitals (LCAO) method as

$$A(\boldsymbol{R}, E, E - eV) = \sum_{i, i'} \sum_{j, j'} \sum_{p, p'} \sum_{q, q'} G^{\mathrm{S}}_{ipi'p'}(E)$$
$$\times G^{\mathrm{T}}_{jqj'q'}(E - eV) J_{ipj'q'}(\boldsymbol{R}) J^{*}_{i'p'jq}(\boldsymbol{R}) . \tag{5.3}$$

Here, $G^{\mathrm{S}}_{ipi'p'}(E)$ and $G^{\mathrm{T}}_{jqj'q'}(E)$ are given by

$$G^{\mathrm{S}}_{ipi'p'}(E) = \sum_{\mu} C^{\mathrm{S}}_{\mu, ip} C^{\mathrm{S}*}_{\mu, i'p'} \delta(E - E_{\mu}) , \tag{5.4}$$

$$G^{\mathrm{T}}_{jqj'q'}(E) = \sum_{\nu} C^{\mathrm{T}}_{\nu, jq} C^{\mathrm{T}*}_{\nu, j'q'} \delta(E - E_{\nu}) , \tag{5.5}$$

and the tunnel matrix element $J_{ipjq}(\boldsymbol{R})$ is expressed as

$$J_{ipjq}(\boldsymbol{R}) = \int_{\Omega_{\mathrm{T}}} d\boldsymbol{r} \, \chi_p(\boldsymbol{r} - \boldsymbol{R} - \boldsymbol{R}_i) V_{\mathrm{T}}(\boldsymbol{r}) \psi^{*}_q(\boldsymbol{r} - \boldsymbol{R}_j) , \tag{5.6}$$

by the atomic basis functions of the surface $\chi_p(\boldsymbol{r})$ and of the tip $\psi_q(\boldsymbol{r})$.

The derivative of the tunnel current, (5.1), with respect to the bias V reads

$$\frac{dI}{dV} = \frac{2\pi e^2}{\hbar} A(\boldsymbol{R}, E_{\mathrm{F}} + eV, E_{\mathrm{F}}) - \frac{2\pi e^2}{\hbar} \int_{E_{\mathrm{F}}}^{E_{\mathrm{F}} + eV} dE \frac{\partial A(\boldsymbol{R}, E', E + eV)}{\partial E} \Big|_{E' = E} . \tag{5.7}$$

If the energy dependence of the tip electronic state is small enough, the second term in the r.h.s. can be ignored. In this case, if the first-order term of the moment expansion [5.13, 14] is performed, the relation

$$\frac{dI}{dV} \propto \varrho(\boldsymbol{R}, E_{\mathrm{F}} + eV) \tag{5.8}$$

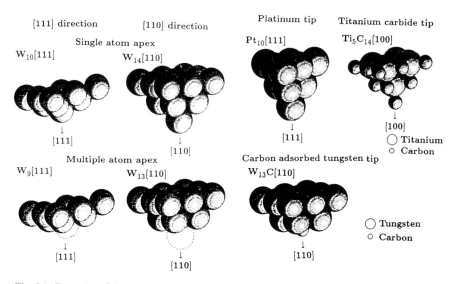

Fig. 5.1. Examples of the cluster models used in the simulation

holds. In our simulation of STM/STS we do not use such an approximation, but apply the whole complete expression of (5.1), (2). For this calculation it is necessary to obtain the Green's function of the tip and the sample surface, respectively. To obtain the tip Green's function, we utilize cluster models made of 10–20 atoms. Examples of cluster models for the tip are shown in Fig. 5.1. For the calculation of the surface electronic state, slab models made up of several atomic layers are conveniently utilized, but sometimes cluster models are used as well.

If the distance between the sample surface and the tip is rather long it is not necessarily appropriate to use the LCAO formalism for the description of the very tail part of the tip. The so-called Connected Vacuum Tail (CVT) method seems reliable and practical in such a case [5.16]. In this method, the space outside the surface is divided by two boundary layers into three regions; (a) far distant vacuum region, (b) intermediate region, and (c) outermost layer and inner region. Then the wave function in the region (a) is described by a linear combination of plane waves multiplied by the decaying exponential function

$$\Psi(k_{\parallel}; r) = \sum_{G_{\parallel}} A(G_{\parallel}) \exp^{[i(k_{\parallel} + G_{\parallel}) \cdot r_{\parallel}]}$$

$$\times \exp\left[-z \sqrt{(k_{\parallel} + G_{\parallel})^2 - \frac{2mE}{\hbar^2}} \right].$$ (5.9)

Here $r_{\parallel}$ and z are the parallel and the perpendicular component of the coordinates, and $G_{\parallel}$ is a reciprocal lattice vector of the surface structure.

On the other hand, in the inner region (c), the electron wave function can be obtained with any desired accuracy by the standard method of LDA, either by the LCAO method or the pseudopotential method. Then we can connect the wave functions of the two regions by solving the Schrödinger equation in the intermediate region.

In the actual calculation this can be performed by the so-called Laue method, as in the dynamical LEED calculation. We expand the wave function as

$$\Psi(k_{\parallel}; r) = \sum_{G_{\parallel}} F(G_{\parallel}; z) \exp^{i(k_{\parallel} + G_{\parallel}) \cdot r_{\parallel}} \tag{5.10}$$

where $k_{\parallel}$ is the crystal momentum of the wave function in the parallel direction to the surface and $F(G_{\parallel}, z)$ are the unknown coefficient functions. These functions can be calculated as the solution of the coupled differential equation obtained by the original Schrödinger equation. By connecting the wave function at the boundaries between the regions (a and b), and between (b and c) as smoothly as possible, we can determine the coefficient $A(G_{\parallel})$ in (5.9). If we fix the eigen-energy of the slab state as a whole, we cannot smoothly connect the wave function at the boundary of regions (b and c), but this causes no serious effect in most of the cases.

5.3 Simulation of STM/STS of the Graphite Surface

5.3.1 Normal Images

In this section, we discuss the effect of the atomic structure of the tip in some detail, taking the example of the graphite surface. We have found that by tips $W_{10}[111]$, $W_{14}[110]$ with a single apex atom the type of STM image, as shown in Fig. 5.2a, d, results [5.17]. In this figure the distance between the tip and the sample surface is fixed at 2.6 Å and the tip bias voltage is chosen to be 0.5 V. The result does not change much with the values of these parameters. This type of STM image agrees well with the experimental observation and can be regarded as a normal image of graphite for the reasons discussed below. The strong current region in these images is located on the B site of graphite, and forms a triangular lattice as a whole. Here the B site denotes one of the two graphite atom sites below which there is no carbon atom in the adjacent layer. The other type of atom site is called the A site. There are carbon atoms on the adjacent layer directly below the A site. The weakest current region is located at the center of the honeycomb structure.

The reason why the tunnel current concentrates around the B site is explained below. According to the band theory, the Bloch waves of the graphite with the energy around the Fermi level have a large amplitude around the B site and not on the A site. At the A site, because of the bonding and the antibonding interaction between the π orbitals of the outermost layer and the underlayer, the LDOS is swept out from the Fermi-level region. It should be pointed out, that at

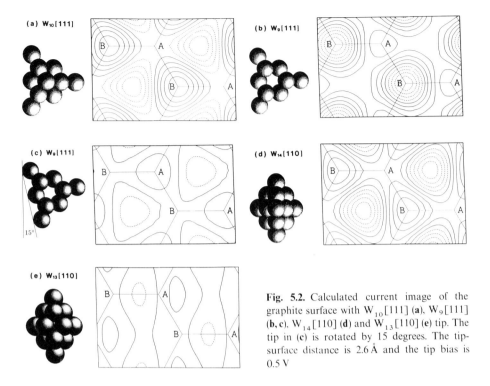

Fig. 5.2. Calculated current image of the graphite surface with $W_{10}[111]$ (**a**), $W_9[111]$ (**b, c**), $W_{14}[110]$ (**d**) and $W_{13}[110]$ (**e**) tip. The tip in (**c**) is rotated by 15 degrees. The tip-surface distance is 2.6 Å and the tip bias is 0.5 V

the center of the hexagon of the honeycomb structure the Bloch wave has the cross point of the nodes. Therefore an axial symmetric orbital of the tip cannot contribute to the tunnel current here. The marked difference between the STM image and the atomic structure of the graphite surface plane is a good lesson that we should bear in mind: the STM does not necessarily image the atomic structure of the surface as a naive ball model, but represents very delicate features of the surface wave functions in the relevant energy range.

5.3.2 Abnormal Images

If we remove the apex W atom from the $W_{10}[111]$ or $W_{14}[110]$ tip mentioned above, the resulting tip, $W_9[111]$ or $W_{13}[110]$, has several atoms on the top plane; the $W_9[111]$ tip has three top atoms while the $W_{13}[110]$ tip has four atoms on the top. The tips with several atoms on the top tend to form abnormal STM images. For example, Fig. 5.2b is obtained by the model $W_9[111]$ tip. Though the image of Fig. 5.2b forms a triangular lattice, the correspondence between the current distribution and the lattice structure is shifted from that for the normal image (Fig. 5.2a, d). In Fig. 5.2b the weakest current region is now located on the B site and not on the center of the hexagon. The strongest current region is moved to the A site. The reason for the shift of tunnel-current distribution is explained as follows. The total tunnel current flowing between the tip and the surface becomes maximum when the current component through each of the top atoms becomes maximum. In the geometry where the triangle formed by the three W atoms and that formed by the B site of graphite are arranged in the same direction, the top W atoms come very close to one of the B sites, if the axis of the tip is located on the A site.

This is the reason why the total tunnel current becomes maximum when the axis of the tip is located on the A site. Such a condition for the maximum of the tunnel current cannot be true if the tip is rotated slightly from the above geometry. Figure 5.2c shows the simulated STM image of graphite when the tip is rotated by 15 degrees from the geometry of Fig. 5.2b. The distribution of the tunnel current becomes significantly changed from that of Fig. 5.2b. Moreover, the magnitude of the variation of the tunnel current over the scan is found to be very much reduced. Therefore the rotation of the tip around its center axis causes a remarkable effect in the case of the tip with several atoms on top. On the other hand, for the tip with a single apex atom, such as $W_{10}[111]$ or $W_{14}[110]$, the STM image hardly changes when the tip is rotated around its axis. This difference in the behavior of the tips with the single apex atom and those tips with several top atoms indicates that the tunnel current is really concentrated on the single apex atom. We will verify this more explicitly by the tunnel current distribution over the tip later.

(a) $W_{10}[111]$

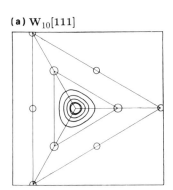

(b) $W_{14}[110]$

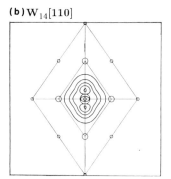

(c) $W_{13}[110]$

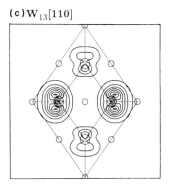

Fig. 5.3. Contour map of the charge density of **(a)** W_{10}, **(b)** W_{14}, and **(c)** W_{13} tip summed up over the energy range from $E_F - 0.5\,eV$ to $E_F + 1.0\,eV$. Charge density is calculated at the plane perpendicular to the symmetry axis and 1 atomic unit below the top

Next we will discuss the STM image shown in Fig. 5.2e obtained with the $W_{13}[110]$ tip which has four W atoms forming a rhombus on the top plane. The STM image of Fig. 5.2e is quite abnormal, flowing to the y direction, and bears little resemblance to the graphite lattice structure. The origin of the abnormal images is ascribed to an interference effect between the tunnel current components flowing to different W atoms on the top of the tip plane. To see it, the LDOS integrated in the energy region from $E_F - 0.5$ eV to $E_F + 1.0$ eV is shown in Fig. 5.3a, b, c for the $W_{10}[111]$, $W_{14}[110]$ and $W_{13}[110]$ tip, respectively. For the $W_{10}[111]$ and $W_{14}[110]$ tips which give the normal STM image, the LDOS is almost concentrated on the single apex W atom. On the other hand, for the $W_{13}[110]$ tip the distribution of the LDOS is split into two peaks on the two W atoms located on the short diagonal of the rhombus. Moreover, according to the detailed analysis, the wave function making the dominant contribution to the LDOS peaks changes its sign on the two atoms. This means that the current contributions by these two atoms almost cancel each other, and the remaining off-diagonal components between the two atoms become dominant, showing up as an abnormal image. This cancellation and the remaining off-diagonal current also depend on the relative geometry of the tip and the sample surface. Thus the abnormal image changes significantly on the rotation of the tip around its axis.

5.3.3 Effect of the Atom Kind of the Tip and the Tunnel Current Distribution

To investigate the effect of the atom kind of the tip, we calculate the STM image of graphite by model tips $Pt_{10}[111]$, $Ti_5C_{14}[001]$, $W_{13}C[110]$ and $W_{10}C_3[111]$ [5.18]. The results of the STM images are shown in Fig. 5.4. The tips $Pt_{10}[111]$, $Ti_5C_{14}[001]$ and $W_{13}C[110]$ have only a single atom on their top, and the resulting STM images with these tips correspond to the normal one. On the other hand, the STM image obtained with the $W_{10}C_3[111]$ tip is found to be abnormal. As shown in Fig. 5.4d, the B site corresponds to the weakest-current region while the center of the hexagon becomes the strongest-current region. From the detailed analysis of the tunnel current, the origin of the abnormal image is explained by the distributed tunnel current over tip atoms. The top plane of the tip is formed by a single W atom at the center and the three surrounding C atoms which are located at almost the same level.

To see the tunnel-current distribution over the atoms on the tip, we write the tunnel current (5.3) as

$$I = \sum_{jj'} I_{jj'} \tag{5.11}$$

$$I_{jj'} = \frac{2\pi e}{h} \int dE \{ f(E) - f(E - eV) \}$$

$$\times \sum_{\substack{ii' pp' \\ qq'}} G^S_{ipi'p'}(E) G^T_{jqj'q'}(E - eV) J_{ipj'q'} J^*_{i'p'jq} . \tag{5.12}$$

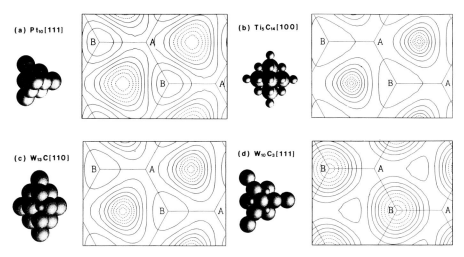

Fig. 5.4. Calculated current image of the graphite surface with (**a**) $Pt_{10}[111]$, (**b**) $Ti_5C_{14}[100]$, (**c**) $W_{13}C[110]$, (**d**) $W_{10}C_3[111]$ tip. The tip-surface distance is 2.6 Å and the tip bias is 0.5 V

Here I_{jj} is the diagonal-current contribution by the atom j and $I_{jj'}$ is the off-diagonal-current contribution over the atom j and j'. The values of the current components are listed in Table 5.1 for the $W_{10}[111]$, $W_{14}[110]$, $W_{13}C[110]$, $W_{10}C_3[111]$ and $Ti_5C_{14}[001]$ tips. For the $W_{10}[111]$, $W_{14}[110]$, $W_{13}C[110]$, and $Ti_5C_{14}[001]$ tips, more than 90% of the tunnel current is concentrated on the single apex atom. This feature does not depend on the atom kind of the apex atom. On the other hand, for the $W_{10}C_3[111]$ tip the tunnel current is distributed over the center W atom and the three C atoms. It is confirmed in this table that for the tip with a single apex atom, the tunnel current is concentrated on this atom as assumed before. This is the reason why we obtain atomic resolution in the STM image irrespective of the kind of atom.

5.4 STM/STS of Si(100) Reconstructed Surfaces

If the Si(100) surface would not be reconstructed, each top atom would have two dangling bonds and such a surface would be quite unstable. That the basic structure unit of the reconstructed Si(100) − 2 × 1 surface is a dimer of the top atoms, has been postulated by *Schlier* and *Farnworth* [5.19] as early as 1959. But firm confirmation had to await the direct observation of the surface structure by STM. The report of STM images of the Si(100) surface has attracted much interest by showing very clear direct images of the dimer rows [5.20, 21]. However, the beautiful STM images of dimer rows also provided a mysterious problem with this reconstructed surface. Other experiments, as well as theoret-

Table 5.1. Relative value of $I_{jj'}$ summed over j-th and j'-th atoms in the same layer when the tips are located just above the A-site of graphite. Other conditions are the same as for Fig. 5.2

(a) $W_{10}[111]$

atoms	top W	2nd W	3rd W
tungsten at the top	1	-0.0096	-0.0201
tungsten in the 2nd layer	-0.0096	0.102	-0.0067
tungsten in the 3rd layer	-0.0201	-0.0067	0.0122

(b) $W_{14}[110]$

atoms	top W	2nd W	3rd W
tungsten at the top	1	-0.0199	-1×10^{-5}
tungsten in the 2nd layer	-0.0199	0.0086	2×10^{-5}
tungsten in the 3rd layer	-1×10^{-5}	2×10^{-5}	4×10^{-7}

(c) $W_{13}C[110]$

atoms	top C	2nd W	3rd W
Carbon at the top	1	0.0018	-0.0003
tungsten in the 2nd layer	-0.0018	0.0044	1×10^{-5}
tungsten in the 3rd layer	-0.0003	1×10^{-5}	7×10^{-6}

(d) $W_{10}C_3[111]$

atoms	top W	top C	2nd W	3rd W
tungsten at the top	1	-0.847	-0.0375	0.0244
carbon at the top	-0.847	6.74	0.187	-0.0974
tungsten in the 2nd layer	-0.0375	0.187	0.132	-0.0209
tungsten in the 3rd layer	0.0244	-0.0974	-0.0209	0.0207

(e) $Ti_5C_{14}[111]$

atoms	top C	2nd Ti	2nd C	3rd Ti	3rd C
carbon at the top	1	-0.0006	-0.0015	-3×10^{-5}	-2×10^{-5}
titanium in the 2nd layer	-0.0006	9×10^{-5}	-2×10^{-6}	4×10^{-9}	2×10^{-8}
carbon in the 2nd layer	-0.0015	-2×10^{-6}	2×10^{-5}	-8×10^{-8}	-2×10^{-7}
titanium in the 3rd layer	-3×10^{-5}	4×10^{-9}	-9×10^{-8}	3×10^{-8}	8×10^{-9}
carbon in the 3rd layer	-2×10^{-5}	2×10^{-8}	-2×10^{-7}	8×10^{-9}	2×10^{-8}

ical calculations, indicated that the most stable structure of this surface must be the asymmetric dimer structure forming a c(4 × 2) lattice. For example, LEED observations [5.22] confirmed the transition of the surface structure from a disordered 2 × 1 to the ordered c(4 × 2) structure below a temperature of around 200 K. The ARUPS experiments [5.23] measured the dispersion of the surface energy band for the single domain c(4 × 2) surface, which agreed excellently with the theoretical prediction [5.24]. On the other hand, the dimer rows observed by STM in a wide defect-free terrace look like symmetric dimers [5.20, 21]. The asymmetric dimers are observed only at steps or around local defects.

What is the reason for the discrepancy between the STM image observation and the other experiments and theory? To investigate this problem, we have performed the theoretical simulation of STM/STS for various reconstructed surfaces of Si(100).

Figure 5.5 depicts the obtained STM image of the Si(100) − 2 × 1 symmetric dimer surface for a tip–surface distance of 2.6 Å and a tip bias of 0.5 V. Figure 5.5a was obtained with the tip model $W_{10}[111]$, while Fig. 5.5b displays the results of the tip model $W_9[111]$, introduced in the previous section [5.25]. The calculated images look rather similar to the ones experimentally found. However, in the simulated STM image with the $W_9[111]$ tip (Fig. 5.5b), a systematic deviation of the tunnel current map from the dimer row structure is found. The image of Fig. 5.5b is, therefore, an abnormal image caused by the distribution of tunnel current over the three W atoms on the top of the tip. The situation is similar to that for the case of graphite. The simulated image with the $W_{10}[111]$ tip (Fig. 5.5a) is considered as a normal image since the image is not significantly changed when the tip is rotated around its axis. However, it is interesting that the image somehow reflects the effect of the microscopic shape of the tip. For example, the contour lines of the current show a more corrugated feature in the left side of the dimer row than in the right hand side of the dimer row. This is

(a) symmetric $W_{10}[111]$ (b) $W_9[111]$

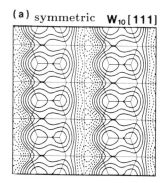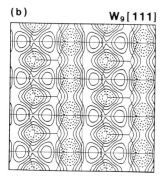

Fig. 5.5. Calculated current image of Si(100)2 × 1 symmetric dimer surface by the model tip (a) $W_{10}[111]$ and (b) $W_9[111]$. One edge of the basal triangle of the tip is placed parallel to the dimer row axis. The tip-surface distance is 2.6 Å and the tip bias is 0.5 V

because there is no mirror plane symmetry of the tip to the plane including the tip axis (z-axis) and the y-axis (the direction of the dimer row). A careful look at the experimental STM images sometimes reveals the similar feature as seen in Fig. 5.5a.

As for the occupied state, the simulation of the symmetric dimer surface seems to reproduce experimental STM images fairly well. However, for the unoccupied states, the correspondence between the theory and the experiment is not satisfactory. Figures 5.6a, b compare the simulated STM images in the gray scale representation for the occupied and the unoccupied state, respectively. Figure 5.6a is essentially the same as Fig. 5.5a. Figure 5.6b was obtained for the tip bias of − 0.5 V. Both images are quite similar to each other. However, the observed STM image is quite different between the occupied and the unoccupied states; individual atoms in the dimer cannot be seen separately for the occupied states, while it can be seen separately for the unoccupied states and, as a result, a very deep valley appears along the middle of the dimer row in the corrugation profile [5.21]. Considering the node lines existing for the antibonding π^* orbital of the dimer, the existence of the very deep lines along the middle of the dimer row for the unoccupied state and its absence for the occupied states would be rather natural.

Then why can our simulation not confirm this expectation? There are two reasons for this. First, the energy-band structure of the symmetric dimer Si(100) − 2 × 1 surface has metallic character. That is to say, the energy band of the bonding π state and that of the antibonding π^* state of the dimer crosses the Fermi level. This can be found in the simulated STS spectrum of the dimer surface, as shown in Fig. 5.7. Therefore, irrespective of the bias polarity, the tunnel current is contributed by both the bonding π and the antibonding π^* band. So the very clear node feature of the antibonding π^* state is smeared out.

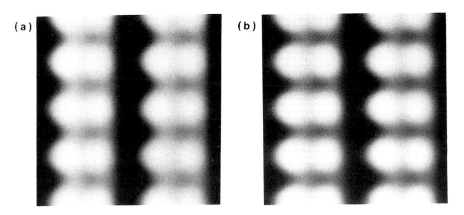

(a)   **(b)**

Fig. 5.6. Gray scale representation of the calculated current image of Si(100)2 × 1 symmetric dimer surface with the model tip W_{10}[111] for the tip bias **(a)** − 0.5 V and **(b)** 0.5 V. The tip-surface distance is 2.6 Å

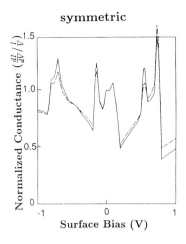

symmetric

Fig. 5.7. Calculated STS spectrum of Si(100)2 × 1 symmetric dimer surface with the model tip $W_{10}[111]$. Full line and dashed line correspond to sites above the dimer atom and the bond center, respectively. The tip-surface distance is 2.6 Å

Another reason is ascribed to the tip electronic state. As discussed in Sect. 5.2, the tip does not probe a pure LDOS of the sample surface. If the tunnel active orbital on the tip is made up of the s or d_{z^2} orbital of the top atom, the surface LDOS is more-or-less faithfully reproduced and the tunnel current becomes very weak at the nodal region of the relevant surface wave function.

However, if the tunnel active orbital of the tip includes p_x, p_y, d_{xz}, d_{xy}, or other types of symmetry components, the tunnel current is dominantly contributed by these components at the nodal region of the relevant surface wave function. Thus the nodal structure of the surface wave function will be smeared out [5.14]. Figures 5.8a, b, c exhibit the contour map of the total tunnel current, and that of the contribution from the 6s orbital component and that from the $5d_{z^2}$ component, respectively. Figure 5.8 was obtained by the CVT method mentioned in Sect. 5.2. Though the node along the dimer row can scarcely be seen for the total current map, the valley feature is recovered for the current map of the 6s or $5d_{z^2}$ orbital component. Note that the spatial resolution with the $5d_{z^2}$ orbital is higher than that with the 6s orbital component. This is coincident with the general theory by *Chen* [5.26].

The electronic structure of the symmetric dimer surface is metallic and not consistent with experiment. The simulated STS spectrum for the symmetric dimer surface (Fig. 5.7) does not reveal any gap at the zero-bias region. On the contrary it shows a peak of the tunnel current. On the other hand, the obtained STS spectrum for the c(4 × 2) asymmetric dimer surface reproduces the experiments as shown in Fig. 5.9a rather well. Figure 5.9a is calculated by the usual LCAO construction for the surface wave function. Almost the same results have been obtained by the calculation with the CVT method [5.16]. In the spectrum of Fig. 5.9a three characteristic peaks are labeled by 1, 2 and 3, respectively. These correspond to those of the experiment with the same label [5.21], as shown in Fig. 5.9b. It should be remarked that the peaks 1, 2 and 3 correspond to the upper band edge of the dimer π bond and the lower and the upper band

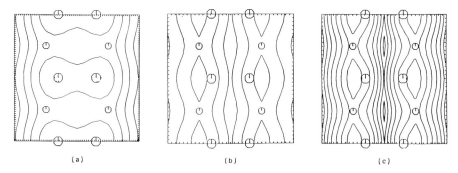

Fig. 5.8. The total current image for Si(100)2 × 1 symmetric dimer surface calculated with the W$_{10}$[111] model tip (**a**) and the current contribution by the 6s (**b**) and the 5d$_{z^2}$ (**c**) orbital of the top W atom. The tip-surface distance is 2.6 Å, and the tip bias is −1.0 V. The CVT method is used for the calculation

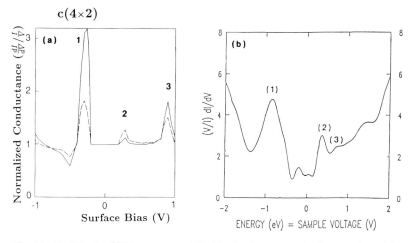

Fig. 5.9. (**a**) Calculated STS spectrum of Si(100)c(4 × 2) asymmetric dimer surface with the model tip W$_{10}$[111]. (**b**) Observed STS for Si(100)2 × 1 symmetric dimer surface by *Hamers* et al.

edge of the dimer π* bond, respectively. The puzzling problem is that the experimental spectrum is taken in the "apparently symmetric dimer" region, while the numerical simulation is performed for the asymmetric dimer surface. The agreement of the spectra may indicate that the "apparently symmetric dimer" is essentially an asymmetric dimer and by some unknown effect, such as the flip-flop oscillation of the dimer axis, is observed as symmetric. Another reason may be the tip-dimer interaction, e.g., the closest tip atom may always raise one of the dimer atoms, when the tip passes over that atom.*

One might ask whether an asymmetric dimer would be observed as a symmetric dimer. However, this possibility can be eliminated as far as we assume

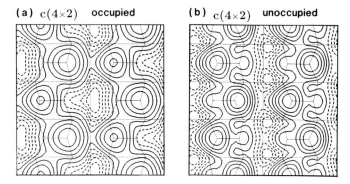

(a) $c(4\times2)$ occupied

(b) $c(4\times2)$ unoccupied

Fig. 5.10. Calculated current image of Si(100)c(4 × 2) asymmetric dimer surface with the model tip $W_{10}[111]$ for the tip bias **(a)** 0.5 V and **(b)** − 0.5 V. The tip-surface distance is 2.6 Å

a fixed structure. To confirm this, the simulated STM image for the asymmetric c(4 × 2) surface is displayed in Fig. 5.10 for the occupied state and the unoccupied state, respectively. As for the tip model, a $W_{10}[111]$ cluster is used. For the surface structure the modified Yin–Cohen model is assumed. The obtained zig-zag pattern is distinctively different from the observed STM image of the "apparently symmetric dimers". Rather these images of the asymmetric dimers reproduce the actually observed zig-zag pattern of the dimer row at a step or around a local defect fairly well.

5.5 The Negative-Differential Resistance Observed on the $Si(111)\sqrt{3}\times\sqrt{3}$-B Surface

The ideal surface of $Si(111)\sqrt{3}\times\sqrt{3}$-B itself shows an exotic structure, which cannot be seen in other chemisorbed systems. For this surface the chemisorbed B atom does not sit on the T_4 adatom site, as in the case of the other group III atoms such as Al, Ga, In. Experimental results as well as theoretical total energy calculations conclude that the B atom occupies the five-fold coordinated site in the third atomic layer (S_5 site) [5.27–29], as shown in Fig. 5.11. There is an adatom Si at the top layer directly above the B site. This structure seems to be more favored by the smallness of the B atomic radius, as compared to the Si atomic radius, than the usual T_4 chemisorption site. The electron originally associated with the adatom dangling bond is transferred to the B atom to complete four tetrahedral covalent bonds around the B atom. Thus the dangling bond of the top atom is deactivated and the Coulomb interaction of the ion pairs of the Si adatom cation and the B negative ion further stabilize the surface. Because of such a unique structure many interesting properties would be expected for this surface.

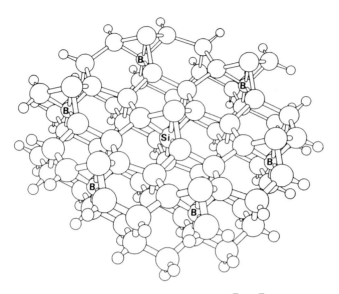

Fig. 5.11. Cluster model $Si_{78}B_6H_{51}$ for $Si(111)\sqrt{3} \times \sqrt{3}$-B-defect surface. The atoms denoted by B and Si are boron atoms at the S_5 sites and a silicon atom at the defect site, respectively. The hydrogen atoms are used to eliminate the dangling bonds of the peripheral atoms

A surprising property of this surface has been found by STS experiments by *Lyo* and *Avouris* [5.9], and *Bedrossian* et al. [5.10]. A negative tunnel resistance has been observed in the STS spectra, at randomly distributed irregular regions, the size of which is of the order of 10 Å. The regular STM image pattern with the $\sqrt{3} \times \sqrt{3}$ periodicity is destroyed in such a region. Therefore, the negative differential resistance is not the feature of the ideal $Si(111)\sqrt{3} \times \sqrt{3}$-B surface, but the feature associated with some defect site of this surface. The negative differential resistance of the tunnel current cannot be explained by the conventional view of STS based on the relation, (5.8), in which dI/dV cannot have a negative sign. If we ignore the structure of the LDOS of the tip and regard it as a constant, then we fail to understand this phenomenon. The appearance of the negative differential resistance is strongly correlated with the electronic structure of the tip.

To clarify the mechanism of the negative differential resistance we have made a theoretical simulation by adopting the following defect model of the $Si(111)\sqrt{3} \times \sqrt{3}$-B surface [5.30]. We introduced the substituted Si for the B atom site in the otherwise perfect $Si(111)\sqrt{3} \times \sqrt{3}$-B structure. To see the electronic state introduced by such a defect, the LDA calculation has been performed for the cluster model $Si_{78}B_6H_{51}$, (Fig. 5.11). For reference, the electronic structure of the cluster $Si_{77}B_7H_{51}$ without a defect has been calculated as well. Moreover, the energy bands have also been determined by slab models. The obtained LDOS of the cluster $Si_{78}B_6H_{51}$ at the adatom above the

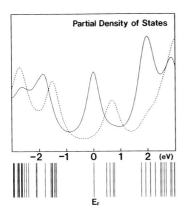

Fig. 5.12. Level structure and the Partial Density Of States (PDOS) of the cluster $Si_{78}B_6H_{51}$. Solid and broken lines are the PDOS at the adatom sites above the defect Si atom and the B atoms, respectively. The energy levels of the cluster around the Fermi level E_F are shown below the abscissa with the same energy scale

B site and above the defect site are shown in Fig. 5.12 by the dashed curve and the solid line curve, respectively. In this figure, the Fermi level position is taken as the energy origin. The conduction band and the valence band are situated 1.7 eV above and -1.1 eV below E_F, respectively. In the energy gap region there is a single peak of the LDOS both for the adatom above the B site and the adatom above the defect site. This peak corresponds to the dangling-bond state of the adatom. For the adatom above the B site, this dangling bond state is empty, but for the adatom above the defect site the dangling bond becomes occupied by a single electron since the peak position of the level is remarkably lowered towards the Fermi level. That this is not an artifact of the cluster model was confirmed by performing similar calculations for the cluster $Si_{77}B_7H_{51}$, whose central B site is not the defect but occupied by a B atom. In this cluster the LDOS of the adatom above the central B site is found to have a peak at the same position as in the LDOS of the other adatom site. The same conclusion is confirmed by the slab calculation. To summarize, there appears to be a special localized state around the adatom over the defect site, which is occupied by a single electron.

Various models of the tip are applied for the simulation of STM/STS of the cluster $Si_{78}B_6H_{51}$ [5.30]. As for the case of the $W_{10}[111]$ tip, the differential tunnel conductance at the adatom above the defect and that above the B atom is shown as a function of the tip bias in Fig. 5.13a and b, respectively. It is found that dI/dV becomes negative around $V = 2.2$ V, for the case of STS above the defect site. On the other hand, dI/dV is positive everywhere for the case over the regular B site. To see the spatial distribution of the values of the differential conductance, the contour map of dI/dV at $V = 2.2$ V is exhibited in Fig. 5.14. It is remarkable that the negative values of dI/dV appear only in a very localized region around the defect site. This feature resembles the experimental observation.

To investigate the mechanism of the negative differential conductance, the contribution of the tunnel current from each of the discrete levels of the tip state

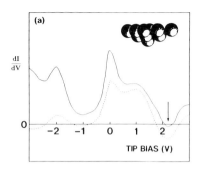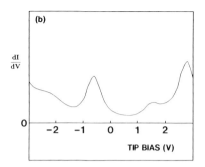

Fig. 5.13. Calculated tunnel conductance of Si(111)$\sqrt{3} \times \sqrt{3}$-B-defect surface using Si$_{78}$B$_6$H$_{51}$ and W$_{10}$[111] cluster models for the surface and the tip, respectively. The tip position is above (**a**) the defect site and (**b**) the boron site. Broken line in figure (**a**) is the contribution to the total conductance by the tip level L. At the upper right corner of (**a**), a W$_{10}$[111] cluster is shown

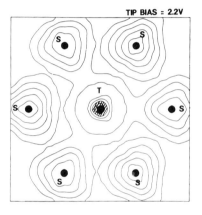

Fig. 5.14. Contour map of the tunnel conductance at the tip bias 2.2 V. Closed circles show the adatom positions. T and S are the adatom positions above the defect Si and the B atoms. Negative conductance region is shaded

is analyzed. It is found that there is one particular level, denoted hereafter, as level L, in the W$_{10}$[111] tip model which carries a large amount of the tunnel current. The contribution to the differential conductance dI/dV from this level is plotted with the dashed line in Fig. 5.13a. One may notice that the negative differential conductance behavior of this component is much larger than that of the total tunnel current. In fact, the negative-differential conductance behavior is due to the contribution by the particular level L alone, and the contributions by all the other levels are positive and tend to cancel the negative contribution from the level L. If the tunnel current is dominated by a single tip level, the tunnel current takes a maximum as a function of the bias voltage when it passes a peak of the LDOS of the surface. In this case, the negative differential resistance is caused by a sort of resonant tunneling between the special tip level and the peak structure of the surface LDOS. In the present system, the special tunnel active level in the tip is the level L, and the LDOS peak of the surface is caused by the localized defect level at the Fermi energy.

The active tunnel state L has an orbital well localized on the top W atom of the tip, which has a lobe extending towards the sample surface. The origin of the occurrence of such a state is not yet completely clarified. However, the existence or non-existence of such a state seems to depend sensitively on the atomic structure and atom kind of the tip. For example, there is no such kind of orbital for $W_{14}[110]$ or $Pt_{10}[111]$ tips.

5.6 The STM Image of the Si(111)$\sqrt{3} \times \sqrt{3}$-Ag Surface and the Effect of the Tip

In the past decade there has been much interest in clarifying the atomic structure of Si(111)$\sqrt{3} \times \sqrt{3}$-Ag surface. Because of the rather complicated structure of this surface, many different models have been proposed, but none of these models of this surface has yet been established. The STM images reported by *van Loenen* et al. [5.31], and *Wilson* and *Chiang* [5.32] were believed to open the way to the final solution. However, as discussed below, the situation is not as simple as at first expected.

A typical observed STM image for a negative tip bias consists of bright spots arranged in a honeycomb lattice (Fig. 5.15). Although these bright spots are often postulated to correspond to atoms located on the top layer, such an assumption is not necessarily correct, as will be shown later. However, even if we admit this assumption, there remains the problem of which kind of atom these bright spots correspond to. Since we cannot identify the atom kind from STM images, two types of models have been proposed for the Si(111)$\sqrt{3} \times \sqrt{3}$-Ag surface on the basis of the observed STM images: In one model [5.32] the bright spots in the observed STM images are assumed to correspond to Ag atoms

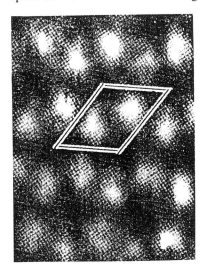

Fig. 5.15. Observed current image of the Si(1 1 1) $\sqrt{3} \times \sqrt{3}$-Ag surface for a bias voltage of $+ 1.0$ V applied to the sample (after van Loenen et al. [5.31])

which form the topmost layer, and in the other one [5.31] the topmost layer is formed by Si atoms with a honeycomb arrangement. However, for both types of models a serious problem is pointed out by *Nagayoshi* [5.33] from the viewpoint of the electronic state theory. He performed LDA calculations for both types of models and found that their surface electronic states come out metallic. This is in qualitative contradiction with the photoemission [5.34] and inverse-photoemission [5.35] experiments, which show that the Si(111)$\sqrt{3} \times \sqrt{3}$-Ag surface have a distinct energy gap around the Fermi level. Slight modification of the models cannot amend this qualitative disagreement.

Recently, *Katayama* et al. [5.36] proposed a new structural model, called the modified Honeycomb-Chained-Trimer (HCT) model, on the basis of experimental results from CoAxial Impact-Collision Ion Scattering Spectroscopy (CAICISS). This model, which is shown in Fig. 5.16, is consistent with almost all experimental results regarding the atomic structure. Therefore, it is a quite challenging and interesting problem to confirm whether this model is also consistent with the reported STM image and whether the surface electronic structure of this model can reproduce the observed insulating feature. *Watanabe* et al. [5.37] have performed LDA calculations of the electronic structure for the modified HCT model with the use of a slab model. The obtained energy-band dispersion (Fig. 5.17) indicates that this model yields an insulating character of the surface; a finite energy gap of the order of 0.6 eV opens up between occupied and unoccupied surface states. Furthermore, the calculated density of states for occupied bands shows good correspondence with the photoemission spectrum reported by *Yokotsuka* et al. [5.34]. In this way, the calculation by *Watanabe* et al. shows that the modified HCT model can reproduce the observed characteristics of the electronic structure of this surface. It should be mentioned that the theoretical total energy calculation performed by *Ding* et al. [5.38] also supports this model.

Since the atomic structure of the modified HCT model (Fig. 5.16) looks at first sight dissimilar to the honeycomb structure seen in Fig. 5.15, it is interesting

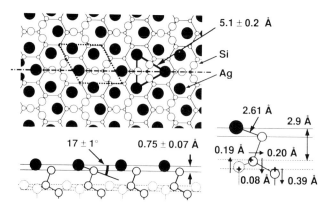

Fig. 5.16. Modified HCT model for the structure of the Si(111)$\sqrt{3} \times \sqrt{3}$-Ag surface

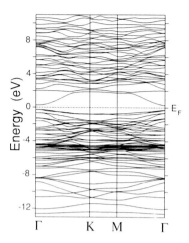

Fig. 5.17. Energy band dispersions of the $Si(111)\sqrt{3} \times \sqrt{3}$-Ag surface calculated for the modified HCT model

to make theoretical simulation of STM images based on the modified HCT model. Figure 5.18 depicts a typical current image for this model for a negative tip bias, which corresponds to the unoccupied surface states, simulated by using the tip model of $W_{10}[111]$ [5.37]. In this figure one can see bright spots with a honeycomb arrangement, which agree excellently with the reported experimental data. Then to what surface feature do the bright spots in the STM image correspond? The answer to this question is given by Fig. 5.19 which shows the calculated contour map of the tunneling current. As seen in this figure, each bright spot is located at the center of the triangle formed by three Ag atoms, and does not correspond to any atom. This is caused by the characteristic of the surface electronic state: It was found that each bright spot corresponds to the maximum amplitude of the lowest unoccupied state.

Thus far, we have discussed only the STM image for the unoccupied surface states. What is the situation for the occupied states? The situation seems complicated since several kinds of STM images have been reported for positive tip bias [5.31]. In one type of the reported images, bright spots form a reticulate pattern. Theoretical simulations using the tip model of either $W_{10}[111]$ or $W_{14}[110]$ can reproduce this type of image quite well [5.39]. By examining the simulated results in detail, it was found that a delicate variation of the brightness in the reticulate pattern is seen in the simulated current image. This fine structure has not been observed in the earlier experiments but has been confirmed by a very recent experiment [5.40]. Although the agreement between the observed and simulated images is quite good in this way, there remains a problem: As far as we use the cluster models of $W_{10}[111]$ and $W_{14}[110]$, we can obtain only this type of image for the occupied states, while other types of images have been observed, as mentioned earlier. The microscopic state of the tip seems to influence the observed STM images. Thus simulations with other types of tip models, $W_9[111]$ and $W_{13}[110]$, have also been performed. As in the case of the graphite surface discussed in Sect. 5.3, theoretical simulations by

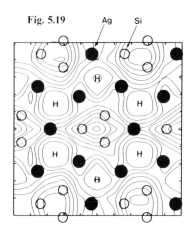

Fig. 5.18. Gray-scale image of the tunneling current in the logarithmic scale calculated with the tip model of $W_{10}[111]$. A bias voltage of -1.0 V was applied to the tip. The distance between Ag layer and the outermost W atom of the tip was 3.7 Å

Fig. 5.19. Contour map of the tunneling current in the logarithmic scale calculated under the same condition as for Fig. 5.18. Filled and open circles denote Ag and Si atoms, respectively, and H indicates the maxima of the tunneling current

using these tip models yield abnormal images [5.39]. Interesting enough, some of the abnormal images of the modified HCT model obtained by using these tip models are quite similar to some of the observed images reported by *van Loenen* et al. [5.31]. It might be worth mentioning that these simulated abnormal images gradually change to the normal one with tilting the tip axis from the surface normal [5.41].

To summarize this section we should say that the modified HCT model of the Si(111)$\sqrt{3} \times \sqrt{3}$-Ag surface explains all the experimental results including STM images. Variations of the STM images for the occupied states may result from the difference of the tip microscopic state.

5.7 Light Emission from a Scanning Tunneling Microscope

Light emission from an STM has been measured by *Coombs* et al. [5.7], *Takeuchi* et al. [5.8] and *Berndt* et al. [5.42] recently. From the viewpoint of application, the possibility of performing Scanning Tunneling Optical Microscopy (STOM) has attracted much interest. In STOM one monitors the intensity of emitted light during the scan of the tip over the surface. Since the spectrum of the emitted light is characteristic of the kind of the surface atom one would expect that STOM performed with different photon energy could provide images of the surface structure discriminating each atom kind. Thus it should be possible to color atoms in STOM. The microscopic mechanism of light emission still seems unsettled. It is quite plausible that there are some different mechanisms depending on the experimental condition and the surface system. Two different mechanisms have been proposed so far, i.e., the inelastic tunneling process and the minority carrier injection process. In the former process, an electron emits a plasmon on tunneling through the vacuum gap between the tip and the surface. The generated plasmon is then converted into light [5.8, 43]. The latter process takes place in the deeper surface region of a semiconductor [5.44]. The minority carriers injected by the STM tip recombine with the majority carriers with conversion of excess energy into photons. Since the possibility of observing the surface atomistic structure is provided by the former inelastic tunneling process, we will focus on it in this section and explore a theoretical prediction for a STOM.

The coupling between the tunneling electron and the surface plasmon is given by [5.43]

$$H' = \sum_\lambda \phi_\lambda^{\mu,\,\nu}(b_\lambda + b_\lambda^\dagger)a_\mu^\dagger c_\nu + \text{h.c.} \ . \tag{5.13}$$

Here, b_λ, a_μ and c_ν are the annihilation operator of plasmon of the λ-th mode, electrons of the μ-th tip state, and electrons of the ν-th sample state, respectively. The coupling matrix element $\phi_\lambda^{\mu,\,\nu}$ can be written as

$$\phi_\lambda^{\mu,\,\nu} = \frac{i\hbar}{e(\varepsilon_\mu - \varepsilon_\nu)} \int d\mathbf{r}\, \mathbf{E}_\lambda(\mathbf{r}) \cdot \mathbf{J}_{\mu\nu}(\mathbf{r}) \ , \tag{5.14}$$

where ε_μ and ε_ν are the eigenenergies corresponding to the wave functions χ_μ and ψ_ν, respectively. $\mathbf{E}_\lambda(\mathbf{r})$ is the field of the λ-th normalized plasmon and $\mathbf{J}_{\mu,\,\nu}(\mathbf{r})$ is the transition current density:

$$\mathbf{J}_{\mu\nu}(\mathbf{r}) = \frac{ie\hbar}{2m} \left[\chi_\mu^*(\mathbf{r})\nabla\psi_\nu(\mathbf{r}) - \psi_\nu(\mathbf{r})\nabla\chi_\mu^*(\mathbf{r})\right] \ , \tag{5.15}$$

If the plasmon field is almost constant, i.e., $\mathbf{E}_\lambda(\mathbf{r}) = \mathbf{E}_\lambda(\mathbf{0})$ in the narrow gap, the coupling constant is written as

$$\phi_\lambda^{\mu,\,\nu} \cong \frac{i\,\mathbf{E}_\lambda(\mathbf{0}) \cdot \mathbf{l}\hbar}{e(\varepsilon_\mu - \varepsilon_\nu)} \times \mathbf{J}_{\mu\nu} \ . \tag{5.16}$$

In the above, $\mathbf{l}$ is the vector of the closest tunneling path between the surface and the tip, and the surface integral $\mathbf{J}_{\mu\nu}$ of the current density $\mathbf{J}_{\mu\nu}(\mathbf{r})$ is performed over the dividing surface between the tip and the surface.

After some algebra we can show that the power spectrum of the emitted light is given by

$$I(\mathbf{R}, V, \omega) = \sum_\lambda \hbar\omega_\lambda \Gamma_\lambda \delta(\omega - \omega_\lambda)$$

$$= A(\omega) \sum_{iji'j'} \int_{E_F + \hbar\omega}^{E_F + eV} \sum dE\, G_{ii'}^S(E) G_{jj'}^T(E - eV + \hbar\omega) J_{ij}(\mathbf{R}) J_{i'j'}^*(\mathbf{R}) \ , \tag{5.17}$$

where $A(\omega)$ is a macroscopic factor

$$A(\omega) = \frac{4\pi}{e^2} \sum_\lambda \delta(\omega^2 - \omega_\lambda^2)|\boldsymbol{l} \cdot \boldsymbol{E}_\lambda(\boldsymbol{0})|^2 \; . \tag{5.18}$$

$G_{ii'}^S$ and $G_{jj'}^T$ are the imaginary parts of the surface and the tip Green's function, respectively, and J_{ij} is the tunneling matrix element, (5.6). The macroscopic

Table 5.2. Table of the contrast parameter. Range of the tip bias V and the photon energy $\hbar\omega$ is from -1.0 V to -3.0 V and from 0.0 eV to 2.71 eV

V_T \ $\hbar\omega$	0.0	0.42	0.71	1.42	1.86	2.14	2.71
-1.0	0.17	0.13	0.14				
-2.0	0.16	0.17	0.18	0.20	0.13		
-3.0	0.09	0.08	0.09	0.06	0.06	0.09	0.13

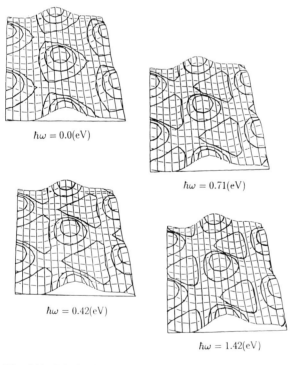

$\hbar\omega = 0.0(\mathrm{eV})$

$\hbar\omega = 0.71(\mathrm{eV})$

$\hbar\omega = 0.42(\mathrm{eV})$

$\hbar\omega = 1.42(\mathrm{eV})$

Fig. 5.20. Calculated Scanning Tunneling Optical Microscopy (STOM) image of the Ag(111) surface for various photon energies. The tip-surface distance is 3.2 Å, and the tip bias is -2.0 V. The model of the tip is $W_{10}[111]$

factor is determined by the coarse grained shape of the tip and the sample surface. On the other hand, the remaining factor in (5.17) is governed by the microscopic structure and the electronic states of the tip and the surface. Since the macroscopic factor changes very slowly during the scan of the tip, we can discriminate the modification of emitted light intensity on an atomic scale. If this variation is relatively large, there is enough possibility that one can obtain an atomistic image of the surface by STOM. In Table 5.2, we list the contrast parameter

$$p(V, \omega) \equiv \frac{I_{max}(V, \omega) - I_{min}(V, \omega)}{I_{max}(V, \omega)} \tag{5.19}$$

for Ag(111) surface for various photon energies and bias conditions. I_{max} and I_{min} are the maximum and the minimum intensity during the scanning of the tip at a fixed bias voltage and photon energy. These values are comparable to the corresponding values of an usual STM. In Fig. 5.20, we display the theoretical image of STOM on the Ag(111) surface for various photon energies. The atomic structure of this surface is well represented by the photon intensity map. Since the electronic structure of the Ag surface has no characteristic structure, we have no prominent structure of the spectra of emitted light as a function of the bias voltage and the photon energy. However, for some adsorbed metal surfaces the spectra would show quite an interesting behavior from which detailed information on the microscopic surface states could be deduced.

5.8 Summary and Future Problems

In the present chapter, we proposed the method of numerical simulations of STM and STS based on first-principles electronic state theory, both for the sample surface and the tip. The STM images of some interesting surface systems have been simulated by this method using realistic cluster models for the tip. The most significant finding is that the tunnel current is dominantly concentrated on a single apex atom of the tip. This might be the reason why an STM image with atomic lateral resolution can be obtained even by a tip with a radius of curvature on the order of several hundreds of Ångstroms. At an atomistic level, the closest point of the tip to the surface is usually formed by a single atom, so it is quite natural to think about the tip formed by a single apex atom. Important findings observed in the results of the theoretical simulation are that abnormal images are obtained if several atoms are located on the top plane of the tip.

The interpretation of the STM image and the STS spectrum is often not trivial, as demonstrated for the case of the graphite and the Si(111)$\sqrt{3} \times \sqrt{3}$-Ag surface. The experimental data provide rich and detailed information of the surface electronic structure in the relevant energy region. To resolve this profound information, theoretical simulation can play an essential role. Moreover, the theory based on realistic first-principles calculations is also indispensable for the analysis and prediction of many kinds of exotic phenomena. In the present article we discussed the example of the negative tunnel conductance localized on an atomic scale and the light emission from the STM. There are many other interesting phenomena observed in STM/STS, which are characteristic of the nano-tunnel system, such as single electron tunneling processes, standing wave excitation, moiré pattern-like STM images and others. Theoretical work on these problems has recently been performed by our group and will be presented elsewhere. An important target for the theory of STM/STS is to provide a predictable basis for the atomic scale manipulation of surfaces. In spite of various difficulties, we might expect a lot of challenging work to appear in the near future.

* Note Added in Proof

Recently observed STM images by *Wolkow* [5.45] at low temperatures showed the zigzag pattern of the asymmetric dimers and confirmed that the flip-flop oscillation is really the reason of the "apparently symmetric dimer".

References

5.1 G. Binnig, H. Rohrer: Helv. Phys. Acta **55**, 726 (1982)
5.2 P. Hansma, J. Tersoff: J. Appl. Phys. **61**, R1 (1987)
5.3 K. Takayanagi, Y. Tanishiro, M. Takahashi, S. Takahashi: Surf. Sci. **164**, 367 (1985)
5.4 R. Wolkow, Ph. Avouris: Phys. Rev. Lett. **60**, 1049 (1988)
5.5 R. M. Feenstra, J.A. Stroscio, A.P. Fein: Surf. Sci. **12**, 95 (1987)
5.6 B. Feuerbach, B. Fitton, R.F. Willis (eds.): *Photoemission and the Electronic Properties of Surfaces* (Wiley, New York 1978)
5.7 J.H. Coombs, J.K. Gimzewski, B. Reihl, J.K. Sass, R.R. Schlittler: J. Microsc. **152**, 325 (1988)
5.8 K. Takeuchi, Y. Uehara, S. Ushioda, N. Morita: J. Vac. Sci. Technol. B **9**, 557 (1991)
5.9 I.W. Lyo, Ph. Avouris: Science **245**, 1369 (1989)
5.10 P. Bedrossian, D.M. Chen, K. Mortensen, J.A. Golovchenko: Nature **342**, 259 (1989).
5.11 R.S. Becker, J.A. Golovchenko, D.R. Hamann, B.S. Swartzentruber: Phys. Rev. Lett. **55**, 2032 (1985)
5.12 W. Krieger, T. Suzuki, M. Völcker, H. Walther: Phys. Rev. B **41**, 10229 (1990)
5.13 M. Tsukada, N. Shima: J. Phys. Soc. Jpn. **56**, 2875 (1987)
5.14 M. Tsukada, K. Kobayashi, N. Isshiki, H. Kageshima: Surf. Sci. Rep. **13**, 265 (1991)
5.15 J. Bardeen: Phys. Rev. Lett. **6**, 57 (1961)
5.16 M. Tsukada, K. Kageshima, N. Isshiki, K. Kobayashi: Surf. Sci. **266**, 253 (1992); H. Kageshima, M. Tsukada: Phys. Rev. B **46**, 6928 (1992)
5.17 N. Isshiki, K. Kobayashi, M. Tsukada: Surf. Sci. **238**, L439 (1990)
5.18 N. Isshiki, K. Kobayashi, M. Tsukada: J. Vac. Sci. Technol. B **9**, 475 (1991)
5.19 R.E. Schlier, H.E. Farnsworth: J. Chem. Phys. **30**, 917 (1959)
5.20 R.J. Hamers, R.M. Tromp, J.E. Demuth: Phys. Rev. B **34**, 5343 (1986)
5.21 R.J. Hamers, Ph. Avouris, F. Bozso: Phys. Rev. Lett. **59**, 2071 (1987)

5.22 T. Tabata, T. Aruga, Y. Murata: Surf. Sci. **179,** L63 (1986)
5.23 T. Enta, S. Suzuki, S. Kono: Phys. Rev. Lett. **53,** 2704 (1990)
5.24 Z. Zhu, N. Shima, M. Tsukada: Phys. Rev. B **40,** 11868 (1989)
5.25 N. Isshiki, H. Kageshima, K. Kobayashi, M. Tsukada: Ultramicroscopy **42–44,** 109 (1992)
5.26 C.J. Chen: J. Vac. Sci. Technol. A **6,** 319 (1988)
5.27 P. Bedrossian, R.D. Meade, K. Mortensen, D.M. Chen, J.A. Golvochenko, D. Vandebilt: Phys. Rev. Lett. **63,** 1257 (1989)
5.28 In-Whan Lyo, E. Kaxiras, Ph. Avouris: Phys. Rev. Lett. **63,** 1261 (1989)
5.29 E. Kaxiras, K.C. Pandey, F.J. Himpsel, R.M. Tromp: Phys. Rev. B **41,** 1262 (1990)
5.30 M. Tsukada, K. Kobayashi, N. Shima, N. Isshiki: Proc. 20th Int'l. Conf. on the Physics of Semiconductors, Ed. by E.M. Anastassaki, J.D. Joannopoulos (World Scientific, Singapour 1990) pp. 171
5.31 E.J. van Loenen, J.E. Demuth, R.M. Tromp, R.J. Hamers: Phys. Rev. Lett. **58,** 373 (1987); J.E. Demuth, E.J. van Loenen, R.M. Tromp, R.J. Hamers: J. Vac. Sci. Technol. B **6,** 18 (1988)
5.32 R.J. Wilson, S. Chiang: Phys. Rev. Lett. **58,** 369 (1987); R.J. Wilson, S. Chiang: Phys. Rev. Lett. **59,** 2329 (1987)
5.33 H. Nagayoshi: J. Phys. Soc. Jpn. **55,** 307 (1986); Solid State Phys. **24,** 371 (1989) (in Japanese)
5.34 T. Yokotsuka, S. Kono, S. Suzuki, T. Sagawa: Surf. Sci. **127,** 35 (1983)
5.35 J.M.. Nicholls, F. Salvan, B. Reihl: Phys. Rev. B **34,** 2945 (1986)
5.36 M. Katayama, R.S. Williams, M. Kato, E. Nomura, M. Aono: Phys. Rev. Lett. **66,** 2762 (1991)
5.37 S. Watanabe, M. Aono, M. Tsukada: Phys. Rev. B **44,** 8330 (1991); S. Watanabe, M. Aono, M. Tsukada: Ultramicroscopy **42–44,** 105 (1992)
5.38 Y.G. Ding, C.T. Chan, K.M. Ho: Phys. Rev. Lett. **67,** 1454 (1991)
5.39 S. Watanabe, M. Aono, M. Tsukada: Appl. Surf. Sci. **60/61,** 437 (1992)
5.40 K.J. Wan, X.F. Lin, J. Nogami: Phys. Rev. B **45,** 9509 (1992)
5.41 S. Watanabe, M. Aono, M. Tsukada: Jpn. J. Appl. Phys. submitted
5.42 R. Berndt, A. Baratoff, J.K. Gimzewski: J. Vac. Sci. Technol. B **9,** 573 (1991)
5.43 T. Schimizu, K. Kobayashi, M. Tsukada: Appl. Surf. Sci. **60/61,** 454 (1992)
5.44 D.L. Abraham, A. Veider, Ch. Schonenberger, H.P. Meier, D.J. Arent: Appl. Phys. Lett. **56,** 1564 (1988)
5.45 R.A. Wolkow, Phys. Rev. Lett. **68,** 2636 (1992)

6. Bohm Trajectories and the Tunneling Time Problem

C.R. Leavens and G.C. Aers

With 21 Figures

Although many approaches based on conventional interpretations of quantum mechanics have been developed for calculating the average time taken for an electron to tunnel through a potential barrier, a satisfactory solution remains elusive. These approaches are discussed very briefly, focussing on the question of whether the concept of 'tunneling time' or, more generally, 'mean transmission time' is a meaningful one. Then it is shown that Bohm's causal or trajectory interpretation provides a well-defined and unambiguous prescription for calculating transmission times that are conceptually meaningful *within that interpretation*. Results of such calculations are presented for single and double rectangular barriers. The time-modulated rectangular barrier is treated in detail to emphasize the importance of considering distributions of transmission and reflection times and not just the mean transmission time. Finally, the possibility of determining tunneling times experimentally is discussed.

6.1 Background

6.1.1 Motivation

In the theoretical analysis of a complex dynamical system, comparison of the various time scales involved often motivates and/or justifies useful approximations. When tunneling through a potential barrier plays a key role it is often assumed that the average time for an electron to tunnel through the barrier, i.e. the tunneling time τ_T, is an important time scale. In the field of scanning tunneling microscopy, such an assumption has been made in investigations of the dynamical image potential [6.1, 2], the generation of d.c. current by laser illumination of the tip–sample junction [6.3, 4], the anomalously low barrier for tunneling through aqueous solutions [6.5], and the effect of atomic vibrations on image resolution [6.6]. The tunneling time of interest in the above context is an intrinsic property, i.e. a property of the undisturbed system. If an external clock were coupled to the system to 'measure' τ_T directly then the observed tunneling time might be so strongly perturbed by interaction with the measuring apparatus as to be of little use in the comparison of relevant time scales for the unperturbed system. Proceeding indirectly by using the dynamics of the undisturbed system as an internal clock, one might infer the order of magnitude of the

Springer Series in Surface Sciences, Vol. 29
Scanning Tunneling Microscopy III Eds.: R. Wiesendanger · H.-J. Güntherodt
© Springer-Verlag Berlin Heidelberg 1993

intrinsic tunneling time by the success of approximations based on its magnitude relative to the other time scales. This would involve comparison of the approximate results with accurate experimental or theoretical ones. However, one should bear in mind that the most basic quantities are the transmission and reflection time *distributions* not just their average values. One should not automatically assume for any phenomenon involving tunneling that the average transmission time is the relevant time scale nor, for that matter, that the entire width of the barrier is the only length scale of importance in determining the temporal characteristics. With this in mind we adopt a more general point of view in the following.

6.1.2 Defining the Problem

To define the quantities of interest in the 'tunneling time problem' [6.7–12] consider an ensemble of scattering experiments in each of which an electron with the same initial wave function $\psi(z, t = 0)$ is incident normally from the left on the potential barrier $V(\mathbf{r}, t) = V(z, t)\theta(z)\theta(d - z)$ which varies only in the z direction.[1] The mean transmission time $\tau_T(z_1, z_2)$ is defined as the average time spent in the region $z_1 \leq z \leq z_2$ subsequent to $t = 0$ by those electrons that are ultimately transmitted. Only a fraction of the ensemble members, equal to the transmission probability $|T|^2$, are involved in the average determining τ_T. The corresponding mean reflection time $\tau_R(z_1, z_2)$ is defined as the average time spent in the region $z_1 \leq z \leq z_2$ by those electrons that are ultimately reflected and involves only the remaining fraction $|R|^2 \equiv 1 - |T|^2$ of the ensemble members. Finally, the mean dwell time $\tau_D(z_1, z_2)$ is the average time spent between z_1 and z_2 by an electron irrespective of whether it is eventually transmitted or reflected. Clearly, since τ_D involves all members of the ensemble, the relation

$$\tau_D(z_1, z_2) = |T|^2 \tau_T(z_1, z_2) + |R|^2 \tau_R(z_1, z_2) \tag{6.1}$$

must hold. Although the positions z_1 and z_2 are arbitrary, except for the restriction $z_1 \leq z_2$, the case of usual interest is $z_1 = 0$ and $z_2 = d$ for the abrupt barriers considered here. The so-called 'tunneling time' or 'traversal time' is then the quantity $\tau_T(0, d)$. Strictly speaking, the former term should be used only when the wave-packet and barrier parameters are such that over-the-barrier propagation makes a negligible contribution to the transmission probability. Moreover, for sloping rather than abrupt barriers the classical turning points depend on energy and the spatial extent of the tunneling region is consequently blurred for a wave packet with finite energy width. Hence, the phrase 'tunneling time' will usually be avoided in the remainder of this chapter.

---

[1] It should be emphasized that in this chapter we do not consider the related and much better understood problem of the escape time for a particle prepared in a metastable state of a potential well.

The mean transmission time $\tau_T(z_1, z_2)$ was defined in words rather than by a mathematical expression, e.g. the expectation value of a Hermitean operator $\hat{\tau}_T(z_1, z_2)$, essentially because in quantum mechanics time is regarded as a parameter or c-number, not as a dynamical observable represented by such an operator. Hence, there is no automatic prescription in the basic formalism for calculating τ_T and the other characteristic times τ_R and τ_D. One might therefore ask if these quantities are meaningful concepts. It has been argued [6.13] that they are not because they imply the existence of microscopically well-defined particle trajectories. The latter concept is, of course, expressly forbidden in conventional interpretations because it is impossible, even in principle, to observe such a trajectory due to the position-momentum uncertainty relation. Another argument [6.13] against transmission and reflection times being meaningful is that, within conventional interpretations, it is impossible to divide the probability density $|\psi(z, t)|^2$ inside the barrier into 'to be transmitted' and 'to be reflected' components. In our opinion, as discussed in the next section, this negative response to the above question is at least a consistent one. On the other hand, within *Bohm*'s causal or trajectory interpretation of quantum mechanics [6.14–16] the notion of a precisely well-defined particle trajectory is not only a meaningful concept but a central one. Moreover, it is possible to divide $|\psi(z, t)|^2$ inside the barrier into 'to be transmitted' and 'to be reflected' components. Consequently, as discussed in Sect. 6.3, Bohm's interpretation leads to a unique, well-defined prescription for calculating meaningful transmission, reflection and dwell times. This prescription is applied to a number of simple systems in Sect. 6.4. Finally, in Sect. 6.5, the problem of experimentally determining mean transmission and reflection times is discussed from the Bohm trajectory and conventional points of view.

6.2 A Brief Discussion of Previous Approaches

Since all of the theoretical results for τ_T and τ_R discussed in this section can be simply related to those obtained with *Feyman* path integral techniques [6.17] by *Sokolovski* and coworkers [6.18–20] we focus on their approach. The starting point is the classical expression

$$t^{cl}[z_1, z_2; z(t)] = \int_0^\infty dt\, \theta[z(t) - z_1]\, \theta[z_2 - z(t)] \tag{6.2}$$

for the time spent subsequent to $t = 0$ in the region $z_1 \leq z \leq z_2$ by a point particle following its classical trajectory $z(t)$. They generalize this expression for the dwell time to the quantum regime by replacing $z(t)$ in the above functional by a Feynman path $z(\bullet)$ and then averaging over all such paths according to the Feynman prescription, i.e.

$$\tau_D(z_1, z_2) = \langle \psi^*(z', t_\infty) t^{cl}[z_1, z_2; z(\bullet)]\, \psi(z, 0)\rangle_F / \langle \psi^*(z', t_\infty)\psi(z, 0)\rangle_F \ . \tag{6.3}$$

Here $\langle \cdots \rangle_F$ denotes an average with weight factor $\exp[iS(z(\bullet))/\hbar]$ over all paths $z(\bullet)$ joining the space-time points $(z, 0)$ and (z', t_∞) followed by an average over all z and z'. $S[z(\bullet)]$ is the classical action and the time t_∞ must be large enough that the scattering process is essentially complete. Evaluation of (6.3) leads to the real, non-negative quantity

$$\tau_D(z_1, z_2) = \int_0^\infty dt \int_{z_1}^{z_2} dz \, |\psi(z, t)|^2 , \qquad (6.4)$$

which many, including the present authors, regard as the only firm result in the field. For the special case of an 'incident' plane-wave $\exp(ikz)$ of wavenumber k it can be shown [6.21] that (6.4) becomes

$$\tau_D(k; z_1, z_2) = \frac{1}{j_k^{(i)}} \int_{z_1}^{z_2} dz \, |\psi_k(z)|^2 , \qquad (6.5)$$

a result first postulated by *Büttiker* [6.22]. Here $j_k^{(i)} = \hbar k/m$ is the incident probability current density and $\psi_k(z)$ is the stationary-state wave function.

At sufficiently large times t_∞ the wave function can be written to a very good approximation as the sum of reflected and transmitted components, i.e. $\psi(z', t_\infty) = \psi_R(z', t_\infty) + \psi_T(z', t_\infty)$. Based on this, *Sokolovski* and *Connor* [6.19] determine the mean transmission and reflection times τ_T and τ_R by replacing $\psi^*(z', t_\infty)$ in (6.3) by $\psi_T^*(z', t_\infty)$ and $\psi_R^*(z', t_\infty)$, respectively. The resulting expressions for τ_T and τ_R are, in general, complex-valued quantities. For an 'incident' plane-wave they take the simple form

$$\tau_T^{SB}(k; z_1, z_2) = i\hbar \int_{z_1}^{z_2} dz \, \frac{\delta \ln T(k)}{\delta V(z)} , \quad \tau_R^{SB}(k; z_1, z_2) = i\hbar \int_{z_1}^{z_2} dz \, \frac{\delta \ln R(k)}{\delta V(z)} , \quad (6.6)$$

derived by *Sokolovski* and *Baskin* [6.18]. Here $T \equiv |T| \exp(i\phi_T)$ and $R \equiv |R| \exp(i\phi_R)$ are the transmission and reflection probability amplitudes. The imaginary part of $|T|^2 \tau_T^{SB} + |R|^2 \tau_R^{SB}$ is exactly zero and the real part is exactly equal to τ_D so that the sum-rule (6.1) is satisfied [6.18]. For the special case of perfect transmission ($|T|^2 = 1$) the transmission probability is stationary with respect to small changes in the barrier potential, i.e. $\delta \ln |T|/\delta V(z) = 0$. Hence, the imaginary part of τ_T^{SB} is zero and τ_T^{SB} is identical to the mean dwell time τ_D. Similarly, for perfect reflection $\tau_R^{SB} = \tau_D$.

The real and (minus) the imaginary parts of $\tau_T^{SB}(k; 0, d)$ are identical to the spin-precession traversal time of *Rybachenko* [6.23] and the spin-rotation traversal time of *Büttiker* [6.22] respectively. These are derived, following *Baz'* [6.24], from an analysis of the effect of an infinitesimal uniform magnetic field, confined to the barrier, on the components of the average spin per transmitted electron in the plane perpendicular to the field and in the field direction respectively. Neither of these traversal times can stand on its own: the former is essentially independent of the width d of an opaque rectangular barrier leading to a mean transmission speed that can exceed the speed of light c even in the corresponding relativistic calculation [6.25]; the latter can be negative for above barrier transmission. Accordingly, Büttiker identified the actual traversal time with the square-root of the sum of their squares, i.e. with $|\tau_{\mathrm{T}}^{\mathrm{SB}}(k; 0, d)|$. This is referred to as the Larmor clock traversal time in the rest of this paper. It is identical to the *Büttiker–Landauer* traversal time derived by considering the sensitivity of T to the instantaneous height of a time-modulated barrier [6.26, 27]. For the special cases of perfect transmission and perfect reflection the spin-precession, Larmor clock and time-modulated barrier approaches all give the correct result (6.5) for τ_{T} and τ_{R} respectively.

When the derivation leading to $\tau_{\mathrm{T}}^{\mathrm{SB}}$ is repeated, ignoring the interference between the incident and reflected components of the wave function, another complex quantity is obtained [6.19] that is closely related to other traversal times in the literature: its real part is identical to the phase time of *Bohm* [6.28] and *Wigner* [6.29] and to the spin-precession time of *Huang* et al. [6.30], while its imaginary part is the spin-rotation traversal time of the latter authors; its modulus is identical to the traversal time obtained by *Büttiker* and *Landauer* [6.31] from an analysis of the coherent transmission of two plane waves with slightly different energies. For the special case of perfect reflection, none of these approaches gives a mean reflection time that agrees with the result (6.5).

It is important to note that all of the above approaches are concerned with *intrinsic* transmission and reflection times because they are ultimately based on (6.2) which represents an ideal clock that 'runs' only when the electron is in the region of interest and does not perturb its motion in any way. Moreover, the fact that they do not all give the same answer to such an apparently simple problem as the dependence of $\tau_{\mathrm{T}}(k; 0, d)$ on the width d of an opaque rectangular barrier[2] should not be swept under the rug by claiming that they are concerned with different quantities from the outset.

It should also be emphasized that, under the right conditions, any one of the above times might be an important parameter. For example, if a tunneling experiment involves a very small constant change in the barrier height $V(z)$ then there is no doubt that the relative sensitivity of the transmission probability amplitude to changes in average barrier height $\bar{V}$, i.e. $\partial \ln T(k)/\partial \bar{V}$, is a relevant quantity. If one chooses to multiply this complex-valued quantity by $i\hbar$ then the resulting 'time' is obviously also relevant (whether it is the real part, imaginary part or modulus which is most important depends on the experimental situation). This however does not necessarily mean that the resulting time, which is identical to $\tau_{\mathrm{T}}^{\mathrm{SB}}(k; 0, d)$ [6.8], should be identified with the actual mean transmission time.

Of more importance than the question of which, if either, of two different approaches is 'better' is the question of whether or not the concepts of transmission and reflection times are meaningful. The point of view that they are not is

---

[2] For example, the time-modulated barrier and spin-precession results for $\tau_{\mathrm{T}}(k; 0, d)$ are linear in d and independent of d, respectively.

consistent with the fact that there is no known approach based on conventional quantum mechanics which leads to real, non-negative transmission and reflection times that satisfy (6.1) and which does not lead to mean transmission speeds in excess of c [6.8]. *Sokolovski* and *Connor* [6.19] claim that, in general, transmission and reflection times must be complex valued and that it is misguided to search for a unique, well-defined prescription for calculating real-valued transmission and reflection times, particularly by invoking non-standard interpretations of quantum mechanics. We do not find their arguments compelling and in the next section explore the problem within an interpretation of quantum mechanics that leads naturally to transmission and reflection times satisfying the above criteria.

6.3 Bohm's Trajectory Interpretation of Quantum Mechanics

6.3.1 A Brief Introduction

In *Bohm*'s interpretation of non-relativistic quantum mechanics [6.14–16], an electron *is* a particle the motion of which is causally determined by an objectively real complex-valued field $\psi(z, t)$ so that it has a well-defined position and velocity at each instant of time. This is diametrically opposed to the fundamental tenets of conventional interpretations. Nevertheless, it is claimed that with three additional postulates, Bohm's interpretation leads to precisely the same results as the conventional ones for all experimentally observable quantities. These three postulates are: (1) the guiding field

$$\psi(z, t) \equiv R(z, t) \exp[iS(z, t)/\hbar] \quad (R \text{ and } S \text{ real}) \tag{6.7}$$

satisfies the time-dependent Schrödinger equation (TDSE) $[i\hbar\partial/\partial t + (\hbar^2/2m)\partial^2/\partial z^2 - V(z, t)]\psi(z, t) = 0$; (2) the velocity of an electron located at the position z at time t is given by

$$v(z, t) = m^{-1}\partial S(z, t)/\partial z \; ; \tag{6.8}$$

(3) $|\psi(z, t)|^2 \, dz$ is the probability of the electron *being* between z and $z + dz$ at time t even in the absence of a position measurement. In conventional interpretations this quantity is the probability of the electron *being found* between z and $z + dz$ at time t by a precise position measurement.

Substitution of (6.7) into the TDSE and separation of the resulting real and imaginary parts gives the modified Hamilton–Jacobi equation

$$\partial S(z, t)/\partial t + (2m)^{-1}[\partial S(z, t)/\partial z]^2 + V(z, t) + Q(z, t) = 0 \tag{6.9}$$

with

$$Q(z, t) \equiv -(\hbar^2/2m) R^{-1}(z, t) \partial^2 R(z, t)/\partial z^2 \; , \tag{6.10}$$

and the continuity equation

$$\partial P(z, t)/\partial t + \partial j(z, t)/\partial z = 0 \tag{6.11}$$

relating the probability and probability current densities

$$P(z, t) \equiv R^2(z, t) \equiv |\psi(z, t)|^2 \ , \tag{6.12}$$

$$j(z, t) \equiv P(z, t)\,v(z, t)$$

$$\equiv (\hbar/2im)\,[\psi^*(z, t)\,\partial \psi(z, t)/\partial z - \psi(z, t)\,\partial \psi^*(z, t)/\partial z] \ , \tag{6.13}$$

respectively. Bohm attributes the differences between quantum and classical mechanics mainly to the 'quantum potential' Q and regards the classical limit as the limit in which Q is completely negligible compared with all other relevant energies. Since the differences between quantum and classical physics can be dramatic it should not come as a surprise that the quantum potential can have correspondingly remarkable properties. For example, since $R(z, t)$ occurs in both the numerator and denominator of (6.10) it is possible for the quantum potential to have an important effect even in a region where $|\psi(z, t)|^2$ is apparently negligible. The quantum potential has even more remarkable properties for many-electron systems: it can be non-local, the quantum potential acting on one electron being determined instantaneously at a distance by others in the system and not just through some unvarying function of particle positions but through the quantum state Ψ of the entire system. Non-locality is a necessary feature, according to *Bell*'s theorem [6.32], of any realistic interpretation of quantum mechanics, such as Bohm's, which reproduces all the experimental consequences of the conventional interpretations.

Bohm's quantum potential approach has provided fresh insight into measurement theory including the notorious collapse of the wave function which is not required in his interpretation [6.33, 34], the Einstein–Podolsky–Rosen thought experiment [6.35] and, returning to one-particle systems for a few more examples, the two-slit interference experiment [6.36], the Aharonov–Bohm effect [6.37], delayed-choice experiments [6.38], and quantum mechanical tunneling [6.39]. The following discussion of transmission, reflection and dwell times within Bohm's interpretation is a natural and long overdue extension of the work of *Dewdney* and *Hiley* [6.39].

6.3.2 Transmission and Reflection Times Within Bohm's Interpretation

Given the initial position $z^{(0)} \equiv z(t = 0)$ of an electron with initial wave function $\psi(z, t = 0)$, its subsequent trajectory $z(z^{(0)}, t)$ is uniquely determined by simultaneous integration of the TDSE and the guidance equation $dz(t)/dt = v(z, t)$ given by (6.8). Alternatively, such trajectories can be obtained by simultaneous integration of the TDSE and Newton's equation of motion with the usual potential energy $V(z, t)$ augmented in the latter by the quantum potential

$Q(z, t)$.[3] Now, suppose that two Bohm trajectories with the same guiding field $\psi(z, t)$ intersect at the space-time point (z_i, t_i). It follows immediately from $v(z, t) = m^{-1} \partial S(z, t)/\partial z$ that both trajectories have the same velocity as well as the same position at time t_i. Since the time evolution of the trajectories is given by a second-order differential equation the two trajectories must then coincide at all times $t > t_i$ in the future and, using time reversal invariance, at all times $t < t_i$ in the past. Hence, Bohm trajectories do not intersect. This fact will be very useful in what follows.

The interpretation of a Bohm trajectory as an actual particle trajectory [6.14–16], rather than an abstract mathematical construct, immediately leads to a unique and well-defined prescription [6.40, 41] for calculating transmission and reflection times. For an electron that *is* at $z = z^{(0)}$ at $t = 0$ the time spent thereafter in the region $z_1 \leq z \leq z_2$ is *unambiguously* given by replacing the classical trajectory in the right-hand-side of (6.2) by a Bohm trajectory, i.e.,

$$t(z^{(0)}; z_1, z_2) = \int_0^\infty dt\, \theta[z(z^{(0)}, t) - z_1]\, \theta[z_2 - z(z^{(0)}, t)] \tag{6.14}$$

In practice, the initial location $z^{(0)}$ of an electron is not known exactly and uncertainty enters Bohm's deterministic theory through the postulated probability distribution $P(z^{(0)}, 0) \equiv |\psi(z^{(0)}, 0)|^2$ for $z^{(0)}$. Proceeding as in classical statistical mechanics, the mean dwell time is then given by

$$\tau_D(z_1, z_2) \equiv \langle t(z^{(0)}; z_1, z_2) \rangle \tag{6.15}$$

where for any function f of $z^{(0)}$

$$\langle f(z^{(0)}) \rangle \equiv \int_{-\infty}^\infty dz^{(0)}\, |\psi(z^{(0)}, 0)|^2 f(z^{(0)}) \ . \tag{6.16}$$

Insertion of an integral over all z of the delta function $\delta[z - z(z^{(0)}, t)]$ into (6.15) immediately gives the equivalent, and perhaps intuitively obvious, result [6.40]

$$\tau_D(z_1, z_2) = \int_0^\infty dt \int_{z_1}^{z_2} dz\, |\psi(z, t)|^2 \tag{6.17}$$

with

$$|\psi(z, t)|^2 \equiv \langle \delta[z - z(z^{(0)}, t)] \rangle \ . \tag{6.18}$$

Equation (6.17) is precisely the result (6.4) derived using standard quantum mechanics by *Sokolovski* and *Baskin* [6.18].

---

[3] Although this method of calculation presumably could be more efficient, trajectories being calculated to order $(\Delta t)^2$ rather than just Δt, it has not been attempted in our work because of possible complications resulting from the classical force $- \partial V(z, t)/\partial z$ being singular at the edges of rectangular barriers.

In the following discussion of mean transmission and reflection times for finite wave packets, it is always assumed that the initial wave packet $\psi(z, t = 0)$ is normalized to unity and is sufficiently far to the left of the barrier region $0 \leq z \leq d$ that the initial probability density $|\psi(z, t = 0)|^2$ is completely negligible for $z \geq 0$. In the calculations presented here, the centroid of the initial wave packet is usually chosen so that $|\psi(z, t = 0)|^2$ integrated over the region $0 \leq z \leq \infty$ is equal to $10^{-4} |T|^2$. Now, since Bohm trajectories do not intersect each other, there is a special starting point $z_c^{(0)}$ given by

$$\int_{z_c^{(0)}}^{\infty} dz^{(0)} |\psi(z^{(0)}, 0)|^2 = |T|^2 \tag{6.19}$$

such that only those trajectories $z(z^{(0)}, t)$ with $z^{(0)} > z_c^{(0)}$ are ultimately transmitted contributing to $|T|^2$ and only those with $z^{(0)} < z_c^{(0)}$ are ultimately reflected contributing to $|R|^2$. Hence the mean transmission and reflection times are uniquely given by

$$\tau_T(z_1, z_2) = \langle t(z^{(0)}; z_1, z_2) \theta(z^{(0)} - z_c^{(0)}) \rangle / \langle \theta(z^{(0)} - z_c^{(0)}) \rangle , \tag{6.20}$$

$$\tau_R(z_1, z_2) = \langle t(z^{(0)}; z_1, z_2) \theta(z_c^{(0)} - z^{(0)}) \rangle / \langle \theta(z_c^{(0)} - z^{(0)}) \rangle , \tag{6.21}$$

where $|T|^2 = \langle \theta(z^{(0)} - z_c^{(0)}) \rangle$ and $|R|^2 = \langle \theta(z_c^{(0)} - z^{(0)}) \rangle$. Obviously τ_T and τ_R are real-valued non-negative quantities and the sum-rule (6.1) is satisfied exactly.

The above prescription should be contrasted with the usual approaches based on the time evolution of the wave function [6.28, 29, 42–46] which assume that the centroid (or peak) of the incident probability density $|\psi(z, 0)|^2$ evolves into the centroid (or peak) of the transmitted component $|\psi_T(z, t_\infty)|^2$. This assumption has no justification in conventional quantum mechanics as has been pointed out by *Büttiker* and *Landauer* [6.31]. Within Bohm's interpretation the assumption is simply incorrect because it is only that part of the initial wave packet to the right of $z_c^{(0)}$ that evolves into the transmitted component. It should be noted that for any wave packet of finite spatial extent, the usual TDSE approaches lead to negative traversal times whenever the initial centroid is too far from the barrier [6.21].

Of more general interest than their mean values are the transmission and reflection time distributions [6.40, 41]

$$P_T[t(z_1, z_2)] \equiv \langle \theta(z^{(0)} - z_c^{(0)}) \delta[t(z_1, z_2) - t(z^{(0)}; z_1, z_2)] \rangle / \langle \theta(z^{(0)} - z_c^{(0)}) \rangle , \tag{6.22}$$

$$P_R[t(z_1, z_2)] \equiv \langle \theta(z_c^{(0)} - z^{(0)}) \delta[t(z_1, z_2) - t(z^{(0)}; z_1, z_2)] \rangle / \langle \theta(z_c^{(0)} - z^{(0)}) \rangle , \tag{6.23}$$

respectively. Often, a very significant contribution to the reflection probability comes from trajectories that do not enter the region $z_1 \leq z \leq z_2$ of interest and it is convenient to split off their contribution to $P_R[t(z_1, z_2)]$ by writing

$$P_R[t(z_1, z_2)] \equiv P_R^{(0)} \delta[t(z_1, z_2)] + P_R^{>}[t(z_1, z_2)] \ . \tag{6.24}$$

The mean transmission and reflection times are obviously given by

$$\tau_T(z_1, z_2) = \int_0^\infty dt \, t \, P_T(t) \ , \quad \tau_R(z_1, z_2) = \int_0^\infty dt \, t \, P_R^{>}(t) \ . \tag{6.25}$$

6.4 Application to Simple Systems

6.4.1 Some Numerical Details

For static barriers the transmission probability for a finite wave packet is given by

$$|T|^2 = \int_0^\infty \frac{dk}{2\pi} |\phi(k)|^2 |T(k)|^2 \tag{6.26}$$

where $|T(k)|^2$ is the stationary-state transmission probability for the 'incident' plane wave $\exp(ikz)$ of wave number k and $\phi(k)$ is the Fourier transform of the initial wave function $\psi(z, 0)$. Hence $|T|^2$ and $z_c^{(0)}$ can be calculated prior to solving the TDSE.

In the following subsections, the numerical method used to solve the TDSE is the fourth order (in time step Δt) symmetrized product formula method developed by *De Raedt* [6.47]. For reasonable wave packet and barrier parameters, the resulting transmission probability

$$|T|^2 = \int_d^\infty dz \, |\psi(z, t_\infty)|^2 \tag{6.27}$$

converges very satisfactorily with decreasing Δt to the exact value, given by (6.26) for static barriers. Although the accuracy of the calculated trajectories is only of order Δt, that of the averaged quantities (6.20–23) is much better due to significant cancellation of errors between the numerators and denominators of the defining equations. Moreover, if one is interested in only τ_T and τ_R, the fact that Bohm trajectories do not intersect enables one to bypass the calculation of trajectories entirely. The alternative expressions, obtained from (6.17), are

$$\tau_T(z_1, z_2) = \frac{1}{|T|^2} \int_0^\infty dt \int_{z_1}^{z_2} dz \, |\psi(z, t)|^2 \, \theta[z - z_c(t)] \ , \tag{6.28}$$

$$\tau_R(z_1, z_2) = \frac{1}{|R|^2} \int_0^\infty dt \int_{z_1}^{z_2} dz \, |\psi(z, t)|^2 \, \theta[z_c(t) - z] \ , \tag{6.29}$$

where $z_c(t)$, the transmission–reflection bifurcation curve, is given by

$$\int_{z_c(t)}^{\infty} dz\, |\psi(z,t)|^2 = |T|^2 \ . \tag{6.30}$$

In most of the following applications of the Bohm trajectory approach to scattering problems the initial wave function is taken to be a Gaussian, i.e.

$$\psi(z, t=0) = \frac{1}{[2\pi(\Delta z)^2]^{1/4}} \exp\left[-\left(\frac{z-z_0}{2\Delta z}\right)^2 + ik_0 z \right]\ . \tag{6.31}$$

The width Δz of this minimum-uncertainty-product wave function is related to the width Δk of its Fourier transform $\phi(k)$ by $\Delta z \Delta k = 1/2$; z_0 is the centroid of $|\psi(z,0)|^2$ and k_0 is the centroid of $|\phi(k)|^2 = [8\pi(\Delta z)^2]^{1/2} \exp[-(k-k_0)^2/2(\Delta k)^2]$. The average energy of the initial wave packet is $\bar{E} = [1 + (\Delta k/k_0)^2] E_0$ where $E_0 \equiv \hbar^2 k_0^2/2m$ is the energy associated with wave number k_0. The prescription discussed above for selecting z_0 takes the explicit form $z_0 = -N^{-1}(1 - 10^{-4}|T|^2)\Delta z$ and the special starting position is given by $z_c^{(0)} = z_0 + N^{-1}(1 - |T|^2)\Delta z$ where $N^{-1}(x)$ is the inverse of the normal distribution function.

6.4.2 Reflection Times for an Infinite Barrier

As a simple application [6.40] of the Bohm trajectory approach the mean reflection time is calculated for a region in front of the perfectly reflecting barrier $V(z) = V_0 \theta(z)$ with $V_0 \to \infty$. For this special case reflection and dwell times are, of course, equivalent.

It is a straightforward exercise to show that the solution of the TDSE is

$$\psi(z, t) = \alpha \exp[\beta_0 + \beta_2 z^2 + i(\gamma_0 + \gamma_2 z^2)]$$
$$\times [\exp(\delta z + i\varepsilon z) - \exp(-\delta z - i\varepsilon z)]\theta(-z) \tag{6.32}$$

where

$$\alpha \equiv [2(\Delta z)^2 \eta/\pi]^{1/4} \exp\{-(\Delta z)^2 k_0^2 - (i/2)\arctan[\hbar t/2m(\Delta z)^2]\}\ ,$$
$$\beta_0 \equiv \eta(\Delta z)^2 [4(\Delta z)^4 k_0^2 - z_0^2 - 2k_0 z_0 \hbar t/m]\ ,$$
$$\beta_2 \equiv -\eta(\Delta z)^2\ ,$$
$$\gamma_0 \equiv -\eta[4(\Delta z)^4 k_0 z_0 + 2(\Delta z)^4 k_0^2 \hbar t/m - z_0^2 \hbar t/2m]\ , \tag{6.33}$$
$$\gamma_2 \equiv \eta(\hbar t/2m)\ ,$$
$$\delta \equiv 2\eta(\Delta z)^2(z_0 + k_0 \hbar t/m)\ ,$$
$$\varepsilon \equiv \eta[4(\Delta z)^4 k_0 - z_0 \hbar t/m]\ ,$$
$$\eta \equiv [4(\Delta z)^4 + (\hbar t/m)^2]^{-1}\ .$$

It should be noted that $\psi(z, t)$ has a string of nodes at the points $z_n \equiv -n\pi/k_0$ $(n = 1, 2, \dots)$ at the instant $t = t_0 \equiv |z_0|/(\hbar k_0/m)$.

After casting $\psi(z, t)$ into the form $R(z, t) \exp[iS(z, t)/\hbar]$ it is easy to obtain the following expressions for the quantum potential and particle velocity:

$$Q(z, t) = \frac{\hbar^2}{2m} \left\{ -2\beta_2 \left[1 + 2\beta_2 z^2 + 2 \left(\frac{\delta \sinh 2\delta z + \varepsilon \sin 2\varepsilon z}{\cosh 2\delta z - \cos 2\varepsilon z} \right) z \right] \right.$$

$$\left. -2 \frac{(\delta^2 \cosh 2\delta z + \varepsilon^2 \cos 2\varepsilon z)}{\cosh 2\delta z - \cos 2\varepsilon z} + \frac{(\delta \sinh 2\delta z + \varepsilon \sin 2\varepsilon z)^2}{(\cosh 2\delta z - \cos 2\varepsilon z)^2} \right\},$$

$$(6.34)$$

$$v(z, t) = \frac{\hbar}{m} \left[2\gamma_2 z + \left(\frac{\varepsilon \tanh \delta z - \delta \sin \varepsilon z \cos \varepsilon z \, \text{sech}^2 \delta z}{\sin^2 \varepsilon z + \tanh^2 \delta z \cos^2 \varepsilon z} \right) \right]. \qquad (6.35)$$

Numerical integration of $v(z, t) = dz(t)/dt$ subject to the initial condition $z(t = 0) = z^{(0)}$ yields the particle trajectory $z(z^{(0)}, t)$. A selection of such trajectories is shown in Fig. 6.1. The bending of the trajectories away from the nodes that appear at the instant $t = t_0$ at the points z_n is quite striking, as is the bunching of trajectories just in front of the infinite barrier. Since $v(z, t_0) < 0$ all the trajectories must turn around before $t = t_0$. For each particle trajectory $z(z^{(0)}, t)$ it is easy to calculate numerically the time $t(z^{(0)}; a, 0)$ spent by the particle in the region $a \leq z \leq 0$. The corresponding mean reflection time $\tau_R(a, 0) = \langle t(z^{(0)}; a, 0) \rangle$ for the ensemble of particles described initially by $\psi(z^{(0)}, 0)$ is obtained by integrating $|\psi(z^{(0)}, 0)|^2 t(z^{(0)}; a, 0)$ over all $z^{(0)}$. Results for $\tau_R(a, 0)$ are compared in Fig. 6.2 with the plane-wave ($\Delta k = 0$) dwell time [6.21]

$$\tau_D(k_0; a, 0) = \tau_R(k_0; a, 0) = \frac{2|a|}{(\hbar k_0/m)} \left(1 - \frac{\sin 2k_0 a}{2k_0 a} \right). \qquad (6.36)$$

While $\tau_R(a, 0)$ for $\Delta k = 0.08 \, \text{Å}^{-1}$ shows significant departures from (6.36) for $a \lesssim -2 \, \text{Å}$ the result for $\Delta k = 0.02 \, \text{Å}^{-1}$ is barely distinguishable from (6.36) over the entire 10 Å range shown in the figure. In this simple case the $\Delta k = 0$ limit is numerically accessible via the Bohm trajectory technique.

6.4.3 Transmission and Reflection Times for Rectangular Barriers

As an application of the trajectory approach involving tunneling [6.41], consider an electron with the initial Gaussian wave function (6.31) incident on the rectangular barrier $V(z) = V_0 \theta(z) \theta(d - z)$ of height $V_0 = 10 \, \text{eV}$ and width d.

Figure 6.3 shows a selection of trajectories for $\Delta k = 0.04$ and $0.08 \, \text{Å}^{-1}$ with starting points $z^{(0)}$ in the vicinity of the transmission–reflection bifurcation point $z_c^{(0)}$. The energy E_0 of the incident wave packet is half of the barrier height V_0. The most striking difference between the two sets of trajectories is the factor of ~ 2 difference in time scales for motion *within* the barrier.

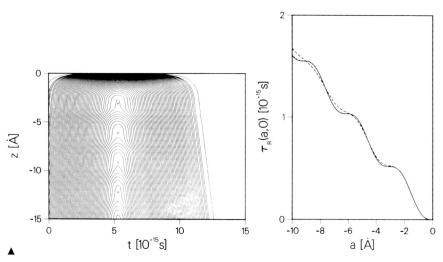

Fig. 6.1. Bohm trajectories $z(z^{(0)}, t)$ for an initial Gaussian wave packet incident on the infinite potential barrier $V(z) = V_0 \theta(z)$ with $V_0 \to \infty$. The wave packet parameters are $E_0 = 4\,\mathrm{eV}$, $\Delta k = 0.04\,\mathrm{\AA}^{-1}$ (i.e., $\Delta z = 12.50\,\mathrm{\AA}$), and $z_0 = -63\,\mathrm{\AA}$. The space-time points (z_n, t_0) at which $\psi(z, t)$ is zero are indicated for $n = 1, 2, 3$ and 4

Fig. 6.2. The average reflection time $\tau_R(a, 0)$ for the region $a \le z \le 0$ in front of the infinite barrier $V(z) = V_0 \theta(z)$ with $V_0 \to \infty$. The initial Gaussian wave packets have $E_0 = 4\,\mathrm{eV}$ and $\Delta k = 0.08$ (*dashed line*) and $0.02\,\mathrm{\AA}^{-1}$ (*dotted line*). The plane-wave dwell time $\tau_D(k_0; a, 0)$ is included for comparison (*solid line*)

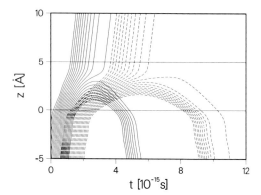

Fig. 6.3. Bohm trajectories for an initial Gaussian wave packet incident on a rectangular barrier of height $V_0 = 10\,\mathrm{eV}$ and width $d = 5\,\mathrm{\AA}$. The location of the barrier is indicated by horizontal lines. The wave packet parameters are: $E_0 = 5\,\mathrm{eV}$, $\Delta k = 0.04\,\mathrm{\AA}^{-1}$, $z_0 = -71.80\,\mathrm{\AA}$ (*dashed lines*); $\Delta k = 0.08\,\mathrm{\AA}^{-1}$, $z_0 = -35.58\,\mathrm{\AA}$ (*solid lines*)

Figure 6.4 shows the dependence of $\tau_T(0, d)$, calculated using the Bohm trajectory and Larmor clock[4] approaches, on the width d of the barrier for wave packets with $E_0 = V_0/2$ and $\Delta k = 0.04, 0.08$ and $0.16\,\mathrm{\AA}^{-1}$. As d approaches zero both sets of calculations merge with the free-particle ($V_0 = 0$) result

---

[4] For finite wavepackets, the Larmor clock results are obtained by averaging $|\tau_T^{SB}(k; 0, d)|$ over k with weight factor $|\phi(k)|^2 |T(k)|^2/(2\pi |T|^2)$.

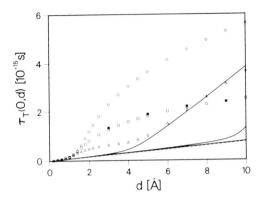

Fig. 6.4. Dependence of $\tau_T(0,d)$ on d for a rectangular barrier of height $V_0 = 10\,\text{eV}$ and incident energy $E_0 = 5\,\text{eV}$. The Bohm trajectory results are shown by circles for $\Delta k = 0.04\,\text{Å}^{-1}$, squares for $\Delta k = 0.08\,\text{Å}^{-1}$ and triangles for $\Delta k = 0.16\,\text{Å}^{-1}$ and the corresponding k-averaged Larmor clock results by the lower, middle and upper solid curves, respectively. For $\Delta k = 0.08\,\text{Å}^{-1}$ and $d = 3, 5, 7$ and $9\,\text{Å}$ the trajectory results recalculated using the value of z_0 appropriate to $\Delta k = 0.04\,\text{Å}^{-1}$ are shown (*filled squares*). The free particle result $d/(\hbar k_0/m)$ is indicated by the dashed line

$\tau_T^{\text{free}}(k_0; 0, d) = d/(\hbar k_0/m)$. The Larmor clock results for $\Delta k = 0.08\,\text{Å}^{-1}$ are influenced by significant above-barrier transmission for $d \geq 6\,\text{Å}$ and τ_T departs significantly from the $\Delta k = 0.04\,\text{Å}^{-1}$ results. For $\Delta k = 0.16\,\text{Å}^{-1}$ this effect is much more dramatic, setting in at $d \sim 3\,\text{Å}$. The behaviour of the mean Bohm trajectory transmission time begins to change qualitatively when d becomes comparable to the characteristic tunneling length $\kappa_0^{-1} = \hbar/[2m(V_0 - E_0)]^{1/2} \sim 0.9\,\text{Å}$, becoming strongly dependent on Δk. For $\Delta k = 0.16\,\text{Å}^{-1}$ the calculated transmission probability for $d \geq 6\,\text{Å}$ has a relatively large above-barrier resonant $(|T(k)|^2 \sim 1)$ contribution and is virtually independent of d; in this region the Larmor clock and Bohm trajectory results are, not surprisingly, in good agreement.

The prescription discussed above for guaranteeing a well-defined transmission probability $|T|^2$ requires that the centroid z_0 of the initial Gaussian wave packet moves further from the barrier as Δz increases. For $d = 3, 5, 7$ and $9\,\text{Å}$ the $\Delta k = 0.08\,\text{Å}^{-1}$ trajectory results recalculated with z_0 values appropriate to $\Delta k = 0.04\,\text{Å}^{-1}$ are also shown in Fig. 6.4. Clearly only a small fraction of the large difference between the $\Delta k = 0.04$ and $0.08\,\text{Å}^{-1}$ curves arises from the dependence on z_0. However, the calculated small changes in τ_T with z_0 are numerically significant and the mean transmission time does depend on z_0 even though $|T|^2$ does not. Since $|\phi(k)|^2$ is independent of z_0 for the initial Gaussian wave packet this means that, within the Bohm trajectory approach, $\tau_T(0, d)$ cannot be obtained by integrating $\tau_T(k; 0, d)|\phi(k)|^2 |T(k)|^2/(2\pi|T|^2)$ over k because the latter quantity is independent of z_0.

In Fig. 6.5 the dependence of $\tau_T(0, d)$ on E_0 and Δk is shown for a relatively narrow ($d = 3\,\text{Å}$) barrier in order to suppress possible above-barrier resonance effects. The Larmor clock and Bohm trajectory results are qualitatively different in both their dependence on E_0 and Δk until E_0 is well above the barrier height V_0. In Fig. 6.6 the same comparison is made for a wider ($d = 5\,\text{Å}$) barrier and an incident wave packet with a relatively small energy width ($\Delta k = 0.02\,\text{Å}^{-1}$) in order to highlight any resonance effects. The Larmor clock result for $\tau_T(0, d)$

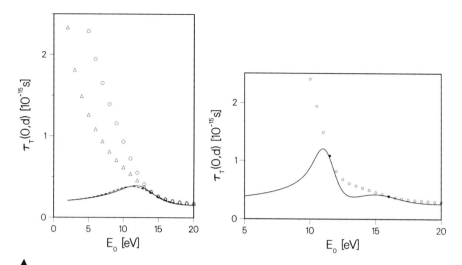

Fig. 6.5. Dependence of $\tau_\mathrm{T}(0,d)$ on the energy E_0 of a Gaussian wave packet incident on a rectangular barrier of height $V_0 = 10\,\mathrm{eV}$ and width $d = 3\,\text{Å}$. The Bohm trajectory results are shown by circles for $\Delta k = 0.04\,\text{Å}^{-1}$ and triangles for $\Delta k = 0.08\,\text{Å}^{-1}$ and the corresponding Larmor clock results by the solid and dashed lines, respectively. The first above barrier resonance $E_\mathrm{r}^{(1)}$ at 14.175 eV is the only one below 20 eV

Fig. 6.6. Dependence of $\tau_\mathrm{T}(0, d)$ on the energy E_0 of an initial Gaussian wave packet with $\Delta k = 0.02\,\text{Å}^{-1}$ incident on a rectangular barrier of height $V_0 = 10\,\mathrm{eV}$ and width $d = 5\,\text{Å}$. The Larmor clock results are shown by the solid curve and the trajectory results by circles (those for the above-barrier resonances at $E_\mathrm{r}^{(1)} = 11.503\,\mathrm{eV}$ and $E_\mathrm{r}^{(2)} = 16.012\,\mathrm{eV}$ by filled circles)

shows considerable structure, peaking fairly close to the above-barrier resonances at $E_\mathrm{r}^{(1)}$ and $E_\mathrm{r}^{(2)}$. The trajectory result, on the other hand, decreases monotonically exhibiting no apparent resonant structure; however, the contribution to the mean dwell time from transmitted particles, i.e. $|T|^2\,\tau_\mathrm{T}(0, d)$, does have well-defined resonance peaks. As expected, the agreement between the Bohm trajectory and Larmor clock results is best very close to the resonances where $|T|^2 \sim 1$.

Bohm trajectory transmission time distributions $P_\mathrm{T}[t(0, d)]$ are shown in histogram form in Fig. 6.7 for barrier widths close to $(d = 2.0\,\text{Å})$ and well above $(d = 5.0\,\text{Å})$ the transparent to opaque barrier 'crossover' evident in Fig. 6.4. For each case there is a well-defined minimum and most probable transmission time. However, each distribution has a long tail on the high side of the peak arising from transmitted trajectories very close to the transmission–reflection bifurcation where the time spent inside the barrier diverges. Transmission times very much in excess of the average, although rare, are possible and could have implications for submicron single-electron devices based on tunneling.

Figure 6.8 shows the reflection time distributions $P_\mathrm{R}[t(0, d)] = P_\mathrm{R}^0\,\delta[t(0, d)] + P_\mathrm{R}^>[t(0, d)]$ corresponding to the $(d = 5\,\text{Å})$ transmission time distributions $P_\mathrm{T}[t(0, d)]$ in the bottom panel of Fig. 6.7. The component

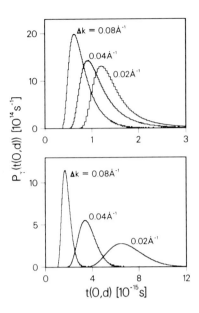

Fig. 6.7. Transmission time distributions $P_T[t(0,d)]$ calculated using the Bohm trajectory approach for a barrier of height $V_0 = 10\,\text{eV}$ and width d of 2.0 Å (*top*) and 5.0 Å (*bottom*). In all cases the incident energy $E_0 = 5\,\text{eV}$

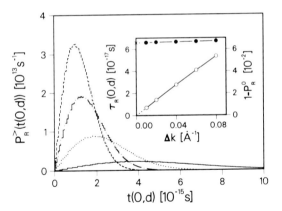

Fig. 6.8. Reflection time distributions $P_R^>[t(0,d)]$ for a rectangular barrier of height $V_0 = 10\,\text{eV}$ and width $d = 5$ Å. The wave packet parameters are $E_0 = 5\,\text{eV}$ and $\Delta k = 0.02$ (*solid line*), 0.04 (*dotted line*) 0.06 (*chained line*) and 0.08 Å^{-1} (*dashed line*). $P_R^>[t(0,d)]$ is normalized to $(1 - P_R^0)$ which is shown in the inset along with the mean reflection time $\tau_R(0,d)$ as open and filled circles, respectively. The plane-wave dwell time $\tau_D(k_0; 0,d)$ is indicated by X. The solid and dotted lines in the inset are guides to the eye

$P_R^>[t(0, d)]$ for electrons which actually enter the barrier region before being reflected is normalized to $(1 - P_R^0)$ which is shown in the inset along with $\tau_R(0, d)$ as a function of Δk. Over the range $0.01\,\text{Å}^{-1} \leq \Delta k \leq 0.08\,\text{Å}^{-1}$ at least, as Δk decreases towards the plane-wave limit ($\Delta k = 0$) a smaller and smaller fraction $(1 - P_R^0)$ of the reflected electrons spends longer and longer in the barrier region in such a way that the mean reflection time $\tau_R(0, d)$ has relatively little dependence on Δk. It is tempting to extrapolate the calculated results for $\tau_R(0, d)$ to $\Delta k = 0$ in order to obtain $\tau_R(k_0; 0, d)$ which, at least for this particular case, would be very close to the plane-wave dwell time $\tau_D(k_0; 0, d)$ given by (6.5), as shown in the figure. However, if the stationary-state Bohm trajectory result, $\tau_T(k_0; 0, d) = |T(k_0)|^{-2} \tau_D(k_0; 0, d)$, of *Spiller* et al. [6.48] is correct then $\tau_R(k_0; 0, d)$ must be zero (for $|R(k_0)|^2 \neq 0$) because $|T(k_0)|^2 \tau_T(k_0; 0, d) = \tau_D(k_0; 0, d)$ exhausts the sum rule (6.1). Unfortunately, it would be necessary to extend the time-dependent wave packet calculations of $\tau_R(0, d)$ to much smaller values of Δk in order to decide whether it remains finite or eventually plummets to zero in the plane-wave limit. We believe that the former alternative is more likely to be correct because it is not obvious that one can determine the temporal characteristics of a scattering process with $0 < |T|^2 < 1$ by studying only the stationary-state ($\Delta k = 0$) case. For example, *Spiller* et al. use the time-indepen-dent particle velocity $v(z) = (1/m)\partial S(z)/\partial z$ which is positive for all z and appar-ently assume that it is a property of transmitted electrons only rather than an average over transmitted and reflected electrons. Hence, all their calculated trajectories are transmitted ones, *even if* $|T(k_0)|^2 \ll 1$. We believe that for $|T|^2 < 1$ the plane-wave limit should be approached using time-dependent calculations with initial wave packets of greater and greater spatial extent, as in Sect. 6.4.2. Unfortunately, it is clear that calculations for initial wave packets with much smaller Δk than those considered in Figs. 6.4, 8 will be required to obtain the plane-wave limit of $\tau_T(0, d)$ and $\tau_R(0, d)$ for finite opaque barriers. This is expected to be computationally demanding.

An interesting question [6.7, 25] is "What is the average time $\tau_R(z_1, z_2)$ spent in a region $z_1 \leq z \leq z_2$ on the far side ($z_1 > d$) of a rectangular barrier by electrons that are ultimately *reflected?*". Although the potential $V(z)$ is zero for $z > d$ and there is apparently nothing to scatter an electron back through the barrier the Larmor clock approach leads to a non-zero result for $\tau_R(z_1 > d, z_2)$. On the other hand, although the quantum potential $Q(z, t)$ for $z > d$ could conceivably launch an electron back through the barrier this has never hap-pened in any of our calculations using the trajectory approach and hence $\tau_R(z_1 > d, z_2)$ appears to be zero.

6.4.4 Coherent Two-Component Incident Wave Packet

The non-intersecting property of Bohm trajectories is graphically illustrated by the scattering of an initial incident wave packet which is the *coherent* sum of two widely separated but otherwise identical Gaussians, i.e.

$$\psi_\gamma(z, t = 0) = \frac{\left[\exp\left(\frac{-(z - z_0)^2}{4(\Delta z)^2}\right) + \exp\left(\frac{-(z - z_0 + \gamma\Delta z)^2}{4(\Delta z)^2}\right)\right]\exp(ik_0 z)}{2^{1/2}\left[1 + \exp\left(-\frac{\gamma^2}{8}\right)\right]^{1/2}[2\pi(\Delta z)^2]^{1/4}},$$

(6.37)

with γ very much greater than 1 (the centroid of the second Gaussian is located $\gamma\Delta z$ further from the barrier than the centroid z_0 of the first). It readily follows from (6.26) that the transmission probability is given by

$$|T_\gamma|^2 = \frac{1}{1 + \exp(-\gamma^2/8)} \int_0^{+\infty} \frac{dk}{2\pi} [1 + \cos\gamma(k - k_0)\Delta z] |\phi_{\gamma=0}(k)|^2 |T(k)|^2$$

(6.38)

where $\phi_{\gamma=0}(k)$ is the Fourier transform of a single (normalized) Gaussian wave function of width Δz and $|T(k)|^2$ is the corresponding plane-wave transmission probability. Using the fact that $|T_{\gamma=\infty}|^2 = |T_{\gamma=0}|^2$ we restrict our attention to the case $\gamma \gg 1$ and for definiteness consider an opaque barrier so that the transmission–reflection bifurcation point $z_{c,\gamma}^{(0)}$ is to the right of z_0. Now, for $z \geq z_0$ and t sufficiently small that there is still negligible overlap of the two components of the wave packet, $|\psi_\gamma(z, t)|^2$ is equal to $(1/2)|\psi_{\gamma=0}(z, t)|^2$ and $Q_\gamma(z, t)$ is equal to $Q_{\gamma=0}(z, t)$ to a very good approximation (recall that the quantum potential Q depends only on the form of $|\psi|^2$ so that the factor of $1/2$ is irrelevant). This means that, for sufficiently large γ, electrons with starting points $z^{(0)}$ to the right of but not too close to the bifurcation point $z_{c,\gamma=0}^{(0)}$ of the single Gaussian wave packet are transmitted following trajectories very similar to those for the single Gaussian case, as can be seen in Fig. 6.9. However, it is clear from the defining equation, (6.19), that $z_{c,\gamma}^{(0)}$ must be further from the barrier than $z_{c,\gamma=0}^{(0)}$ because $|\psi_\gamma(z, 0)|^2$ is smaller than $|\psi_{\gamma=0}(z, 0)|^2$ by a factor of 2 for $z \gtrsim z_0$. Hence, some of the trajectories that were initially reflected by the barrier, namely

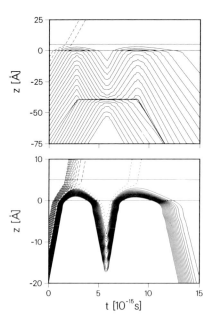

Fig. 6.9. Two selections of Bohm trajectories $z(z^{(0)}, t)$ for an initial wave packet which is the coherent sum of two widely separated but otherwise identical Gaussians incident on a rectangular barrier ($V_0 = 10\,\text{eV}$, $d = 5\,\text{Å}$, $E_0 = 5\,\text{eV}$ and $\Delta k = 0.08\,\text{Å}^{-1}$). There are two distinct types of transmitted trajectories: those that are transmitted directly (*long-dashed lines*) and those that are transmitted after first being reflected by the barrier (*short-dashed lines*)

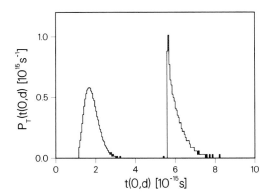

Fig. 6.10. The distribution $P_T[t(0, d)]$ of transmission times $t(0, d)$ for a double-Gaussian wave packet incident on a rectangular barrier. The wave packet and barrier parameters are those of Fig. 6.9

those with $z_{c, \gamma}^{(0)} \leq z^{(0)} \leq z_{c, \gamma = 0}^{(0)}$, must ultimately be transmitted. What happens, of course, is that eventually the (initially) reflected part of the first component of the wave packet significantly overlaps and interferes with the incoming second component. The resulting strong fluctuations in the quantum potential launch the initially 'unsuccessful' trajectories back towards the barrier for another 'attempt'. Those with $z^{(0)} \geq z_{c, \gamma}^{(0)}$ are transmitted this time (see the bottom panel of Fig. 6.9).

The Bohm trajectories of Fig. 6.9 are for the initial wave packet $\psi_\gamma(z, t = 0)$ with $E_0 = 5$ eV, $z_0 = -35.6$ Å, $\Delta z = 6.25$ Å ($\Delta k = 0.08$ Å^{-1}) and $\gamma = 13$ incident on the rectangular barrier $V(z) = V_0 \theta(z) \theta(d - z)$ with $V_0 = 10$ eV and $d = 5$ Å. The prominent feature in the middle of the top panel arises when the two trajectories originating at the centroids of the two Gaussian components of the initial wave packet come within about 10 Å of each other during the collision of the reflected component with the incoming component of $\psi_\gamma(z, t)$, sandwiching all of the trajectories with $z_0 - \gamma \Delta z < z^{(0)} < z_0$ between them. It is clear from the bottom panel of the figure that, on average, those trajectories that are transmitted without being initially reflected involve considerably shorter transmission times than those which are initially reflected. This is apparent in the calculated transmission time distribution shown in Fig. 6.10. Clearly, using the fact that Bohm trajectories do not intersect can allow one to predict important qualitative features of both trajectory maps and transmission time distributions.

6.4.5 Transmission Times for Time-Modulated Barriers

The time-modulated barrier approach of *Büttiker* and *Landauer* [6.26, 27] was motivated by the belief that adding a small oscillatory component $V_1 \cos \omega t$ to a static barrier leads to distinct low ($\omega \ll \tau_T^{-1}$) and high ($\omega \gg \tau_T^{-1}$) frequency regimes in the resulting tunneling behaviour (τ_T is the mean transmission time for the unperturbed static barrier). In particular, it is argued that for $\omega \ll \tau_T^{-1}$ the tunneling electron 'sees' a static barrier to a very good approximation while for $\omega \gg \tau_T^{-1}$ it 'sees' the time-averaged barrier but, in addition, now has a significant probability of absorbing or emitting modulation quanta leading to transmission and reflection side-bands. In their analysis of this appealing physical picture *Büttiker* and *Landauer* considered a plane wave $\exp(ikz)$ of energy $E = \hbar^2 k^2/2m$ 'incident' on the time-modulated rectangular barrier $V(z, t) = (V_0 + V_1 \cos \omega t)\,\theta(z)\,\theta(d - z)$ with $V_1 \ll \hbar\omega \ll E$, $V_0 - E$. In the low frequency limit they showed that the probability of transmission at the first-order side-band energies $E \pm \hbar\omega$ is given by

$$|T_\pm|^2 = [V_1\,\tau_{\mathrm{T}}^{\mathrm{BL}}(k)/2\hbar]^2\,|T(k)|^2 \quad (\omega \to 0) \tag{6.39}$$

where

$$\tau_{\mathrm{T}}^{\mathrm{BL}}(k) = (m/\hbar\kappa)|\partial\ln\,T(k)/\partial\kappa| \quad (\kappa = [2m(V_0 - E)/\hbar^2]^{1/2}) \tag{6.40}$$

is the well-known Büttiker–Landauer traversal time for a rectangular barrier. *Støvneng* and *Hauge* [6.49] and also *Jauho* and *Jonson* [6.50] showed that (6.39) holds for an arbitrary barrier when (6.40) is replaced by

$$\tau_{\mathrm{T}}^{\mathrm{BL}}(k) = \hbar|\partial\ln\,T(k)/\partial\bar{V}| \tag{6.41}$$

where $\bar{V}$ is the average height of the barrier. However, both groups expressed reservations about the identification of $\tau_{\mathrm{T}}^{\mathrm{BL}}(k)$ with the actual traversal time. It should be recalled that (6.41) is identical to the modulus of the mean transmission time of Sokolovski and Baskin, $|\tau_{\mathrm{T}}^{\mathrm{SB}}(k)|$.

Recently *De Raedt* et al. [6.51] numerically studied the scattering of Gaussian wave packets by time-modulated rectangular barriers but were unable to extract τ_{T} from the frequency dependence of the calculated transmission probability. Our calculations for a significantly different range of parameters [6.52] support this conclusion. Figure 6.11 shows the dependence of $|T|^2$ on modulation frequency ω for an opaque rectangular barrier. The modulation amplitude V_1 is sufficiently small that only the first-order side-bands are important. The transmission probability is strongly enhanced when the upper side-band at $E_0 + \hbar\omega$ coincides with the first ($n = 1$) and second ($n = 2$) above-barrier resonances of the unmodulated barrier at $E_r^{(n)} = V_0 + (\hbar^2/2m)(n\pi/d)^2$. The oscillatory structure at low frequencies is due to the finite spatial width Δz of the wave packet. The frequency range over which such oscillations are of non-negligible amplitude decreases with increasing Δz and is expected to shrink to zero in the plane-wave limit ($\Delta z \to \infty$) leaving a smoothly varying featureless curve in the vicinity of $\tau_{\mathrm{T}}^{\mathrm{BL}}(k_0)^{-1}$ ($= 0.26 \times 10^{16}$ s^{-1}). This trend is confirmed by calculations with larger Δz than those shown. The figure shows $|T|^2$ for both $V_1 \sin \omega t$ and $V_1 \cos \omega t$ modulations. We prefer the former because in the static limit $\omega \to 0$ the (total) barrier height is then the original unmodulated barrier height V_0 rather than $V_0 + V_1$. Only results for the $V_1 \sin \omega t$ modulations are shown in the remaining figures of this subsection.

We now consider the time-modulated barrier from the Bohm trajectory point of view which clearly shows that there is no reason to expect a signature of

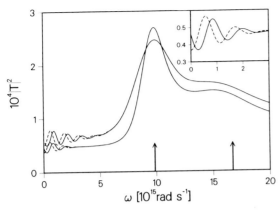

Fig. 6.11. The frequency dependence of the transmission probability $|T|^2$ for a Gaussian wave packet incident on a time-modulated rectangular barrier of unmodulated height $V_0 = 10$ eV and width $d = 5$ Å. The results for the modulation $V_1\theta(z)\theta(d - z)\sin\omega t$ are shown by the solid curves and for $V_1\theta(z)\theta(d - z)\cos\omega t$ by the dashed curves with $V_1 = V_0/50 = 0.2$ eV. The incident energy $E_0 = 5$ eV, $z_0 = -71.8$ Å and $\Delta k = 0.08$ Å^{-1} for the upper curves (near $\omega = 0$) and $\Delta k = 0.04$ Å^{-1} for the lower curves. The resonance condition $\hbar\omega = E_r^{(n)} - E_0$ for the first-order sideband is indicated by an arrow for $n = 1$ and 2. The inset shows the low frequency behaviour for $\Delta k = 0.04$ Å^{-1}

τ_T in the frequency dependence of $|T|^2$. In the limit $t \to \infty$ when the scattering of the wave packet is complete each trajectory $z(z^{(0)}, t)$, with the exception of the one with $z^{(0)} = z_c^{(0)}$, can be labelled as either transmitted or reflected

$$\theta_T(z^{(0)}) \equiv \theta(z^{(0)} - z_c^{(0)}) = \begin{cases} 1 \text{ (transmitted trajectory)} \\ 0 \text{ (reflected trajectory)} \end{cases}. \tag{6.42}$$

Now, for the trajectory starting at a particular $z^{(0)}$, transmission is 'all or nothing', i.e. $\theta_T(z^{(0)}) = 1$ or 0. Hence, if the original barrier is perturbed in some way there are four distinct possibilities for the effect of the perturbation on θ_T, only three of which can occur for a given perturbation because Bohm trajectories do not intersect. If the transmission probability $|T|^2 = \langle\theta_T(z^{(0)})\rangle$ increases then one possibility, denoted by the shorthand $0 \to 1$, is that the perturbation causes $\theta_T(z^{(0)})$ to change from 0 to 1; for the two other possibilities, $0 \to 0$ and $1 \to 1$, $\theta_T(z^{(0)})$ is unaffected by the perturbation. The important $0 \to 1$ possibility occurs for and *only* for $z_c^{(0)} \leq z^{(0)} \leq z_{c,0}^{(0)}$ where $z_{c,0}^{(0)}$ denotes $z_c^{(0)}$ for the unperturbed system. If $|T|^2$ decreases because of the perturbation the three possibilities are $1 \to 0$, $0 \to 0$ and $1 \to 1$, the first occurring only for $z_{c,0}^{(0)} \leq z^{(0)} \leq z_c^{(0)}$. Any change in the transmission probability arises solely from those particle trajectories for which the perturbation changes $\theta_T(z^{(0)})$, namely from 0 to 1 if $|T|^2$ increases or from 1 to 0 if $|T|^2$ decreases. For example, if the relative change in $|T|^2$ due to the perturbation is a very small decrease then the crucial $1 \to 0$ trajectories have $z^{(0)}$ very close to and to the right of $z_{c,0}^{(0)}$. The corresponding trajectories of the unperturbed system are characterized by anomalously long transmission times far out in the tail of the transmission time distribution (see the $\omega = 0$ distribution in Fig. 6.14 below). Similarly, if the relative change in $|T|^2$ is a very small increase then the trajectories of the unperturbed system that are most dramatically changed by the perturbation (i.e. $0 \rightarrow 1$) are characterized by anomalously large *reflection* times. These can be especially atypical in this case particularly for an opaque barrier for which the typical reflected trajectory does not even reach the barrier and has a reflection time $t(z^{(0)}; 0, d)$ of zero. Clearly, at least within the Bohm interpretation, there is no reason whatsoever to expect anything special to happen to $|T|^2$ when the modulation frequency ω passes through τ_T^{-1}.

Figure 6.12 compares a selection of Bohm trajectories for an unmodulated rectangular barrier with the corresponding trajectories when a small modulation $V_1 \sin \omega t$ is applied to the barrier with V_1 and $\hbar\omega$ satisfying the conditions $V_1 \ll \hbar\omega \ll E_0$, $V_0 - E_0$ of the Büttiker–Landauer papers. Although in this case the perturbation leads to a decrease in $|T|^2$, none of the pairs of trajectories shown here have $z^{(0)}$ close enough to $z_{c,0}^{(0)}$ to be of the $1 \rightarrow 0$ type associated with the decrease, but the trend is clear. Increasing the density of trajectories would, of course, eventually reveal such trajectories.

Figure 6.13 shows a selection of trajectories for the special case in which the modulation frequency satisfies the resonance condition $E_0 + \hbar\omega = E_r^{(1)}$ (we are ignoring the very small difference between E_0 and $\bar{E}$). Those trajectories that spend a significant length of time inside the barrier show well-defined oscillations of period $2\pi/\omega$. Since the classical force $-\partial V(z, t)/\partial z$ is zero for $0 < z < d$ these oscillations arise indirectly from the $V_1 \sin \omega t$ modulation through the quantum force $-\partial Q(z, t)/\partial z$. Although most of the transmitted trajectories enter the barrier at fairly regular intervals they leave in bunches. This combined with the general increase in $t(z^{(0)}; 0, d)$ as $z^{(0)}$ approaches $z_c^{(0)}$ leads to the multi-peaked transmission time distribution shown in Fig. 6.14 where it is compared with the distribution for a far-from-resonance modulation frequency and for the static case.

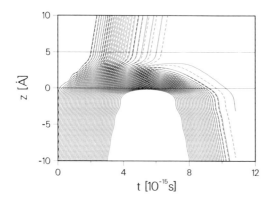

Fig. 6.12. Bohm trajectories for a Gaussian wave packet with $E_0 = 5$ eV, $\Delta k = 0.04$ Å^{-1} and $z_0 = -71.8$ Å incident on the static barrier $V_0\theta(z)\theta(d-z)$ and on the time-modulated barrier $(V_0 + V_1 \sin \omega t)\theta(z)\theta(d-z)$ are shown by the solid and dashed curves, respectively. The barrier parameters are $V_0 = 10$ eV, $d = 5$ Å, $V_1 = 0.2$ eV and $\hbar\omega = 1.0$ eV

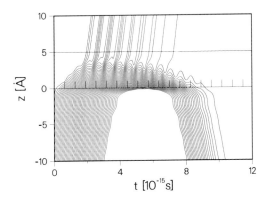

Fig. 6.13. Bohm trajectories for a Gaussian wave packet with $E_0 = 5\,\text{eV}$, $\Delta k = 0.04\,\text{Å}^{-1}$ and $z_0 = -71.8\,\text{Å}$ incident on the time-modulated barrier $(V_0 + V_1 \sin \omega t)\theta(z)\theta(d - z)$ with $V_0 = 10\,\text{eV}$, $d = 5\,\text{Å}$, $V_1 = 0.2\,\text{eV}$ and $\hbar\omega = E_r^{(1)} - E_0$. The vertical lines along the time axis indicate integral multiples of the modulation period $2\pi/\omega$

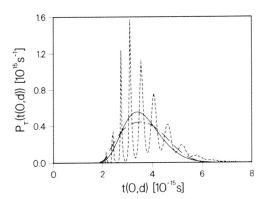

Fig. 6.14. Transmission time distributions $P_T[t(0, d)]$ for $\hbar\omega = 0$ (*solid line*), $\hbar\omega = 1\,\text{eV}$ (*chained line*) and $\hbar\omega = E_r^{(1)} - E_0$ (*dashed line*) for a Gaussian wave packet with $E_0 = 5\,\text{eV}$, $\Delta k = 0.04\,\text{Å}^{-1}$ and $z_0 = -71.8\,\text{Å}$ incident on the time-modulated barrier $(V_0 + V_1 \sin \omega t)\theta(z)\theta(d - z)$ with $V_0 = 10\,\text{eV}$, $d = 5\,\text{Å}$ and $V_1 = 0.2\,\text{eV}$

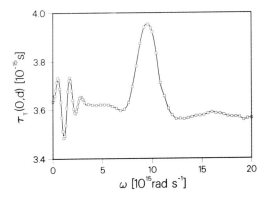

Fig. 6.15. The frequency dependence of the mean transmission time for the wave packet and barrier parameters of Fig. 6.14. The solid line is a guide to the eye

Figure 6.15 shows the dependence of the mean transmission time $\tau_T(0, d)$ on modulation frequency ω for one of the cases considered in Fig. 6.11. Although $\tau_T(0, d)$ is significantly enhanced at the $n = 1$ resonance it is considerably less sensitive to ω than is $|T|^2$.

The distribution of energy for a wave packet is usually considered in terms of the free-particle energy $E(k) \equiv \hbar^2 k^2/2m$, i.e.

$$P(E, t) = (2\pi)^{-1} \int_{-\infty}^{+\infty} dk \, |\phi(k, t)|^2 \, \delta(E - E(k)) \tag{6.43}$$

where $\phi(k, t)$ is the Fourier transform of $\psi(z, t)$. Now, in Bohm's interpretation an electron with position z at time t has a well-defined energy

$$E(z, t) = (m/2)v(z, t)^2 + V(z, t) + Q(z, t) \ . \tag{6.44}$$

Hence, an alternative way of distributing the energy of a wave packet is in terms of $E(z, t)$, i.e.

$$P^B(E, t) = \int_{-\infty}^{+\infty} dz \, |\psi(z, t)|^2 \, \delta(E - E(z, t)) \tag{6.45}$$

$$= \int_{-\infty}^{+\infty} dz^{(0)} |\psi(z^{(0)}, 0)|^2 \, \delta(E - E(z(z^{(0)}, t), t)) \ . \tag{6.46}$$

Both distributions (6.43 and 6.46) lead to the same expectation value, $\bar{E}(t)$, for the energy [6.14].

In the standard interpretations of quantum mechanics one has no way of knowing where, on average, a transmitted electron associated with the upper side-band absorbed a modulation quantum of energy $\hbar\omega$. Despite this, it has been suggested [6.26] that for an opaque rectangular barrier the absorption occurs at the leading edge of the barrier because the tunneling electron then 'sees' an effective barrier that is lower by $\hbar\omega$ for all of its journey through the classically forbidden region. This sounds very plausible. However, it seems inconsistent with the idea that it is only when the time spent by an electron inside the barrier is much greater than ω^{-1} that there is a significant probability for the absorption of a modulation quantum [6.26]. This inconsistency arises at least partly from attempting to picture an electron in a state extending over all space as a localized particle during the tunneling process within an interpretation which does not allow such pictures.

In the Bohm interpretation, on the other hand, one can follow the energy distribution for the 'to be transmitted part' of the wave packet as a function of time, right from $t = 0$, simply by replacing the lower limit on the integral over $z^{(0)}$ in (6.46) by $z_c^{(0)}$. Figure 6.16 shows the evolution of the particle energy $E(z, t)$, with t implicit, for a selection of Bohm trajectories $z(z^{(0)}, t)$ calculated for the wave packet and barrier parameters of Fig. 6.13 where $\hbar\omega = E_r^{(1)} - E_0$. For this resonant situation the reflected trajectories shown all have $E(z, t) \sim E_0$ and the transmitted ones $E(z, t) \sim E_0 + \hbar\omega$ at sufficiently large t. It should be noted that the transmitted electrons associated with the upper side-band experience large fluctuations in energy that continue well beyond the barrier region before their energy finally settles down to $E_0 + \hbar\omega$.

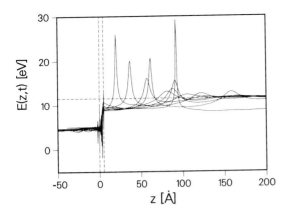

Fig. 6.16. The particle energy $E(z, t)$ along a selection of Bohm trajectories $z(z^{(0)}, t)$ with t implicit and -39.0 Å $\leq z^{(0)} \leq -21.0$ Å in steps of 1 Å for the wave packet and barrier parameters of Fig. 6.13 ($z_c^{(0)} = -28.54$ Å). The resonance energy $E_r^{(1)}$ of the static rectangular barrier is indicated by the horizontal dashed line and the position of the barrier by the vertical dashed lines

6.4.6 Transmission Times for Symmetric Double Rectangular Barriers

Larmor clock results [6.25] for $\tau_T(k; 0, z)$ with $0 \leq z \leq d$ have been calculated for a plane-wave electron of wave number k 'incident' on the symmetric double rectangular barrier $V_0[\theta(z)\theta(a - z) + \theta(z - d + a)\theta(d - z)]$. The detailed results for the special case of resonant transmission with $|T(k = k_r)|^2 = 1$ are directly relevant to this chapter because the Larmor clock and Bohm trajectory approaches give identical results for the limiting cases of perfect transmission and perfect reflection.[5] Rather than simply reproduce these $|T(k_r)|^2 = 1$ results we instead discuss how the Bohm trajectory approach illuminates an interesting puzzle connected with the Larmor clock approach.

For the special case $|T(k_r)|^2 = 1$ the average 'transmission speed' defined in [6.25] as

$$\bar{v}_T(k; z) \equiv [\partial \tau_T(k; 0, z)/\partial z]^{-1} \tag{6.47}$$

is given by

$$\bar{v}_T(k_r; z) = j_{k_r}(z)/P_{k_r}(z) = (\hbar k_r/m)/|\psi_{k_r}(z)|^2 \tag{6.48}$$

where the plane-wave probability current density $j_k(z) = \hbar k/m$ is independent of z. Now, a glance at (6.13) reveals that an alternative expression for the Bohm particle velocity is $v(z, t) = j(z, t)/P(z, t)$. This definition is preferable to the familiar $v(z, t) \equiv m^{-1}\partial S(z, t)/\partial z$ because it is readily generalized [6.53] to particles described by the Pauli or Dirac equations simply by inserting the appropriate expressions for j and P. Clearly, $\bar{v}_T(k_r; z)$ is identical to the plane-wave limit of $v(z, t)$ for the special case $|T(k = k_r)|^2 = 1$ considered here. This is not true in general, not even for resonant transmission with $|T(k = k_r)|^2 < 1$ as is the case when the transmission probabilities for the individual barriers are unequal for $k = k_r$.

---

[5] *Leavens* and *Aers* were not familiar with Bohm's trajectory interpretation when [6.25] was written.

For $k = k_r^{(n)}$ with $n > 1$ it is possible to have quasinodes of the probability density in the well region $a \leq z \leq d - a$ where $|\psi_k(z)|^2$ is so small that (6.48) leads to $\bar{v}_T(z) \gg c$, the speed of light. However, if the Schrödinger expressions for the plane-wave probability and probability current densities $P_k(z)$ and j_k are replaced in (6.48) by the corresponding expressions [6.54] for positive energy spin-up (or down) Dirac electrons then one can readily prove [6.25] that $\bar{v}_T(k_r; z) \leq c$.[6] This is a very satisfactory result because it would be difficult to regard $v_T(k_r; z)$ as a meaningful quantity if it could exceed the speed of light. However, an electron trapped at resonance in the potential well between identical single barriers, each having transmission probability $|T(k_r)|^2_{single} \ll 1$ when acting alone, is often pictured as oscillating back and forth in the well very many (of order $|T(k)|^{-2}_{single}$) times before finally escaping through the second barrier. If this picture bears any relation to reality then we cannot use the fact that $\bar{v}_T(k_r; z) \leq c$ to argue that $\bar{v}_T(k_r; z)$ is a meaningful quantity because the mean transmission time $\tau_T(k; 0, d)$ from which it is derived via (6.47) is a cumulative quantity. Hence, if $\bar{v}_T(k_r; z) \sim c$ and the trapped electron passes the position z more than once before escaping then on at least one of these occasions its speed must have exceeded c. Within the Larmor clock approach an obvious, but perhaps not consistent, way out of this dilemma is simply to enforce a basic tenet of the Copenhagen interpretation and abandon the picture of an electron oscillating back and forth in the well as completely meaningless. Within Bohm's interpretation this escape route is not available. Instead one must show explicitly that such oscillations do not occur.

Bohm [6.14] showed that an electron in an eigenstate $\psi_n(z)$ of a potential well with perfectly reflecting walls has zero velocity, the kinetic energy of the usual picture having been completely converted into quantum potential energy Q. Hence, it is at least plausible that for steady-state resonant tunneling $[|T(k_r^{(n)})|^2 = 1]$ enough of the kinetic energy $E_r^{(n)} \equiv \hbar^2 k_r^{(n)2}/2m$ of an 'incident' electron is converted into quantum potential energy within the well region that the above deleterious oscillations do not occur. Figure 6.17 shows the stationary-state (non-relativistic) kinetic energy $[\partial S(z)/\partial z]^2/2m \equiv E_r - V(z) - Q(z)$ of Bohm's interpretation as a function of position for a double rectangular barrier system with five quasinodes of the probability density $|\psi_{k_r}(z)|^2$ in the well region. Except in the immediate vicinity of the quasinodes the kinetic energy in the well region is reduced, as required, to a very tiny fraction of its incident value. Since $|T(k_r)|^2 = 1$ an incident electron cannot avoid altogether the regions where the probability density is very small, but must propagate through them very quickly, much faster than the initial speed $\hbar k_r/m$ for the case shown. In fact, for quite ordinary parameters, using the Schrödinger equation can lead to calculated speeds very much in excess of c, as mentioned above. Within a relativistic

---

[6] We are not concerned here with barriers of enormous height, $\gtrsim mc^2$, for which spontaneous electron–positron creation must be taken into account. Strictly speaking, the barrier height V_0 of Sects. 6.4.2 and 6.5.1 should not be allowed to become infinite: $V_0 \to \infty$ should be interpreted as $E \ll V_0 \ll mc^2$.

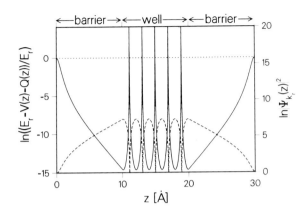

Fig. 6.17. Variation of the stationary-state kinetic energy $E_r - V(z) - Q(z)$ (*solid line*) and probability density $|\psi_{k_r}(z)|^2$ (*dashed line*) across the symmetric double rectangular barrier $V(z) = V_0[\theta(z)\theta(a - z) + \theta(z - d + a)\theta(d - z)]$ with $V_0 = 10$ eV, $a = 10$ Å and $d = 30$ Å. The resonance energy $E_r = E_r^{(6)} = 9.768$ eV

extension of the Bohm interpretation, as shown unwittingly in [6.25], this unphysical result is eliminated in a consistent way.

The stationary-state results of Fig. 6.17 which were generated analytically are now supplemented by numerical results for the time-dependent resonant scattering ($|T|^2 \sim 1$) of an initial Gaussian wave packet. For obvious reasons of numerical accuracy the parameters of the wave packet and the symmetric double rectangular barrier have been chosen so that the time taken for the scattering process to be essentially completed is not too large. None of the Bohm trajectories of Fig. 6.18 which were calculated with $\bar{E} = E_r$ shows an electron oscillating back and forth in the well region, even though the time spent in the well for most of the trajectories is sufficiently long for several oscillations to occur for a particle having a speed of $\sim \hbar k_r/m$.

Our calculations indicate that although the mean dwell time $\tau_D(0, d)$ peaks at $\bar{E} = E_r$ when the average energy $\bar{E}$ of the wave packet is increased through the resonance, the mean transmission time $\tau_T(0, d)$ shows only a monotonic decrease. However, the transmission component $|T|^2 \tau_T(0, d)$ of $\tau_D(0, d)$ does peak at resonance. Figure 6.19 shows the transmission time distributions

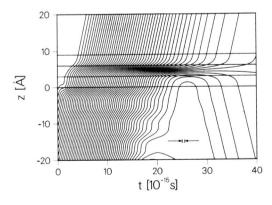

Fig. 6.18. Bohm trajectories for an initial Gaussian wave packet with $\bar{E} = E_r^{(1)} = 1.020$ eV, $\Delta k = 0.02$ Å^{-1} and $z_0 = -96.6$ Å incident on the symmetric double rectangular barrier $V(z) = V_0[\theta(z)\theta(a - z) + \theta(z - d + a)\theta(d - z)]$ with $V_0 = 2$ eV, $a = 3$ Å and $d = 9$ Å. The time taken for a completely free ($V_0 = 0$) electron to cross the well region is indicated. $|T|^2$ for the wave packet is 0.556

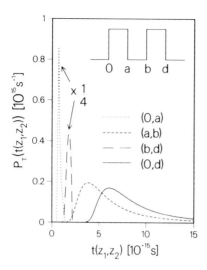

Fig. 6.19. Transmission time distributions for a Gaussian wave packet incident on a symmetric double rectangular barrier. The parameters are the same as those of Fig. 6.18

$P_T[t(z_1, z_2)]$ for the individual barriers, the well and the entire structure calculated for $\bar{E} = E_r$. The distributions for the two barrier regions both show well-defined minimum and maximum transmission times, while that for the well region has a very long tail. Although the two barriers are identical there is no left–right symmetry in the problem because the wave packet is incident from the left. Hence, it should not be a complete surprise that the mean transmission time for the second barrier is considerably larger than that for the first.

6.5 Discussion

6.5.1 'Measurement' of Particle Momentum

Before discussing the possibility of determining τ_T and τ_R experimentally, we consider the much simpler problem of 'measuring' the instantaneous momentum of a particle.

Heisenberg [6.55] considered a thought experiment in which the instantaneous value of the momentum of an electron in a hydrogen atom could be precisely determined, in principle, by instantaneously switching off the interaction between the electron and proton at the time of interest and doing a time-of-flight experiment on the liberated electron. This assumes, of course, that the momentum of the electron does not change, until it triggers the detector, from the value it had the instant before the potential was turned off and that the initial uncertainty in its position is negligible compared to the distance to the detector. Bohm discussed the same problem for an electron trapped in the potential well formed by the confining potential $V(z) = V_0[\theta(-z) + \theta(z-a)]$ with $V_0 \to \infty$

The stationary-state wave function $\psi_n(z)$ is equal to $(2/a)^{1/2} \sin k_n z$ with $k_n \equiv n\pi/a\,(n=1,2,\ldots)$ for $0 \leq z \leq a$ and to zero elsewhere. This standing wave is conventionally interpreted in terms of an electron with kinetic energy $E_n = \hbar^2 k_n^2/2m$ bouncing back and forth between the perfectly reflecting walls at $z=0$ and $z=a$. In Bohm's interpretation, because the stationary-state wave function $\psi_n(z)$ is real, the electron momentum $p(z) = \partial S(z)/\partial z$ is zero for all z and the electron is at rest at some unknown position z, with probability density $|\psi_n(z)|^2$, inside the well. Bohm claimed that these very different pictures of an electron trapped in a potential well lead to the same result for a time-of-flight determination of the momentum 'observable' [6.14]. In the first picture, the experimentally determined momentum distribution is identified with the momentum distribution immediately prior to the instantaneous collapse of the confining potential. In the second picture, the quantum potential evolves in time following this collapse and guides an ensemble of electrons, all initially at rest in positions distributed according to $|\psi(z,0)|^2 = |\psi_n(z)|^2$, in such a way that the resulting large-time velocity distribution is identical to the experimental one. According to this latter point of view, what is actually 'measured' in the time-of-flight experiment has no simple relation to the actual property of interest, the instantaneous momentum of an electron while it is confined in a potential well. The following analysis illustrates this important point. Evaluating the Fourier transform $\phi(k)$ of $\psi_n(z) = (2/a)^{1/2}\theta(z)\theta(a-z)\sin k_n z$ gives the *time-independent* velocity distribution

$$P(v) = \frac{m}{\hbar}|\phi(k)|^2 = \frac{4mk_n^2[1-(-1)^n\cos ka]}{ha(k_n^2-k^2)^2}\quad\left(v=\frac{\hbar k}{m}\right),\qquad (6.49)$$

according to the first version of the time-of-flight experiment[7]. In the second version, based on Bohm's interpretation, the particle velocity is given by (6.8). Hence, the velocity distribution depends on time: at $t=0$ it is given by $P(v;t=0)=\delta(v)$ and if the two pictures are to make identical predictions for the time-of-flight experiment it must evolve into the time-independent distribution $P(v)$ of the other picture for $t \gg t_n \equiv a/v_n$ with $v_n \equiv \hbar k_n/m$. That this is indeed the case is clearly indicated in Fig. 6.20 by the snapshots of the velocity distribution

$$P(v;t) = \int_{-\infty}^{+\infty} dz^{(0)}|\psi(z^{(0)},t=0)|^2\,\delta(v-v(z(z^{(0)},t),t))$$

$$= \int_0^a dz^{(0)}|\psi_n(z^{(0)})|^2\,\delta(v-v(z(z^{(0)},t),t))\qquad (6.50)$$

---

[7] It readily follows from (6.49) that the mean speed, mean-square velocity, and relative width of either half of the symmetrical distribution $P(v)$ are $|\bar{v}| = (2/\pi)v_n\{\mathrm{Si}(n\pi)-[1-(-1)^n]/n\pi\}$, $\overline{v^2}=v_n^2$, and $\Delta v/v_n = (\overline{v^2}-|\bar{v}|^2)^{1/2}/v_n \approx (2/\pi)n^{-1/2}$, respectively. The latter is close to $1/2$ for $n=2$ and decreases slowly with n. Hence, the relative deviation of $|v|$ from v_n is very small only for large n. For $n=1$ the distribution, far from being concentrated very close to $\pm v_{n=1}$, is actually largest at $v=0$.

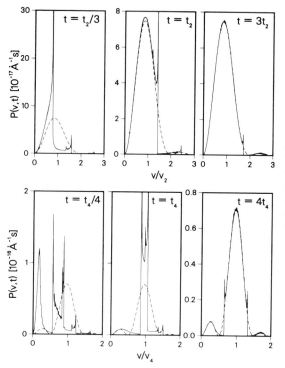

Fig. 6.20. Evolution of the time-dependent electron velocity distribution $P(v; t)$ (*solid line*) in the Bohm picture towards the time-independent distribution of the usual picture (*dashed line*) following the instantaneous collapse at $t = 0$ of the confining potential $V(z) = V_0[\theta(-z) + \theta(z-a)]$ with $a = 10\,\text{Å}$ and $V_0 \to \infty$. The initial velocity distribution is $P(v; t = 0) = \delta(v)$. The characteristic velocities and times are $v_n \equiv \hbar k_n/m$ with $k_n \equiv n\pi/a$ and $t_n \equiv a/v_n$. For the top panel $n = 2$ and for the bottom panel $n = 4$. Since $P(-v; t) = P(v; t)$ results are shown only for $v \geq 0$

for $v > 0$ calculated at $t/t_n = 1/3$, 1 and 3 for $n = 2$ and $t/t_n = 1/4$, 1 and 4 for $n = 4$. The large number of Bohm trajectories $z(z^{(0)}, t)$ needed for the evaluation of (6.50) were generated by simultaneously integrating the TDSE and $v(z, t) = dz/dt$ numerically starting from the initial wave function $\psi(z, t = 0) = \psi_n(z)$.[8]

The evolution of the time-dependent velocity distribution of the Bohm interpretation towards the time-independent distribution of the usual one following the instantaneous collapse of a confining potential is most clearly illustrated with the ground state of the harmonic oscillator potential. The wave function at the instant of collapse is

$$\psi(z, 0) = [2\pi(\Delta z)^2]^{-1/4} \exp[-z^2/4(\Delta z)^2] . \tag{6.51}$$

Casting the text-book expression [6.56] for the resulting time-dependent wave function into the form $\psi(z, t) = R(z, t) \exp[iS(z, t)/\hbar]$ and using (6.8) immedi-

---

[8] The finite-difference approximation to $\partial^2\psi(z, t)/\partial z^2$ fails at $t = 0$ for $z = 0$ and $z = a$ because it cannot reproduce the necessary singular behaviour of $\partial^2\psi_n(z)/\partial z^2$ at these points. The resulting spurious oscillations in $P(v, t)$ have been reduced to an acceptable level in the above examples, primarily by shifting the mesh so that $z = 0$ and $z = a$ are both located halfway between adjacent mesh points.

ately gives

$$v(z, t) = (z/t)/[1 + (t_0/t)^2] \quad (t_0 \equiv 2\Delta z/v_0; v_0 \equiv \hbar/m\Delta z) \tag{6.52}$$

for the Bohm velocity of the electron. Since a precise time-of-flight determination of the velocity distribution at the detector requires that $z(t)$ be proportional to t over most of the flight path, it is clear that the detection times must be much greater than the characteristic time t_0 associated with the harmonic oscillator potential. In the conventional picture where the electron's velocity does not change from the value it had at $t = 0$ the requirement for an accurate experimental result is that the detector distance be much greater than Δz, the uncertainty in the initial position of the electron. Clearly, since $t_0 = 2\Delta z/v_0$ the two pictures involve essentially the same experimental requirements.

Integrating $dz/dt = v(z, t)$ over time from 0 to t and over position from $z^{(0)}$ to z gives

$$z(z^{(0)}, t) = z^{(0)}[1 + (t/t_0)^2]^{1/2} \tag{6.53}$$

for the Bohm trajectory starting from $z^{(0)}$ at $t = 0$. It follows from (6.52, 53) that

$$v[z(z^{(0)}, t), t] = (z^{(0)}/t_0)/[1 + (t_0/t)^2]^{1/2} . \tag{6.54}$$

Substituting this into (6.50) and integrating gives

$$P(v; t) = \frac{(2/\pi)^{1/2}}{v_0}\left[1 + \left(\frac{t_0}{t}\right)^2\right]^{1/2} \exp\left\{-2\left[1 + \left(\frac{t_0}{t}\right)^2\right]\left(\frac{v}{v_0}\right)^2\right\}. \tag{6.55}$$

This is $\delta(v)$ at $t = 0$ and for $t \gg t_0$ converges to

$$P(v; t = \infty) = \left(\frac{2}{\pi}\right)^{1/2}\frac{m\Delta z}{\hbar}\exp\left[-2\left(\frac{mv\Delta z}{\hbar}\right)^2\right] = \frac{m}{\hbar}|\phi(k)|^2 \equiv P(v) \tag{6.56}$$

where $\phi(k) = [8\pi(\Delta z)^2]^{1/4}\exp[-k^2(\Delta z)^2]$ is the well-known Fourier trans-

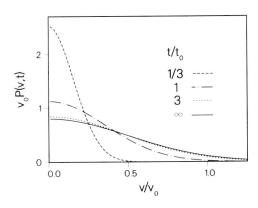

Fig. 6.21. Evolution of the time-dependent electron velocity distribution $P(v; t)$ for $v > 0$ of the Bohm picture towards the time-independent distribution of the usual picture (solid line) following the instantaneous collapse at $t = 0$ of the harmonic oscillator confining potential. The initial wave function is the ground state of the harmonic oscillator potential and the corresponding initial velocity distribution is $P(v; t = 0) = \delta(v)$. The characteristic velocity and time are $v_0 \equiv \hbar/m\Delta z$ and $t_0 \equiv 2\Delta z/v_0$

form of the Gaussian wave function (6.51). Figure 6.21 shows $v_0 P(v, t)$, a universal function of v/v_0 and t/t_0, for $t/t_0 = 1/3, 1, 3$ and ∞. Clearly, in both pictures the hypothetical time-of-flight experiment performed on a suitably large ensemble will yield precisely the same velocity distribution. However, only in the conventional picture would this distribution be identified with the electron velocity distribution immediately prior to the collapse of the confining harmonic potential.

6.5.2 'Measurement' of Mean Transmission and Reflection Times

It is obviously pointless to discuss the possibility of experimentally determining the average transmission and reflection times if they are meaningless concepts. This is definitely not the case within Bohm's interpretation and one can envisage measuring these quantities, at least in principle. However, an experimentally-derived quantity that is identified as the mean transmission time within this interpretation may not be accepted as such by adherents of other interpretations, not even by those who believe that the concept is a meaningful one. To illustrate this important point we consider an indirect method for determining mean transmission and reflection times based on *Lamb*'s operational approach to state preparation and measurement [6.57].

Lamb's admittedly *idealized* experiments are based on the assumption that 'all classically describable potentials $U(z, t)$ are available experimentally'. For example, any desired initial one-electron state $\psi(z, 0) \equiv R(z, 0) \exp[iS(z, 0)/\hbar]$ is prepared as follows: With the potential $V(z, t)$ of interest switched off, the potential $U_1(z)$ given by $[-(\hbar^2/2m)\partial^2/\partial z^2 + U_1(z) - E]R(z, 0) = 0$ is applied to prepare a state which has the real-valued wave function $R(z, 0)$ when the energy of an electron captured by the potential has the correct eigenenergy E. Then, at $t = 0^-$, $U_1(z)$ is switched off instantaneously, the pulse potential $U_2(z, t) = -S(z, 0)\delta(t)$ is applied to generate the desired initial wave function $\psi(z, 0)$, and at $t = 0^+$ the potential $V(z, t)$ is switched on instantaneously to complete the preparation. It should be noted that the quantity $E - U_1(z)$ is identical to the quantum potential $Q(z, 0)$ and that integrating the 'acceleration' $(1/m)[-\partial U_2(z,t)/\partial z]$ over all t gives $(1/m)\partial S(z, 0)/\partial z$ which is precisely the particle velocity $v(z, 0)$ of Bohm's approach. This helps to elucidate the dual role of $\psi(z, t)$ as a field guiding the motion of an individual electron and as the position probability amplitude for an ensemble of identically prepared one-electron systems. Lamb has also shown how the probability density $|\psi(z, t)|^2$ could be precisely measured, in principle at least, by switching on a sufficiently short-ranged potential trap at the time and position of interest and simultaneously switching off the potential $V(z, t)$. Given sufficiently accurate measured values of $|\psi(z, t)|^2$ for the scattering problem of interest, the mean Bohm trajectory transmission and reflection times $\tau_T(z_1, z_2)$ and $\tau_R(z_1, z_2)$ are readily obtained by carrying out the integrations appearing in (6.28–30). However, the physical meaning attached to (6.28, 29) is based on a set of postulates unique to the Bohm interpretation of quantum mechanics. Hence, to a proponent of any conventional interpretation, there is no reason to identify the experimental 'times' based on these equations with actual mean transmission and reflection times. The analogy with the time-of-flight thought experiment is clear. However, in that case it is the proponent of the Bohm interpretation who cannot accept the measured momentum distribution as the one of actual interest, i.e. the distribution immediately prior to the collapse of the confining potential.

There is no consensus among proponents of conventional interpretations on whether or not transmission and reflection times are meaningful concepts. The point of view that they are not meaningful appears to be a consistent one *within conventional interpretations*. Despite this, several physical phenomena involving tunneling have been analyzed assuming that the mean transmission time $\tau_T(0, d)$ is an important parameter [6.1–6]. This has led to suggestions for practical experiments to measure τ_T, some of which have been attempted [6.1–4]. These suggestions are based on the assumption that there will be a qualitative change in behaviour of some measured property when some experimentally-controlled 'time' parameter t_{exp}, usually the reciprocal of a characteristic frequency, is swept through a value equal to τ_T [6.26, 27]. Again the problem is one of interpretation: Does a qualitative change in behaviour at $t_{exp} = t_{exp}^0$ necessarily correspond to $t_{exp}^0 \sim \tau_T$ or is some other time scale involved? In this connection, it should be emphasized that one should think about such experiments in terms of distributions of transmission *and* reflection times $P_T[t(z_1, z_2)]$ and $P_R[t(z_1, z_2)]$ rather than just the mean transmission time $\tau_T(0, d)$. It is quite possible that $\tau_T(0, d)$ is not a relevant time scale. This is certainly the case for the frequency dependence of the transmission probability for the time-modulated barrier considered above.

6.5.3 Concluding Remarks

In this chapter a long standing problem which has not been satisfactorily resolved within standard interpretations of quantum mechanics has been explored within Bohm's trajectory interpretation. The criteria that had previously led us to have serious reservations about all approaches to the 'tunneling time problem' then known to us [6.8] are satisfied by the Bohm trajectory approach. It thus appears to us that there are currently two *consistent* answers to this problem: (1) within Bohm's interpretation the basic postulates provide a unique, well-defined prescription for calculating meaningful transmission times with physically reasonable properties; (2) within conventional interpretations the concept of transmission time is not a meaningful one and forcing it upon the basic formalism ultimately leads to unphysical results. Unfortunately, as illustrated above, an experimentally-determined quantity identified with a mean transmission time within Bohm's interpretation might not be accepted as such within conventional ones (or vice versa). Hence, it may not be possible to obtain universal agreement as to what would constitute a definitive experiment to resolve the issue.

The following statements by *Feynman* et al. [6.58] should be borne in mind in connection with the tunneling time problem: "Just because we cannot *measure* position and momentum precisely does not *a priori* mean that we *cannot* talk about them. It only means that we *need* not talk about them. A concept or an idea which cannot be measured or cannot be referred directly to an experiment may or may not be useful." The beauty of the Bohm interpretation, in this context, is that the concepts of transmission and reflection times follow directly from the basic postulates and consequently one can talk about them in an internally consistent way. Furthermore, due to the nature of these postulates, one can use the language of classical mechanics without apology (e.g., the above discussion of the time-modulated barrier is not peppered with the phrase "crudely speaking"). Moreover, the averages discussed in the Bohm trajectory parts of this chapter are averages in the classical sense and the transmission times are real quantities. In standard interpretations on the other hand, to quote from *Feynman* and *Hibbs* [6.17], " ... the weighting function in quantum mechanics is a complex function. Thus the result is not an 'average' in the ordinary sense." This leads, for example, to the result that even the simple correlation function $\langle z(t) z(t') \rangle_F$ is complex-valued [6.17]. For reasons such as these, which are largely ones of taste, we prefer the Bohm trajectory solution of the tunneling time problem.

This chapter is dedicated to the late David Bohm.

References

6.1   P. Guéret: Dynamic polarization effects in tunneling, in *Electronic Properties of Multilayers and Low-Dimensional Semiconductor Structures*, ed. by J.M. Chamberlain, L. Eaves, J.C. Portal (Plenum, New York 1990) pp. 317–329

6.2   B.N.J. Persson, A. Baratoff: Phys. Rev. B **38**, 9616 (1988)

6.3   A.A. Lucas, P.H. Cutler, T.E. Feuchtwang, T.T. Tsong, T.E. Sullivan, Y. Yuk, H. Nguyen, P.J. Silverman: J. Vac. Sci. Technol. A **6**, 461 (1988)

6.4   R. Möller: to be published

6.5   J.K. Gimzewski, J.K. Sass: to be published

6.6   C.R. Leavens, G.C. Aers: Effect of lattice vibrations in scanning tunneling microscopy, in *Scanning Tunneling Microscopy '86*, ed. by N. García (North-Holland, Amsterdam 1987)

6.7   E.H. Hauge, J.A. Støvneng: Rev. Mod. Phys. **61**, 917 (1989)

6.8   C.R. Leavens, G.C. Aers: Tunneling times for one-dimensional barriers, in *Scanning Tunneling Microscopy and Related Methods*, ed. by R.J. Behm, N. García, H. Rohrer (Kluwer, Dordrecht 1990) pp. 59–76

6.9   M. Büttiker: Traversal, reflection and dwell times for quantum tunneling, in *Electronic Properties of Multilayers and Low-Dimensional Semiconductor Structures*, ed. by J.M. Chamberlain, L. Eaves, J.C. Portal (Plenum, New York 1990) pp. 297–315

6.10 R. Landauer: Ber. Bunsenges. Phys. Chem. **95**, 404 (1991)

6.11 A.P. Jauho: Tunneling times in semiconductor heterostructures: A critical review, in *Hot Carriers in Semiconductor Nanostructures—Physics and Applications*, ed. by J. Shah (Academic, New York 1992) pp. 121–151

6.12 M. Jonson: Tunneling times in quantum mechanical tunneling, in *Quantum Transport in Semiconductors*, ed. by D.K. Ferry, C. Jacoboni (Plenum, New York 1992) pp. 193–238.

6.13 Private Communications with several individuals

6.14 D. Bohm: Phys. Rev. **85**, 166 (1952); 180 (1952)

6.15 D. Bohm, B.J. Hiley, P.N. Kaloyerou: Physics Rep. **144**, 323 (1987)
6.16 J.S. Bell: *Speakable and Unspeakable in Quantum Mechanics* (Cambridge Univ. Press, Cambridge 1987) Chaps. 4, 11, 14, 15, 17
6.17 R.P. Feynman, A.R. Hibbs: *Quantum Mechanics and Path Integrals* (McGraw-Hill, New York 1965) Chap. 7
6.18 D. Sokolovski, L.M. Baskin: Phys. Rev. A **36**, 4604 (1987)
6.19 D. Sokolovski, J.N.L. Connor: Phys. Rev. A **42**, 6512 (1990)
6.20 D. Sokolovski, J.N.L. Connor: Phys. Rev. A **44**, 1500 (1991)
6.21 C.R. Leavens, G.C. Aers: Phys. Rev. B **39**, 1202 (1989)
6.22 M. Büttiker: Phys. Rev. B **27**, 6178 (1983)
6.23 V.F. Rybachenko: Sov. J. Nucl. Phys. **5**, 635 (1967)
6.24 A.I. Baz': Sov. J. Nucl. Phys. **4**, 182 (1967)
6.25 C.R. Leavens, G.C. Aers: Phys. Rev. B **40**, 5387 (1989)
6.26 M. Büttiker, R. Landauer: Phys. Rev. Lett. **49**, 1739 (1982)
6.27 M. Büttiker, R. Landauer: Phys. Scr. **32**, 429 (1985)
6.28 D. Bohm: *Quantum Theory* (Prentice-Hall, New York 1951)
6.29 E.P. Wigner: Phys. Rev. **98**, 145 (1955)
6.30 Z. Huang, P.H. Cutler, T.E. Feuchtwang, R.H. Good Jr., E. Kazes, H.Q. Nguyen, S.K. Park: J. Physique **C6**, 17 (1988)
6.31 M. Büttiker, R. Landauer: IBM J. Res. Develop. **30**, 451 (1986)
6.32 J.S. Bell: Physics **1**, 195 (1965)
6.33 D. Bohm, B.J. Hiley: Found. Phys. **14**, 255 (1984)
6.34 B.J. Hiley: The role of the quantum potential in determining particle trajectories and the resolution of the measurement problem, in *Open Questions in Quantum Physics*, ed. by G. Tarozzi, A. van der Merwe (Reidel, Dordrecht 1985) pp. 237–256
6.35 D. Bohm, B.J. Hiley: Nonlocality and the Einstein–Podolsky–Rosen experiment as understood through the quantum-potential approach, in *Quantum Mechanics versus Local Realism*, ed. by F. Selleri (Plenum, New York 1988) pp. 235–256
6.36 C. Philippidis, C. Dewdney, B.J. Hiley: Nuovo Cim. B **52**, 15 (1979)
6.37 C. Philippidis, D. Bohm, R.D. Kaye: Nuovo Cim. B **71**, 75 (1982)
6.38 D. Bohm, C. Dewdney, B.J. Hiley: Nature **315**, 294 (1985)
6.39 C. Dewdney, B.J. Hiley: Found. Phys. **12**, 27 (1982)
6.40 C.R. Leavens: Solid State Commun. **74**, 923 (1990)
6.41 C.R. Leavens: Solid State Commun. **76**, 253 (1990)
6.42 A.V. Dmitriev: J. Phys. Cond. Matt. **1**, 7033 (1989)
6.43 T.E. Hartman: J. Appl. Phys. **33**, 3427 (1962)
6.44 S. Collins, D. Lowe, J.R. Barker: J. Phys. C **20**, 6213 (1987)
6.45 E.H. Hauge, J.P. Falck, T.A. Fjeldly: Phys. Rev. B **36**, 4203 (1987)
6.46 J.P. Falck, E.H. Hauge: Phys. Rev. B **38**, 3287 (1988)
6.47 H. De Raedt: Comp. Phys. Rep. **7**, 1 (1987)
6.48 T.P. Spiller, T.D. Clark, R.J. Prance, H. Prance: Europhys. Lett. **12**, 1 (1990)
6.49 J.A. Støvneng, E.H. Hauge: J Stat. Phys. **57**, 841 (1989)
6.50 A.P. Jauho, M. Jonson: J. Phys. Cond. Matt. **1**, 9027 (1989)
6.51 H. De Raedt, N. García, J. Huyghebaert: Solid State Commun. **76**, 847 (1990)
6.52 C.R. Leavens, G.C. Aers: Solid State Commun. **78**, 1015 (1991)
6.53 J.S. Bell: *Speakable and Unspeakable in Quantum Mechanics* (Cambridge Univ. Press, Cambridge 1987) pp. 127–133
6.54 J.D. Bjorken, S.D. Drell: *Relativistic Quantum Mechanics* (McGraw-Hill, New-York 1968) Chap. 3
6.55 W. Heisenberg: *Physical Principles of the Quantum Theory* (Univ. of Chicago Press, Chicago 1930) p. 21
6.56 L.I. Schiff: *Quantum Mechanics* (McGraw-Hill, New York 1968) Chap. 3
6.57 W.E. Lamb Jr.: Physics Today **22**, 23 (April 1969)
6.58 R.P. Feynman, R.B. Leighton, M. Sands: *The Feynman Lectures* (Addison-Wesley, Reading 1965) Vol. III.

Additional References with Titles

The following two books on the de Broglie–Bohm interpretation of quantum mechanics are scheduled for publication in 1993:

D. Bohm, B.J. Hiley: *The Undivided Universe: An Ontological Interpretation of Quantum Mechanics* (Routledge, London 1993)

P.R. Holland: *The Quantum Theory of Motion* (Cambridge University Press, Cambridge 1993)

7. Unified Perturbation Theory for STM and SFM

C.J. Chen

With 16 Figures

In this chapter, we describe perturbation theory for STM and SFM. To understand the influence of tip electronic states and the tip–sample interactions in the imaging process, and to interpret the observed images, perturbation theory provides a simple and straightforward picture. First, besides STM, there are a number of experimental methods which contribute to the understanding of the sample surface as well as the tip. From an experimental point of view, the perturbation theory can provide insights for the understanding of the images from the properties of the bare tip and the bare sample, thus to achieve a conceptual understanding of the relations among different experimental measurements. Second, first-principles numerical calculations of the electronic structures of the free sample and the free tip have reached a high level of sophistication that a detailed comparison with those experimental measurements has become everday practice. Perturbation theory can bring the results of those calculations together and make predictions to the STM and SFM images. A consistency among the results from different approaches is a sign of true understanding. In words, a natural way of bringing various theoretical and experimental fields into a unified picture is through a proper perturbation theory. Finally, the perturbation theory provides a natural linkage between the tunneling phenomenon and attractive atomic forces through the concept of *resonance*, which is the foundation of Pauling's theory of the chemical bond. As we shall see, this brings about a unified perturbation theory of STM and AFM.

Two crucial questions must be answered concerning a perturbation-theory-approach, namely, its accuracy and the form of its final results. We will show that by modifying the Bardeen perturbation theory to include the polarization effect (i.e., van der Waals effect), the final results can be sufficiently accurate. We will also show that the tunneling matrix elements relevant to STM and SFM have extremely simple forms, which can result in simple analytic expressions for the images.

7.1 Background

7.1.1 A Brief Summary of Experimental Facts

It has been ten years since *Binnig* and *Rohrer* [7.1] invented the STM. A large body of experimental facts have been accumulated, as detailed in the first two

Springer Series in Surface Sciences, Vol. 29
Scanning Tunneling Microscopy III   Eds.: R. Wiesendanger · H.-J. Güntherodt
© Springer-Verlag Berlin Heidelberg 1993

volumes of this treatment [7.2]. Modern physics is an empirical science, as Heike Kamerlingh Onnes said, *meten is weten* (to measure is to know). The mission of a theory is to explain the experimental facts as comprehensively and comprehensibly as possible, and the validity of a theory is further tested by ongoing measurements. The main experimental facts which a theory of STM must explain are: atomic resolution; the dependence of STM images on bias voltage, set current, and tip structure; the simultaneous observation of force between the tip and the sample and the effect of such a force on STM images. The last fact actually motivated the invention of Scanning Force Microscopy (SFM). The following is a brief description of those facts:

Atomic Resolution. STM has resolved nearest-neighbor atoms and the details of the vicinity of individual atoms on a wide variety of surfaces of metals, semiconductors, and layered materials. For example, various semiconductor surfaces [7.3], with nearest-neighbor atomic distance 2.2 to 3.8 Å; low Miller index metal surfaces, with atomic distance 2.5 to 2.9 Å [7.4–7], and layered materials, 1.5 to 3.5 Å [7.8].

Voltage Dependence of STM Images. In general, the STM image depends on the polarity and magnitude of the bias, especially on semiconductors. For example, on Si(111)2 × 1 and Ge(111)2 × 1 [7.3], along the (011) direction, the corrugation of the STM image at $+1V$ bias and at $-1V$ is almost reversed. The local tunneling spectra contain rich information on the local electronic structure of the sample and the tip [7.3].

Distance Dependence of Corrugation Amplitudes. The corrugation amplitude of STM images of clean surfaces shows an exponential dependence on tip–sample distance. On most metals, the observed atomic corrugation decays one order of magnitude per 1–2 Å [7.5–7]. Experimentally, the tip–sample distance is adjusted through the reference tunneling conductance, i.e., the set current.

Very Small Tunneling Gaps. As a consequence of the rapid decay of atomic information, the normal tunneling gaps enabling atomic resolution should be very small. Direct measurements show that under normal operational conditions of STM, the tip–sample distance is 1–4 Å from a mechanical contact [7.9–11]. Because of the image potential, the top of the potential barrier is often close to or lower than the Fermi level.

Effects of Tip Structure. The corrugation of the STM image depends dramatically on tip structure. The corrugation amplitude for the same surface at the same tunneling conditions can differ by one to two orders of magnitude, depending upon the tip condition [7.5–7]. Almost all reported atom-resolved STM imaging uses W or Pt-Ir tips. The apex atoms are either a component of the tip material, or sample atoms (Si, C, etc.), picked up by the tip.

Effects of Attractive and Repulsive Forces. Direct measurements also show that under normal operational conditions of STM, the attractive force as well as the repulsive force are often as large as 0.1 to 10 nanonewton [7.9–11]. The deformation of the tip end and the sample near the gap region is large enough to become measurable. The force and sample deformation play an essential role in the imaging process on graphite and many other layered materials [7.10]. Actually, the necessity of considering the effect of attractive force in the interpretation of the metal-vacuum-metal tunneling experiments was already emphasized by *Teague* in 1978 [7.12], several years before the invention of STM.

7.1.2 The Bardeen Approach for Tunneling Phenomena

In the early years of STM, Bardeen's approach to tunneling phenomena [7.13] was the basis of much of the theoretical understanding of STM, as described in Chap. 2. Before presenting our new tunneling theory, we make a brief remark on the origin and spirit of the Bardeen approach.

Three years after Bardeen, Cooper, and Schrieffer (BCS) published their theory of superconductivity in 1957 [7.14], *Giaever* [7.15] did a tunneling experiment by applying a voltage V on a sandwich of a normal metal (Al), a thin layer of insulator (Al_2O_3), and a superconductor (Pb, Sn, In, etc.), as shown in Fig. 7.1. The measured dynamic conductance dI/dV strikingly resembled the *theoretical* Density Of States (DOS) ϱ_s of the superconductor [7.15] predicted by BCS [7.14],

$$\varrho_s = \varrho_n \frac{|E|}{\sqrt{E^2 - \Delta^2}}, \quad |E| \geq \Delta \ ,$$

$$\varrho_s = 0, \qquad\qquad |E| < \Delta \ , \tag{7.1}$$

where Δ is the BCS energy gap, and ϱ_n is the DOS of normal metal, a constant. If the dynamic tunneling conductance dI/dV is indeed proportional to the DOS of the superconductor, then Giaever's experiment is a direct verification of the BCS theory. In Giaever's first paper, this is an assumption . To verify the assumption, *Bardeen* [7.13] considered the tunneling process from the point of view of time-dependent perturbation theory [7.16]. In general, the transition probability follows the Fermi golden rule

$$w = \frac{2\pi}{\hbar} |M|^2 \varrho_f \ , \tag{7.2}$$

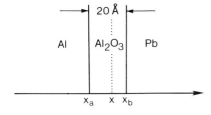

Fig. 7.1. Giaever's tunneling experiment and Bardeen's treatment

where M is the transition matrix element from the initial state to the final state, and ϱ_f is the final-state DOS. Therefore, the issue is reduced to whether the matrix element M is approximately constant in the energy range of interest.

To resolve this problem, Bardeen derived an explicit expression for the matrix element. From Fig. 7.1, it is seen that the thickness of the tunneling barrier, typically 20 Å, is much greater than the typical decay length of the wave function, 1 Å. Thus, the two electrodes are weakly linked. Under this condition, *Bardeen* showed that the matrix element can be expressed as a surface integral [7.13],

$$M = \frac{\hbar}{2m} \int \left(\chi^* \frac{\partial \psi}{\partial x} - \psi \frac{\partial \chi^*}{\partial x} \right) dS \; , \tag{7.3}$$

where ψ is the wave function of the left electrode, assuming that the insulator extends from x_a to $+\infty$; and χ is the wavefunction of the right electrode, assuming that the insulator extends from $-\infty$ to x_b (Fig. 7.1). The surface integral is performed on any plane in the insulator, i.e., $x_a < x < x_b$. If the temperature is low enough, that is, when $k_B T$ is much smaller than the feature size in the energy spectra, then the rate of electron transfer is determined by

$$I = \frac{4\pi e}{\hbar} \int_0^{eV} \varrho_L(E_F - eV + \varepsilon) \, \varrho_R(E_F + \varepsilon) |M|^2 \, d\varepsilon \; , \tag{7.4}$$

where ϱ_L and ϱ_R are the DOSs of the left and right electrodes, respectively. Using the explicit expression for the matrix element, Bardeen showed that M is indeed roughly independent of energy. Thus, if the left electrode is a normal metal, then (7.4) implies that the tunneling conductance in Gaiever's experiment is *directly proportional* to the DOS of the superconductor, which is the right electrode. Giaever's tunneling experiment was an important factor which led to the universal recognition of the BCS theory of superconductivity in 1972.

In *Bardeen's* original paper [7.13], the tunneling problem is formulated as the time evolution of the many-body wave function Ψ of the entire system (which consists of three components). *Duke* [7.17] showed that *Bardeen's* treatment is equivalent to *Oppenheimer's* treatment of the field-ionization problem [7.18], cast in the independent-particle formalism. Oppenheimer even gave an explicit expression for an error estimation of the first-order time-dependent perturbation theory [7.18]. Furthermore, as a fact known already in the sixties [7.17], the Bardeen approach is not valid when the interaction between the two electrodes is strong. In fact, in *Bardeen's* original paper, his justification is that his method is consistent with the WKB results [7.13]. If the tunneling barrier becomes closer to or even lower than the Fermi level, both the WKB theory and the Bardeen theory collapse. In the seventies, several groups developed more accurate formalisms for treating tunneling phenomena, for example, *Caroli* et al. [7.19], and *Feuchtwang* et al. [7.20].

In 1983, *Tersoff* and *Hamann* applied Bardeen's approach to the theory of STM [7.21]. As described in Chap. 2, the Tersoff–Hamann theory is equivalent to idealizing the tip by a point current source (or a point current drain), which has no interaction with the sample. The STM image in that idealized model is independent of the tip structure. In other words, the STM image in the Tersoff–Hamann theory is a property of the bare sample surface, without reference to the tip–sample system. Because of the weak-interaction restriction of the Bardeen approach, the Tersoff–Hamann theory is only applicable when the tip-sample distance is large, e.g., $> 6 \text{Å}$. At such a distance, the atomic corrugation has already faded away. In the original papers of *Tersoff* and *Hamann* [7.21], their theory is applied to the profiles of large superstructures. Also, the Tersoff–Hamann theory is applicable to the case of a single large atom adsorbed on a metal surface (with its atomic structure neglected), which is discussed extensively in Chap. 2. In order to understand the atom-scale ($\simeq 2 \text{Å}$) phenomena in STM, the starting point must go beyond Bardeen's approach [7.22]. In the two previous chapters of this book, Noguera presented the Green's function method, and Doyen presented the scattering-theory method. Both have gone beyond the Bardeen's approach and have the potential to account for many important experimental observations.

7.1.3 Perturbation Approach for STM and SFM

From a theoretical point of view, it is preferable to interpret most of the phenomena observed by STM from a single and natural starting point. It is also preferable to correlate STM images with other experimental results regarding the free sample and the free tip, as well as first-principles numerical calculations of the electronic structures of the free sample and the free tip. In fact, for a large number of surfaces studied by STM, there are numerous other experimental methods which have provided a plethora of data, for example, X-Ray diffraction methods, electron microscopies, LEED, photoemission and inverse photo-emission, and, notably, atom beam scattering. First-principles numerical calculations of the electronic structure of surfaces have reached such a high level of sophistication that a detailed comparison with those experimental data has become everyday practice. On the other hand, the metal tip, similar to those used in STM, has been the subject of intense experimental studies for several decades, using field-ion microscopy, as well as field-emission microscopy and spectroscopy. First-principles theoretical calculations have also achieved general agreement with those experimental results regarding the tip. An extensive summary of the first-principles studies of the tip electronic structure and its effects are given in Chap. 5. Finally, since the Tersoff–Hamann theory is the correct limit for large structures on metals, it is preferable to have a theory which explicitly spells out the physics beyond the Tersoff–Hamann theory, but also explicitly reduces to the Tersoff–Hamann theory for large structures on metal surfaces at a low bias.

A natural way of bringing various fields into a unified picture is through a proper perturbation theory. Consider the following four regimes of interest, where d denotes the tip–sample distance with respect to a mechanical contact:

1) When the tip–sample distance is large, for example, $d > 100$ Å, the mutual interaction is negligible. By applying a large electrical field between them, field emission may occur. However, the phenomena can be described as the interaction of one electrode with the electrical field, without considering any interactions from the other electrode.

2) At intermediate distance, for example, 10 Å $< d < 100$ Å, a long-range interaction between them takes place. The wave functions of both the tip and the sample are distorted, and a van der Waals force arises. The van der Waals interaction follows a power law, with an order of magnitude of a few meV per atom.

3) At shorter distance, for example, 1 Å $< d < 10$ Å, the electrons may transfer from one side to another. The exchange of electrons gives rise to an attractive interaction, that is, the *resonance energy* of Heisenberg [7.30] and Pauling [7.31]. If a bias is applied between them, a *tunneling current* occurs. The resonance energy follows an exponential law, and can be much larger than the van der Waals interaction, e.g., of the order of eV per atom.

4) At extremely short distances, for example, $d < 1$ Å, the repulsive force starts to dominate. It has a very steep distance dependence. The tip-sample distance is virtually determined by the repulsive force. By pushing the tip further towards the sample surface, the tip and sample deform accordingly.

It is important to note that the polarization interaction or the van der Waals interaction starts at a much larger distance than the exchange interaction. The former can be accounted for by a stationary-state perturbation theory, effectively and accurately, without concerning the exchange interaction. The exchange interaction or tunneling can be treated by time-dependent perturbation theory, following the method of *Oppenheimer* [7.19] and *Bardeen* [7.13]. However, since the polarization or the van der Waals interaction is still in effect, the time-dependent perturbation theory directly based on wave functions of free sample and free tip would not provide the correct answer. Most of the STM experiments are performed in the region where both wave function distortion and electron exchange are present. Hence, both perturbations must be considered simultaneously.

The goal of this chapter is to describe a perturbation theory to account for wave function distortion and electron exchange effects simultaneously. The new perturbation theory is essentially the Oppenheimer–Bardeen time-dependent perturbation theory combined with a stationary-state perturbation treatment of the van der Waals interaction. We will call it a *Modified Bardeen Approach*, acronymized MBA. On the one hand, it provides a formula for the tunneling current which is *explicit* but much more accurate than the Bardeen approach. On the other hand, it provides the formulas describing the force arising from the exchange interaction between the tip and the sample, which is the dominant force under normal operational conditions of STM.

As a justification of the Modified Bardeen Approach (MBA), we will compare its results with analytically soluble cases. One is the elementary square-barrier problem. We will show that the MBA result is accurate even if the top of the barrier is lower than the energy level. Another case is the hydrogen molecular ion, which is one of the most important analytically soluble problems in quantum mechanics. The numerical results are well tabulated. We will show that the MBA can reproduce its *potential energy curve* with very high accuracy when the distance between the two protons is greater than 3 Å. The hydrogen molecular ion problem is the basis of the understanding of a covalent chemical bond and, in general, forces between atoms. Therefore, the MBA accounts for both tunneling conductance and attractive atomic force. Because the MBA is valid even when the top of the potential barrier becomes lower than the Fermi level, we will use the term *quantum transmission* to cover the transport phenomena across both classically-forbidden and classically-allowed barriers. The details of derivations can be found in the original papers [7.23–28]. A pedagogic presentation, understandable from an undergraduate-senior quantum mechanics background, can be found in a recent book [7.29].

7.2 The Modified Bardeen Approach

7.2.1 General Derivation

In this section, we outline a general derivation for the MBA. We will show that by evaluating Bardeen's tunneling matrix element from properly *modified* wave functions, the transmission matrix element becomes accurate even if the barrier *collapses*.

Figure 7.2 is the energy schematics of an STM, in the independent-electron picture. As the tip and the sample approach each other with a finite bias V, the potential U in the barrier region becomes different from the potentials of the free tip and the free sample. To make perturbation calculations, we draw a separation surface between the tip and the sample, then define a pair of sub-systems with potential surfaces U_S and U_T, respectively. As we show later on, the exact position of the separation surface is not important. As shown in Fig. 7.2, we *define* that the sum of the two potentials of the individual systems is equal to the potential of the entire system, that is,

$$U_S + U_T = U \ . \tag{7.5}$$

In other words, in the tip region, $U_T = U$, and $U_S = 0$; and in the sample region, $U_S = U$, and $U_T = 0$. Clearly, there is another relation between them: the product of the two potentials is zero throughout the entire space,

$$U_S U_T = 0 \ . \tag{7.6}$$

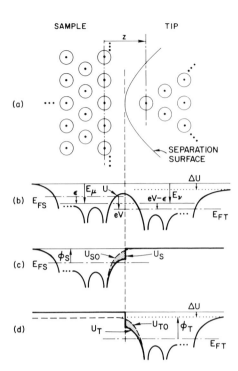

SAMPLE   TIP

Fig. 7.2. Perturbation approach for quantum transmission. **(a)** a separation surface is drawn between the two subsystems. The precise location is not critical. **(b)** the potential surface of the system. **(c)** and **(d)**, the potential surfaces of the subsystems [7.26]

The reference point of energy, the vacuum level, is well defined in the STM problem. The entire system is neutral. Therefore, at infinity, there is a well-defined vacuum potential. In the vicinity of the apex of the tip, the potential barrier in the gap is substantially lowered. However, the barrier lowering is confined in a small region near the tip end. Outside the interaction region, the potential in the space equals to the vacuum level. The second relation, (7.6) is also a necessary condition to reduce the error of perturbation treatment. As shown below, it minimizes *Oppenheimer's* error estimation term [7.18].

As seen from Fig. 7.2, the potentials U_S and U_T are different from the potentials of the free tip, U_{T0}, and the free sample, U_{S0}, respectively. The difference, i.e., $(U_{S0} - U_S)$, is then used as a perturbation to the bare wave functions. Moreover, it is the origin of the van der Waals force.

At $t < 0$, U_T is turned off. Schrödinger's equation gives the stationary states of the sample:

$$(T + U_S)\,\psi_\mu = E_\mu \psi_\mu \;,\tag{7.7}$$

where $T = -(\hbar^2/2m)\nabla^2$ is the kinetic energy. At $t > 0$, the potential in the tip body is turned on. Following the standard time-dependent perturbation theory [7.16], the wave function is expanded in terms of the modified wavefunctions of the tip χ_v, defined by the following Schrödinger equation,

$$(T + U_T)\chi_v = E_v\chi_v \ , \tag{7.8}$$

the transition probability of an electron from ψ_μ to χ_v in the first order perturbation theory is then given by the Fermi golden rule

$$w_{\mu v}^{(1)} = \frac{2\pi}{\hbar}|M_{\mu v}|^2\,\delta(E_v - E_\mu) \ , \tag{7.9}$$

where the matrix element is

$$M_{\mu v} = \int_{\Omega_T} \chi_v^* \, U_T \, \psi_\mu \, d\tau \ . \tag{7.10}$$

Since U_T is non-vanishing only in the tip body, the matrix element $M_{\mu v}$ is evaluated only in the volume of the tip, Ω_T. Using Green's theorem, the transition matrix element can be converted into a surface integral similar to Bardeen's, in terms of *modified wave functions*:

$$M_{\mu v} = -\frac{\hbar^2}{2m}\int_\Sigma (\chi_v^* \nabla\psi_\mu - \psi_\mu \nabla\chi_v^*)\cdot dS \ . \tag{7.11}$$

The matrix element has the dimension of *energy*. In Sect. 7.2.3, we will show that the physical meaning of Bardeen's matrix element is the energy lowering due to the overlap of the two wave functions, or the exchange interactions. In other words, it is the *resonance energy* of *Heisenberg* [7.30] and *Pauling* [7.31].

The error of the first-order time-dependent perturbation theory can be estimated by evaluating the second-order term explicitly, following the method of *Oppenheimer* [7.18]. In our case, the error estimation term is

$$\begin{aligned}
w_{\mu v}^{(2)} &\simeq \frac{2\pi}{\hbar}\left|\sum_\lambda \frac{(\chi_v, U_S\chi_\lambda)(\chi_\lambda, U_T\psi_\mu)}{\bar{E} - E_v}\right|^2 \delta(E_\mu - E_v) \\
&= \frac{2\pi}{\hbar}\left|\frac{(\chi_v, U_S U_T\psi_\mu)}{\bar{E} - E_v}\right|^2 \delta(E_\mu - E_v) = 0 \ ,
\end{aligned} \tag{7.12}$$

where $\bar{E}$ is a certain energy value [7.18]. The error estimation term vanishes because of (7.6). Thus, as a consequence of (7.5 and 6) and, the accuracy of the first-order perturbation is optimized.

As we have discussed previously, the wave functions in (7.7, 8) are different from the wave functions of the free tip and free sample. The effect of the distortion potential, e.g., $V \equiv U_S - U_{S0}$, can be evaluated through a Green's function method. Using Dyson's equation, to the first order, the distorted wavefunction ψ is related to the bare one ψ_0 through

$$\psi(r) = \psi_0(r) + \int G(r, r')\, V(r')\, \psi_0(r')\, d^3r' \ . \tag{7.13}$$

The Green's function is defined by

$$\left(-\frac{\hbar^2}{2m} \nabla^2 + U_{SO} - E \right) G(r, r') = - \delta(r - r') \ . \tag{7.14}$$

For a one-dimensional case, the corrected transmission probability is

$$T = T_0 \left[1 - \frac{m}{\kappa \hbar^2} \int_{-\infty}^{z_0} V(z) \, dz \right]^2 \left[1 - \frac{m}{\kappa \hbar^2} \int_{z_0}^{\infty} V(z) \, dz \right]^2 \ . \tag{7.15}$$

In the following, we test the accuracy of the MBA by comparing its results with two solvable problems. We first apply it to a one-dimensional tunneling problem, as shown in Fig. 7.3. The modified Bardeen integral can be evaluated analytically. By comparing it with its exact solution, we find that the MBA is accurate even if the barrier collapses, as shown in Fig. 7.4. If the barrier lowering follows the law of the classical image potential, the correction is essentially a constant. This provides a simple explanation of the observed constant apparent barrier height. Second, we show that the MBA provides an accurate analytic formula for the potential curve in the attractive-force regime of the hydrogen molecular ion problem, as shown in Figs. 7.6–8. The wave-function modification for the simplest molecule, the hydrogen molecular ion, is again a constant multiplier independent of internuclear distance [7.27].

Although to the first order, in the two simple cases, the wave-function modification is a constant, such a correction factor is indispensable. First, it is the source of the van der Waals force. Second, without this correction, even asymptotically, the Bardeen formula is inaccurate. By comparing with the exact solution of the H_2^+ problem [7.33], we show that in the original Bardeen approach, even asymptotically, the exchange interaction energy is off by a factor

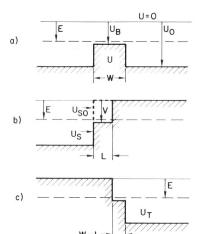

Fig. 7.3. Perturbation treatment of the square barrier problem (a) for original problem. (b) and (c), the potentials of the subsystems for a perturbation treatment [7.26]

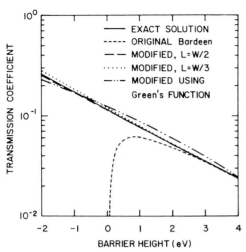

Fig. 7.4. Different approximate methods for the square barrier problem. (Parameters used: $W = 2$ Å; $E = 4$ eV; $U_0 = 16$ eV.) The original Bardeen theory breaks down when the barrier top comes close to the energy level. The modified Bardeen tunneling theory is accurate with separation surfaces either centered ($L = W/2$) or off-centered ($L = W/3$). By approximating the distortion of wavefunctions using Green's-function, the error in the entire region is only a few percents [7.26]

of $(4/e) \simeq 1.471$. The tunneling probability would differ by a factor of $(4/e)^2 \simeq 2.165$. In the square-barrier problem, with the inclusion of the classical image potential, the original Bardeen approach can be off by a factor of about 7.

7.2.2 The Square-Barrier Problem

A schematics of the perturbation treatment of the square-barrier problem is depicted in Fig. 7.3. Figure 7.4 presents a comparison of the exact solution and the perturbation results.

As shown in Fig. 7.4, the approximate transmission coefficient of the MBA remains accurate even with a barrier 2 eV below the energy level. Figure 7.4 also exhibits that by taking a different separation surface, the accuracy of the MBA remains intact. The transmission coefficient of the original BA can be obtained by letting $\kappa_0 = \kappa$. As shown in Fig. 7.4, the BA becomes inaccurate when the barrier top is close to or lower than the energy level. The accuracy of semi-classical theory is even worse than the BA, also displayed in Fig. 7.4. The result of corrections with Green's function is also presented. For the case of Fig. 7.4, the values of the integrals are $[- U_B L]$ and $[- U_B(W - L)]$, respectively. As shown, the result is fairly accurate, which verifies the adequacy of the Green's function method.

Using the above results, we can explain why the barrier lowering by the image force is not observable experimentally. The definition of the apparent barrier height, (7.16), in units of Å and eV, is

$$\phi = 0.952 \left(\frac{d \ln I}{dz} \right)^2 . \tag{7.16}$$

In an STM, as the tip is moving towards the sample, the barrier height and the

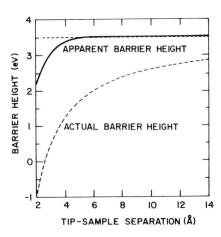

Fig. 7.5. Apparent barrier height calculated from the exact solution. Variation of the apparent barrier height $0.952 \, (\mathrm{d} \ln I/\mathrm{d}z)^2$ with barrier thickness, as calculated from the exact solution of the square potential barrier problem. The actual barrier height, dashed curve, drops dramatically because of the image potential. The apparent barrier height, solid curve, almost equals the nominal value of the barrier height in the entire region. (Parameters used: $U = 3.5 \, \mathrm{eV}$, $C = 10 \, \mathrm{eV} - \text{Å}$; $E_k = 7.5 \, \mathrm{eV}$.) [7.29]

tip-sample distance are changing at the same time. Therefore, the definition of the apparent barrier height is no longer equal to the nominal barrier height in the energy diagram. The actual barrier, with all the exchange and correlation interactions included, can be well described by the classical image potential [7.12, 27]

$$ U = U_0 - \frac{9.97}{W} \, [\mathrm{eV \, Å}] \; , \tag{7.17} $$

then the apparent barrier height should be approximately a constant even if the barrier collapses.

By definition, for square barriers, the apparent barrier height is

$$ \phi_{\mathrm{ap}} = 0.952 \left\{ \frac{\partial}{\partial W} \log \left[1 + \frac{1}{2} \left(\frac{\kappa}{q} + \frac{q}{\kappa} \right)^2 \sinh^2 (2 \kappa W) \right] \right\}^2 . \tag{7.18} $$

In units of Å and eV, $\kappa = 0.513 \sqrt{U}$, $q = 0.513 \sqrt{E_k}$. As depicted in Fig. 7.5, the apparent barrier height is almost a constant down to a negative barrier.

The result has a simple explanation using the MBA. Using the modified Bardeen integral and the Green's function method, we immediately reach the conclusion that the apparent barrier height should be constant under very general conditions: According to (7.15), as long as the integral of the distortion potential is a constant, (i.e., the shaded area in Fig. 7.2 remains constant while varying the tip–sample distance,) the effect of barrier lowering is not observable.

7.2.3 The Hydrogen Molecular Ion

The square-barrier problem is an idealized model which cannot be found in nature. Real tunneling problems are almost always too complicated to have an exact solution. But, there is *one exception*: the hydrogen molecular ion. In the

Born–Oppenheimer approximation, the Schrödinger equation is separable and analytically soluble. The numerical results are extensively tabulated [7.28]. It is the cornerstone of Pauling's theory of chemical bonds [7.26], and the starting point for the understanding of condensed-matter physics [7.29]. We will show that the MBA provides an accurate analytic expression of its potential energy curve. Naturally, it will shed light on the understanding of STM and SFM, as branches of condensed-matter physics.

Three Regimes of Interaction. Figure 7.6 exhibits three regimes of interaction in a hydrogen molecular ion. At large distances, the system can be considered as a neutral hydrogen atom plus a proton. The electrical field of the proton polarizes the hydrogen atom. As a result, a van der Waals force is induced. The van der Waals force dominates as the proton–proton distance $R > 8\,\text{Å}$. At a distance $R < 8\,\text{Å}$, the 1s electron at the vicinity of the right proton has an appreciable probability to tunnel into the 1s state of the left proton, and *vice versa*. This tunneling phenomenon gives rise to a *resonance* and results in a lowering of the total energy 7.26. The resonance gives rise to a bonding state, with a lower total energy (attractive force); and an antibonding state, with a higher total energy (repulsive force). At even shorter distances, e.g., $R < 2.5\,\text{Å}$, the repulsive force between the protons becomes important, and the net force becomes repulsive regardless of the type of the state.

Van der Waals Force. The quantum mechanical treatment of the van der Waals force was resolved in the early thirties. The interaction energy for the hydrogen molecular ion is

$$E = -\frac{9}{4\,R^4}\,. \tag{7.19}$$

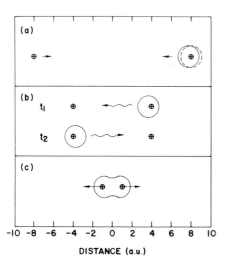

Fig. 7.6. Three regimes of interaction of a hydrogen atom with a proton. (a) at large distances, $R > 16$ a.u., the system can be considered as a neutral hydrogen atom plus a proton. The polarization of the hydrogen atom due to the field of the proton generates a van der Waals force. (b) at intermediate distances, 16 a.u. $> R > 4$ a.u. the electron can tunnel to the vicinity of another proton, and *vice versa*. A resonance force is generated, which is either attractive or repulsive. (c) at short distances, $R < 4$ a.u., proton-proton repulsion becomes important [7.28]

DISTANCE (a.u.)

Throughout this section, the atomic units (a.u.) are used, that is, by setting $e = \hbar = m_e = 1$. In the atomic units, length is in bohrs (0.529 Å), energy is in hartrees (27.21 eV).

Resonance Energy as Tunneling Matrix Element. As shown in Fig. 7.6, at a shorter proton-proton separation ($R < 16$ a.u. or $R < 8$ Å), the electron in the 1s state in the vicinity of one proton has an appreciable probability to tunnel to the 1s state in the vicinity of another proton. The tunneling matrix element can be evaluated using the MBA as we presented in the previous section. A schematic of this problem is shown in Fig. 7.7. Usually, the hydrogen molecular ion problem is treated as a stationary-state problem. Here, we start with the time-dependent Schrödinger equation,

$$i\frac{\partial \Psi(r,t)}{\partial t} = \left(-\frac{1}{2}\nabla^2 + U \right) \Psi(r,t) \ . \tag{7.20}$$

By defining a pair of single-well potentials, U_L and U_R, we define the time-dependent right-hand-side states and the left-hand-side states by the following Schrödinger equations,

$$i\frac{\partial \Psi_L(r,t)}{\partial t} = \left(-\frac{1}{2}\nabla^2 + U_L \right) \Psi_L(r,t) \ , \tag{7.21}$$

$$i\frac{\partial \Psi_R(r,t)}{\partial t} = \left(-\frac{1}{2}\nabla^2 + U_R \right) \Psi_R(r,t) \ . \tag{7.22}$$

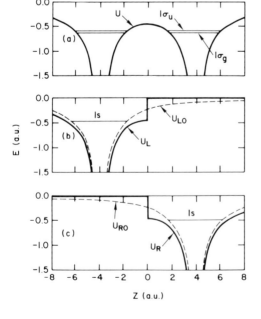

Fig. 7.7. Perturbation treatment of the hydrogen molecular ion. (a) the exact potential curve and the exact energy levels of the problem. (b) solid curve, the left-hand-side potential for a perturbation treatment; dotted curve, the potential for a free hydrogen atom. (c) solid curve, the right-hand-side potential for a perturbation treatment; dotted curve, the potential for a free hydrogen atom. Distance is in atomic units, or bohrs (1 a.u. = 0.539 Å), and energy is also in atomic units, or hartrees (1 a.u. = 27.21 eV) [7.28]

Because the potential U_L is different from the potential of a free proton, U_{L0}, the wave function ψ_L and the energy level E_0 are different from the 1s state of a free hydrogen atom. (The same is true for U_R and ψ_R.) The distortion is actually the same as the origin of the van der Waals force and was evaluated explicitly in 1955 by *Holstein* [7.32]. (Holstein's internal report, incomplete and full of typos, was never published. A reproduction of his derivation is included in the Appendix.)

A specific solution of (7.20) now depends on the initial condition. If at $t = 0$, the electron is in the left-hand-side state, the solution is:

$$\Psi_1(r, t) = [\cos Mt\, \psi_L(r) + i \sin Mt\, \psi_R(r)]\, e^{-iE_0 t} \ . \tag{7.23}$$

This solution describes a back-and-forth migration of the electron between the two protons at a frequency $v = |M|/h$. Similarly, we have another solution

$$\Psi_2(r, t) = [\cos Mt\, \psi_R(r) + i \sin Mt\, \psi_L(r)]\, e^{-iE_0 t} \ , \tag{7.24}$$

which starts with a right-hand-side state at $t = 0$. The transition matrix element M in the above equations is the modified Bardeen integral, (7.11). In atomic units, it is

$$M = \frac{1}{2} \int (\psi_R \nabla \psi_L - \psi_L \nabla \psi_R) \cdot dS \ , \tag{7.25}$$

which is evaluated on the separation surface, i.e., the median plane (Fig. 7.7).

The linear combinations of the solutions, (7.23, 24), are also good solutions of the time-dependent Schrödinger equation, (7.20). In particular, there is a state symmetric with respect to the median plane

$$\Psi_g(r, t) = (\Psi_1 + \Psi_2) = [\psi_L(r) + \psi_R(r)]\, e^{-i(E_0 + M)t} \ . \tag{7.26}$$

as well as an antisymmetric state

$$\Psi_u(r, t) = (\Psi_1 + \Psi_2) = [\psi_L(r) + \psi_R(r)]\, e^{-i(E_0 + M)t} \ . \tag{7.27}$$

Obviously, these solutions are *stationary states* of (7.20) with energy eigenvalues $(E_0 + M)$, and $(E_0 - M)$, respectively. Because both E_0 and M are negative, the symmetric state has a lower energy, which means an attractive force (Fig. 7.8).

The above discussion is a quantitative formulation of the concept of *resonance* introduced by *Heisenberg* [7.30] for treating many body problems in quantum mechanics.

The integral in (7.25) has been evaluated explicitly by *Holstein* [7.34] (Appendix):

$$M = -\frac{2}{e} Re^{-R} \ . \tag{7.28}$$

The total coupling energy is the sum of the van der Waals energy, (7.19), and the

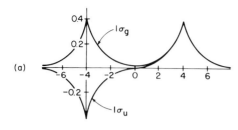

Fig. 7.8. Wave functions of the hydrogen molecular ion. (a) The exact wavefunctions of the
hydrogen molecular ion. The lowest states are
shown. The two exact solutions can be considered as symmetric and antisymmetric linear
combinations of the solutions of the left-hand-
side and right-hand-side problems, (b) and (c),
defined by the potential curves in Fig. 7.7. For
brevity, the normalization constant is omitted
[7.28]

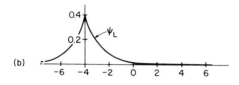

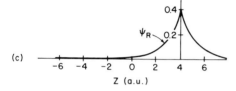

Z (a.u.)

resonance energy, (7.28). For the $1\sigma_g$ state, it is

$$\Delta E(1\sigma_g) = -\frac{9}{4R^4} - \frac{2}{e}Re^{-R} ,\tag{7.29}$$

and for the $1\sigma_u$ state,

$$\Delta E(1\sigma_u) = -\frac{9}{4R^4} + \frac{2}{e}Re^{-R} .\tag{7.30}$$

A comparison between the above equations with the exact solution of the
H_2^+ problem, published by *Bates* et al. [7.33] is shown in Table 7.1. In the range
$R > 6$ a.u., or $R > 3$ Å, the agreement is excellent. It should be emphasized that
(7.29, 30) are based on pure theoretical reasoning with no adjustable parameters.
Therefore, the interpretation of the resonance energy in terms of tunneling is
verified quantitatively also in the case of the hydrogen molecular ion.

Repulsive Force. As shown in Fig. 7.8, as the proton–proton separation becomes even smaller, the picture of resonance becomes obscured, and the
proton–proton repulsion is no longer screened by the electron. *Slater* [7.34]
demonstrated that the Morse curve matches the exact potential curve very
precisely:

$$f = C\{\exp[-2\kappa(s - s_0)] - \exp[-\kappa(s - s_0)]\} ,\tag{7.31}$$

Table 7.1. Energy eigenvalues of hydrogen molecule ion: a comparison between the exact solutions tabulated in [7.33] and the analytic expression from the MBA. Energy in Rydbergs, (1 Rydberg = 0.5 Hartree $\simeq$ 13.6 eV), distance in bohrs (1 bohr = 0.529 Å) [7.28]

R [bohr]	$-E(1\sigma_g)$ [7.33]	[Rydberg] MBA	Difference	$-E(1\sigma_u)$ [7.33]	[Rydberg] MBA	Difference
6.0	1.35726	1.35869	0.00143	1.31462	1.31492	0.00030
6.5	1.32412	1.32459	0.00047	1.29581	1.29583	0.00002
7.0	1.29690	1.29698	0.00008	1.27826	1.27820	− 0.00006
7.5	1.27426	1.27419	− 0.00007	1.26206	1.26198	− 0.00008
8.0	1.25514	1.25505	− 0.00009	1.24721	1.24715	− 0.00006
8.5	1.23878	1.23870	− 0.00008	1.23365	1.23361	− 0.00004
9.0	1.22461	1.22454	− 0.00007	1.22131	1.22127	− 0.00004

where s_0 is the equilibrium point, and C is a constant. As we will discuss later, the sum of the Morse forces on a periodic surface has a finite analytic form.

7.2.4 The Tunneling Time

Tunneling time has been a subject of theoretical discussions for decades, as summarized in Chap. 6. There are various different definitions of the tunneling time for a variety of different models. Regarding a real case existing in nature, the hydrogen molecular ion, there is a unique definition of the tunneling time. It is the time for the electron to migrate from the vicinity of one proton to another proton. It is a well-defined function of the proton–proton distance R. To a high accuracy, for $R > 6$ bohrs, the time is (in the Hartree unit, 2.4189×10^{-17} s),

$$\tau \equiv \frac{\pi}{|M|} = \frac{\pi e}{2R} e^R . \tag{7.32}$$

Table 7.2 lists the dependence of electron-tunneling time between two protons, as a function of the distance between the two protons.

Table 7.2. Tunneling time of an electron between two protons as a function of the proton–proton distance

Distance [Å]	Tunneling time	Distance [Å]	Tunneling time
3	5.28 fs	10	880 ps
4	26.2 fs	11	5.29 ns
5	139 fs	12	32.1 ns
6	765 fs	13	196 ns
7	4.34 ps	14	1.21 µs
8	25.1 ps	15	7.44 µs
9	148 ps	16	46.2 µs

7.2.5 Asymptotic Accuracy of the Bardeen Tunneling Theory

In the two examples we have discussed the perturbing potentials for the modified wave functions are of a long-range nature. Both the image potential and the overlap of the Coulomb potential has a $1/D$ dependence on the distance D. For the case of a square barrier, the asymptotic transmission coefficient is

$$T = \exp\left\{ -2\sqrt{2m_e\left[\phi - \frac{C}{D}\right]\frac{D}{\hbar}}\right\}$$

(7.33)

$$\simeq e^{-2\kappa D} \times \exp\left\{\frac{\kappa C}{\phi}\right\} .$$

The first factor in the second line of (7.33) is the WKB value of the transmission without wave-function correction. The second factor is the correction factor. Taking the typical values for κ, C, and ϕ, the factor is about 7.4. For the case of the hydrogen molecular ion, the wave-function modification makes a correction factor 1.471 for the tunneling matrix element, and a factor 2.165 for the tunneling probability. The necessity of such a correction was extensively discussed by *Herring* [7.35]. As Herring pointed out, without this correction, the traditional Heitler–London treatment for the hydrogen molecular ion is inaccurate even asymptotically.

As we have shown for the cases of simple geometries, the correction is a constant factor independent of energy level and distance. Regarding the application of Bardeen in the interpretation of Giaever's experiment, this constant correction does not matter. However, regarding STM, the correction does sometime matter. The lateral variation of the correction factor is reflected in the STM image's and force distributions, as shown in Chap. 8.

Certainly, the modification does not affect Bardeen's interpretation of Giaever's experiment, because the correction factor is independent of the energy level. In the interpretation of local tunneling spectra with STM, Bardeen's approximation, i.e., the matrix element is roughly independent of energy, is still a useful starting point if the structure in the tip DOS or sample DOS or both is pronounced. We will discuss this in Sect. 7.6.

7.2.6 Tunneling Conductance and Attractive Atomic Force

The equality of the resonance energy and the tunneling matrix element has a measurable consequence. As shown in [7.23], there is a universal equation between the tunneling conductance G and the attractive atomic force F

$$F = -f\kappa\varepsilon\sqrt{GR_K}$$

(7.34)

$$\simeq -0.257 f\kappa\varepsilon\sqrt{G(\mu S)}\, nN ,$$

where f is a dimensionless constant of the order unity, which depends on the geometry of the tip. R_K is von Klitzing's constant, $h/e^2 \simeq 25812.8 \, \Omega$. The second line gives the value in terms of commonly used units: length in Å, energy ε in eV, conductance G in microsiemens (μS), or $10^{-6}\,\Omega^{-1}$ and force in nanonewtons (nN). This relation is found to be in quantitative agreement with the simultaneous measurement of tunneling conductance and attractive atomic forces [7.9].

7.3 Explicit Expressions for Tunneling Matrix Elements

From (7.36), the transmission matrix elements are determined by the wave functions of the tip and the sample at the separation surface. For both the tip and the sample, the wave functions on and beyond the separation surface satisfy the Schrödinger equation in the vacuum,

$$(\nabla^2 - \kappa^2)\chi(r) = 0 \ , \tag{7.35}$$

where $\kappa = (2m\phi)^{1/2}\hbar^{-1}$ is the decay constant, determined by the work function ϕ. Near the center of the apex atom, (7.35) is not valid for the tip wave function. However, it does not matter. In order to calculate the tunneling matrix element we only need the tip wave function near the separation surface, which always satisfies (7.35).

It is convenient to expand the tip wave function and the sample wave function into spherical-harmonic components, $Y_{lm}(\theta, \phi)$, with the nucleus of the tip apex atom as the origin. Each component is characterized by quantum numbers l and m. In other words, we are looking for solutions of (7.35) in the form

$$\chi(r) = \sum_{l,m} C_{lm} f_{lm}(\kappa \varrho) Y_{lm}(\theta, \phi) \ , \tag{7.36}$$

where $\varrho = |r = r_0|$, and r_0 is the center of the apex atom. Substituting (7.36) into (7.35), we obtain the differential equation for the functions $f_{lm}(u)$,

$$\frac{d}{du}\left(u^2 \frac{df(u)}{du}\right) - [u^2 + l(l+1)]f(u) = 0 \ . \tag{7.37}$$

As seen, the radial functions depend only on l.

There are two standard linear-independent solutions for (7.37): the spherical modified Bessel function of the first kind

$$i_l(u) = \sqrt{\frac{\pi}{2u}} \, I_{l+1/2}(u) \ , \tag{7.38}$$

and of the second kind

$$k_l(u) = \sqrt{\frac{2}{\pi u}} K_{l+1/2}(u) \ . \tag{7.39}$$

These so-called special functions are actually *elementary functions*:

$$i_l(u) = u^l \left(\frac{d}{u\,du} \right)^l \frac{\sinh u}{u} \ , \tag{7.40}$$

and

$$k_l(u) = (-1)^l u^l \left(\frac{d}{u\,du} \right)^l \frac{\exp(-u)}{u} \ . \tag{7.41}$$

Obviously, the functions $i_l(u)$ diverge at large u, which are not appropriate to represent tip wave functions. The functions $k_l(u)$ are regular at large u, which satisfies the desired boundary condition. Therefore, a component of tip wave function with quantum numbers l and m has the general form

$$\chi_{lm}(r) = C k_l(\kappa \varrho) Y_{lm}(\theta, \phi) \ . \tag{7.42}$$

Table 7.3. Vacuum tails of tip wave functions

State	Wave function
s	$C \dfrac{1}{\kappa\varrho} e^{-\kappa\varrho}$
p_z	$C \left(\dfrac{1}{\kappa\varrho} + \dfrac{1}{(\kappa\varrho)^2} \right) e^{-\kappa\varrho} \cos\theta$
p_x	$C \left(\dfrac{1}{\kappa\varrho} + \dfrac{1}{(\kappa\varrho)^2} \right) e^{-\kappa\varrho} \sin\theta \cos\phi$
p_y	$C \left(\dfrac{1}{\kappa\varrho} + \dfrac{1}{(\kappa\varrho)^2} \right) e^{-\kappa\varrho} \sin\theta \sin\phi$
d_{xz}	$C \left(\dfrac{1}{\kappa\varrho} + \dfrac{3}{(\kappa\varrho)^2} + \dfrac{3}{(\kappa\varrho)^3} \right) e^{-\kappa\varrho} \sin 2\theta \cos\phi$
d_{yz}	$C \left(\dfrac{1}{\kappa\varrho} + \dfrac{3}{(\kappa\varrho)^2} + \dfrac{3}{(\kappa\varrho)^3} \right) e^{-\kappa\varrho} \sin 2\theta \sin\phi$
d_{xy}	$C \left(\dfrac{1}{\kappa\varrho} + \dfrac{3}{(\kappa\varrho)^2} + \dfrac{3}{(\kappa\varrho)^3} \right) e^{-\kappa\varrho} \sin^2\theta \sin 2\phi$
d_{z^2}	$C \left(\dfrac{1}{\kappa\varrho} + \dfrac{3}{(\kappa\varrho)^2} + \dfrac{3}{(\kappa\varrho)^3} \right) e^{-\kappa\varrho} \left(\cos^2\theta - \dfrac{1}{3} \right)$
$d_{x^2-y^2}$	$C \left(\dfrac{1}{\kappa\varrho} + \dfrac{3}{(\kappa\varrho)^2} + \dfrac{3}{(\kappa\varrho)^3} \right) e^{-\kappa\varrho} \sin^2\theta \cos 2\phi$

In the absence of magnetic field, it is convenient to write those tip wave functions in real form, as listed in Table 7.3. The normalization constant C is to be determined by comparing with the results of first-principle calculations of actual tip states, for example, from the results of Tsukada et al. in Chap. 5. Up to $l = 2$, we define the coefficients of the expansion by the following expression:

$$
\chi = \beta_{00} k_0(\kappa\varrho) + \left(\beta_{10} \frac{z}{\varrho} + \beta_{11} \frac{x}{\varrho} + \beta_{12} \frac{y}{\varrho} \right) k_1(\kappa\varrho)
$$

$$
+ \left[\beta_{20} \left(\frac{z^2}{\varrho^2} - \frac{1}{3} \right) + \beta_{21} \frac{xz}{\varrho^2} + \beta_{22} \frac{yz}{\varrho^2} \right.
$$

$$
\left. + \beta_{23} \frac{xy}{\varrho^2} + \beta_{24} \left(\frac{x^2}{\varrho^2} - \frac{y^2}{\varrho^2} \right) \right] k_2(\kappa\varrho) . \tag{7.43}
$$

By expanding the sample wave function ψ into spherical-harmonic components, and using the properties of spherical harmonics and the modified spherical Bessel functions, the tunneling matrix element for an arbitrary tip state, up to $l = 2$, can be written as a linear combination of the derivatives of the

Table 7.4. Tunneling matrix elements

Tip state	Matrix element
s	$\dfrac{2\pi C\hbar^2}{\kappa m} \psi(\mathbf{r}_0)$
p_z	$\dfrac{2\pi C\hbar^2}{\kappa m} \dfrac{\partial \psi}{\partial z}(\mathbf{r}_0)$
p_x	$\dfrac{2\pi C\hbar^2}{\kappa m} \dfrac{\partial \psi}{\partial x}(\mathbf{r}_0)$
p_y	$\dfrac{2\pi C\hbar^2}{\kappa m} \dfrac{\partial \psi}{\partial y}(\mathbf{r}_0)$
d_{zx}	$\dfrac{2\pi C\hbar^2}{\kappa m} \dfrac{\partial^2 \psi}{\partial z \partial x}(\mathbf{r}_0)$
d_{zy}	$\dfrac{2\pi C\hbar^2}{\kappa m} \dfrac{\partial^2 \psi}{\partial z \partial y}(\mathbf{r}_0)$
d_{xy}	$\dfrac{2\pi C\hbar^2}{\kappa m} \dfrac{\partial^2 \psi}{\partial x \partial y}(\mathbf{r}_0)$
d_{z^2}	$\dfrac{2\pi C\hbar^2}{\kappa m} \left(\dfrac{\partial^2 \psi}{\partial z^2} - \dfrac{1}{3} \kappa^2 \psi \right)(\mathbf{r}_0)$
$d_{x^2-y^2}$	$\dfrac{2\pi C\hbar^2}{\kappa m} \left(\dfrac{\partial^2 \psi}{\partial x^2} - \dfrac{\partial^2 \psi}{\partial y^2} \right)(\mathbf{r}_0)$

sample wave function at the nucleus of the apex atom of the tip [7.23, 24],

$$
\begin{aligned}
M = \frac{2\pi\hbar^2}{m_e\kappa} \Bigg[&\beta_{00}\psi + {} + \beta_{10}\frac{\partial\psi}{\kappa\partial z} + \beta_{11}\frac{\partial\psi}{\kappa\partial x} + \beta_{12}\frac{\partial\psi}{\kappa\partial y} \\
&+ \beta_{20}\left(\frac{\partial^2\psi}{\kappa^2\partial z^2} - \frac{1}{3}\psi\right) + \beta_{21}\frac{\partial^2\psi}{\kappa^2\partial x\partial z} + \beta_{22}\frac{\partial^2\psi}{\kappa^2\partial y\partial z} \\
&+ \beta_{23}\frac{\partial^2\psi}{\kappa^2\partial x\partial y} + \beta_{24}\left(\frac{\partial^2\psi}{\kappa^2\partial x^2} - \frac{\partial^2\psi}{\kappa^2\partial y^2}\right) \Bigg] .
\end{aligned}
\tag{7.44}
$$

These results can be summarized as the *derivative rule*: Write the angle dependence of the tip wave function in terms of x, y, and z. Replace them with the simple rule

$$
x \to \frac{\partial}{\kappa\partial x}; \qquad y \to \frac{\partial}{\kappa\partial y}; \qquad z \to \frac{\partial}{\kappa\partial z} ;
\tag{7.45}
$$

and acting on the sample wave function. The tunneling matrix element is obtained. For individual tip states, the tunneling matrix elements are listed in Table 7.4.

7.4 Theoretical STM Images

The derivative rule can be directly used for STM image interpretation. In principle, to obtain the complete information, we should start with the details of all surface wave functions. However, on real surfaces, those are usually very complicated. Hereby we describe two simplified methods which may provide enough details to compare with experimental STM images. The first one is appropriate for simple and defect-free surfaces, where the STM images are periodic functions of x and y, and can be described by a few lowest Fourier components. In order to predict those lowest Fourier components, it is sufficient to consider the surface Bloch waves at several special points on the surface Brillouin zone. This method was first introduced by *Harris* and *Liebsch* in the treatment of helium atom scattering [7.36]. The second method considers the surface wave function as an aggregation of independent atomic states, and the total tunneling current is the sum of the tunneling current between a tip state and those atomic states on the sample surface. This method can treat non-periodic phenomena, and follows well the surface chemists' intuition. If the atomic wave functions on the sample surface is approximated by the Slater function, then for periodic surfaces, the Fourier coefficients of the corrugation function can be written in a finite analytic form. For simple surfaces, the result from the second method matches well with the result from the leading-Bloch-waves method.

7.4.1 The Method of Leading Bloch Waves

A number of cases have been analyzed in [7.23, 24]. As an example, we treated
the STM imaging of a hexagonal close-packed metal surface. The method is
similar to the treatment of the helium scattering problem on graphite [7.36]. As
shown in Fig. 7.9, the top layer atoms exhibit a hexagonal symmetry, $p6mm$. For
$m = 0$ tip states, the image should have a hexagonal symmetry, i.e., invariant
with respect to plane group $p6mm$. Up to the lowest non-trivial Fourier
components, the most general form of surface charge density with a hexagonal
symmetry is

$$\varrho(r) = \sum_{E_F - \Delta E}^{E_F} |\psi(r)|^2 \simeq a_0(z) + a_1(z)\, \phi^{(6)}(kx) \ , \tag{7.46}$$

where $x = (x, y)$, and $k = 4\pi/\sqrt{3}a$ is the length of a primitive reciprocal lattice
vector. A hexagonal cosine function is defined for convenience,

$$\phi^{(6)}(X) \equiv \frac{1}{3} + \frac{2}{9}\sum_{n=0}^{2} \cos u_n \cdot X \ , \tag{7.47}$$

where $u_0 = (0, 1)$, $u_1 = (-\frac{1}{2}\sqrt{3}, -\frac{1}{2})$, and $u_2 = (\frac{1}{2}\sqrt{3}, -\frac{1}{2})$, respectively. It is
easy to show that the function $\phi^{(6)}(kx)$ has maximum value 1 at each atomic site,
and minimum value 0 at the center of each atomic triangle.

The functional dependence of the coefficients a_0 and a_1 can be obtained
using the leading Bloch-wave method. The first term (7.46) comes from the
Bloch waves at the center of the Brillouin zone. It has the form

$$a_0(z) = C_0 e^{-2\kappa z} \ . \tag{7.48}$$

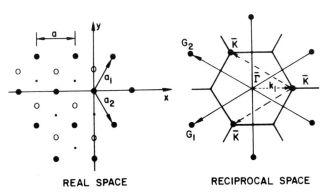

REAL SPACE   RECIPROCAL SPACE

Fig. 7.9. Geometrical structure of a close-packed metal surface. The second-layer atoms (*circles*) and
third-layer atoms (*small dots*) have little influence on the surface charge density, which is dominated
by the top-layer atoms (*large dots*). The top layer exhibits a sixfold symmetry, which is invariant with
respect to plane group $p6mm$, i.e., point group C_{6v} together with the translational symmetry. b, the
corresponding surface Brillouin zone. The lowest non-trivial Fourier components of LDOS arise
from Bloch functions near the $\bar{\Gamma}$ and $\bar{K}$ points [7.24]

The contribution to the second term in (7.46) comes from Bloch functions near the six corners of the Brillouin zone. It has the form

$$a_1(z) = C_1 e^{-2\alpha z} , \qquad (7.49)$$

where $\alpha = (\kappa^2 + q^2)^{1/2}$ is the corresponding decay constant, and $q = k/\sqrt{3}$ is the length of the $\mathbf{k}$ vector at the corners of the first Brillouin zone. The corrugation amplitude of charge-density contour, Δz, as a function of z, can be obtained from the above equations,

$$\Delta z = \frac{C_1}{2\kappa C_0} e^{-2(\alpha-\kappa)z} . \qquad (7.50)$$

The ratio (C_1/C_0) can be determined by comparing (7.50) with the corrugation amplitudes of the charge-density contours obtained from first-principle calculations, or helium-scattering experiments. It is also the STM image for an s-wave tip state. For a p_z tip state, the STM image is

$$\Delta z(\mathbf{x}) = \frac{C_1}{2\kappa C_0}\left(1 + \frac{q^2}{\kappa^2}\right) e^{-2(\alpha-\kappa)z} \phi^{(6)}(k\mathbf{x}) , \qquad (7.51)$$

and the STM image for the d_{z^2} tip state is

$$\Delta z(\mathbf{x}) = \frac{C_1}{2\kappa C_0}\left(1 + \frac{3q^2}{2\kappa^2}\right)^2 e^{-2(\alpha-\kappa)z} \phi^{(6)}(k\mathbf{x}) , \qquad (7.52)$$

A comparison of the theory with experiments is shown in Fig. 7.10. For Al(111), $a = 2.88$ Å, $\phi = 3.5$ eV, it follows that $\kappa = 0.96$ Å^{-1}, $\alpha = 1.74$ Å^{-1}. The slope of the $\ln \Delta z \sim z$ curve from (7.50) through (7.52) fits well with experimental data. The absolute tip–sample distance is obtained from curve fitting, which gives the shortest average tip–sample distance at $I = 40$ nA (with bias 50 mV) to

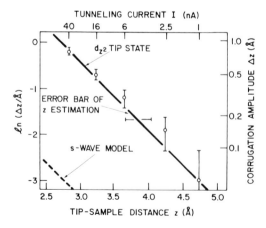

Fig. 7.10. Interpretation of the STM corrugation observed on Al(111). The predicted corrugation amplitude with a d_{z^2} tip state, solid curve, agrees well with the experimental data from Wintterlin et al. (1989), circles with error bars. The parameters of the theoretical curve are taken from a first-principle calculation of Al(111) surface, Wang et al. (1981). The tip-sample distance is defined as the distance from the plane of the top-layer nuclei of the sample to the center of the apex atom of the tip [7.24]

be about $2.9\,\text{Å}$, which agrees well with first-principles calculations and direct measurements.

7.4.2 The Method of Independent Atomic Orbitals

For the image of a single orbital with axial symmetry, the apparent radius R (or its inverse, the apparent curvature K) is a measurable quantity. We start with the case of the image of an s-wave tip state and an s-wave sample state. From simple symmetry considerations, it is clear that the apparent radius of the image is equal to the nominal distance z between the centers of the tip state and the sample state

$$K \equiv \frac{1}{R} = \frac{1}{z} \; . \tag{7.53}$$

In general, the apparent radius R can be obtained from the tunneling conductance distribution by

$$K \equiv \frac{1}{R} = \left(\frac{\partial g(r)}{\partial z} \right)^{-1} \frac{\partial^2 g(r)}{\partial x^2} \; , \tag{7.54}$$

which is evaluated at $(0,0,z)$. To illustrate the meaning of (7.54), we make an explicit calculation for a spherical tunneling-conductance distribution,

$$g(\mathbf{r}) \equiv g(r) \; , \tag{7.55}$$

where $r^2 = x^2 + y^2 + z^2$. With

$$\left[\frac{\partial g}{\partial z} \right]_{z=r} = \left[g' \frac{z}{r} \right]_{z=r} = g' \; , \tag{7.56}$$

and

$$\left[\frac{\partial^2 g}{\partial x^2} \right]_{x=0} = \left[g'' \frac{x^2}{r^2} + \frac{g'}{r} - g' \frac{x^2}{3r^3} \right]_{x=0} = \frac{g'}{r} \; , \tag{7.57}$$

we find that the definition of the radius R, (7.54), coincides with the nominal distance between the two nuclei (Fig. 7.11).

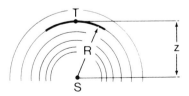

Fig. 7.11. Apparent radius for a spherical conductance distribution. For a spherical tunneling conductance distribution, the apparent radius equals the nominal tip-sample distance, regardless of the specific functional form of the distribution [7.29]

Intuitively, we will expect that for p_z or d_{z^2} states on the tip and on the sample, the images should be sharper, i.e., the apparent radius should be smaller. This is indeed true. The tunneling conductance distribution of atomic states, in the form of Slater functions, can be obtained using the derivative rule [7.23, 24]. For example, for a d_{z^2} tip state and an s-wave sample state, the tunneling-conductance distribution is

$$g(r) = \left(\frac{3}{2} \frac{z^2}{r^2} - \frac{1}{2} \right)^2 e^{-2\kappa r} . \tag{7.58}$$

Using (7.54), we find

$$K \equiv \frac{1}{R} = \frac{1}{z} \left(1 + \frac{3}{\kappa z} \right) . \tag{7.59}$$

The results are listed in Table 7.5, and illustrated in Fig. 7.12. Clearly, the apparent radius is reduced substantially for p_z and d_{z^2} states on the tip as well as on the sample. In other words, with p_z and d_{z^2} states, the images of individual atoms at surfaces look much sharper than those with s states, which is expected.

Table 7.5. Conductance distribution and apparent curvature of atomic images

Tip state	Sample state	Conductance distribution	Apparent curvature
s	s	$e^{-2\kappa r}$	$\dfrac{1}{z}$
s	p_z	$\cos^2 \theta \, e^{-2\kappa r}$	$\dfrac{1}{z} \left(1 + \dfrac{1}{\kappa z} \right)$
s	d_{z^2}	$(3\cos^2 \theta - 1)^2 e^{-2\kappa r}$	$\dfrac{1}{z} \left(1 + \dfrac{3}{\kappa z} \right)$
p_z	s	$\cos^2 \theta \, e^{-2\kappa r}$	$\dfrac{1}{z} \left(1 + \dfrac{1}{\kappa z} \right)$
p_z	p_z	$\cos^4 \theta \, e^{-2\kappa r}$	$\dfrac{1}{z} \left(1 + \dfrac{2}{\kappa z} \right)$
p_z	d_{z^2}	$(3\cos^2 \theta - 1)^2 \cos^2 \theta \, e^{-2\kappa r}$	$\dfrac{1}{z} \left(1 + \dfrac{4}{\kappa z} \right)$
d_{z^2}	s	$(3\cos^2 \theta - 1)^2 e^{-2\kappa r}$	$\dfrac{1}{z} \left(1 + \dfrac{3}{\kappa z} \right)$
d_{z^2}	p_z	$(3\cos^2 \theta - 1)^2 \cos^2 \theta \, e^{-2\kappa r}$	$\dfrac{1}{z} \left(1 + \dfrac{4}{\kappa z} \right)$
d_{z^2}	d_{z^2}	$(3\cos^2 \theta - 1)^4 e^{-2\kappa r}$	$\dfrac{1}{z} \left(1 + \dfrac{6}{\kappa z} \right)$

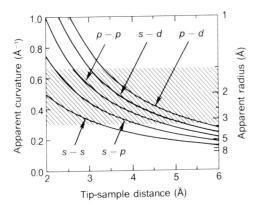

Fig. 7.12. Apparent radius as a function of tip–sample distance. For different combinations of tip states and sample states, the apparent radii are given by Table 7.5. The shaded area indicates the condition for achieving atomic resolution, i.e., having an apparent radius comparable to the actual radius of an atom. The minimum tip–sample distance is limited by the mechanical contact, i.e., about 2.5 Å. Therefore, most of the atom-resolved images are obtained with p_z or d_{z^2} tip states [7.29]

It has for several decades been known that if the charge density of single atoms can be written as a Slater function, then the Fourier coefficients of the charge density of a periodic surface have analytic forms [7.37]. For simple surfaces, to account for STM images, it is sufficient to keep a few low order Fourier components. If the symmetry of the surface is higher than $p1$, then those Fourier components are related by additional equations. For example, for hexagonal close-packed surfaces, to describe the basic features of the STM image, two independent parameters are sufficient. Table 7.6 lists the corrugation amplitudes of a hexagonal close-packed surface. The parameters are: a the atomic distance; $b = 4\pi/3a$ the primitive reciprocal vector; and $\gamma = \sqrt{4\kappa^2 + b^2}$ is the decay constant of the corrugated component of the charge density near the Fermi level. In the last column of Table 7.6, the ratios of the corrugation amplitudes with respect to the s-wave case are given for a surface with atomic distance $a = 2.88$ Å and work function $\phi = 3.5$ eV. The relevant quantities are: $\kappa = 0.96\,\text{Å}^{-1}$, $b = 2.52\,\text{Å}^{-1}$, and $\gamma = 3.17\,\text{Å}^{-1}$. For many cases, the enhancements can be greater than one order of magnitude.

For the STM images of the Al(111) surface, the sample states can be approximated by s-states. The Fermi-level LDOS corrugation is too small (< 0.03 Å) to be observable. However, if at the apex of the W or Pt–Ir tip, there is a d_{z^2} state, the corrugation can be as large as a large fraction of 1 Å. The prediction from the independent-orbital picture is similar to the prediction from the leading Bloch wave picture, and agrees well with experimental measurements (Fig. 7.13).

For transition metal surfaces, the d states contribute substantially to the surface charge density near the Fermi level. Therefore, the corrugation amplitudes of the charge density on transition metals can be much larger than that of the simple metals, such as Al. This is indeed observed experimentally by helium scattering and verified by first-principle calculations of the He-surface system [7.38]. The corrugation observed by He scattering is approximately proportional to the corrugation of the charge density near the Fermi level. In the cases of transition metals, the corrugation is determined by the amount of

Table 7.6. Corrugation amplitudes in the Independent-atomic-orbital model for a hexagonal close-packed surface

Tip state	Sample state	Corrugation amplitude	Ratio to the s–s case
s	s	$\dfrac{9\kappa}{\gamma^2} e^{-(\gamma - 2\kappa)z}$	1
s	p_z	$\left(\dfrac{\gamma}{2\kappa}\right)^2 \dfrac{9\kappa}{\gamma^2} e^{-(\gamma - 2\kappa)z}$	2.73
s	d_{z^2}	$\left(\dfrac{3}{2}\dfrac{\gamma^2}{4\kappa^2} - \dfrac{1}{2}\right)^2 \dfrac{9\kappa}{\gamma^2} e^{-(\gamma - 2\kappa)z}$	12.9
p_z	s	$\left(\dfrac{\gamma}{2\kappa}\right)^2 \dfrac{9\kappa}{\gamma^2} e^{-(\gamma - 2\kappa)z}$	2.73
p_z	p_z	$\left(\dfrac{\gamma}{2\kappa}\right)^4 \dfrac{9\kappa}{\gamma^2} e^{-(\gamma - 2\kappa)z}$	7.45
p_z	d_{z^2}	$\left(\dfrac{\gamma}{2\kappa}\right)^2 \left(\dfrac{3}{2}\dfrac{\gamma^2}{4\kappa^2} - \dfrac{1}{2}\right)^2 \dfrac{9\kappa}{\gamma^2} e^{-(\gamma - 2\kappa)z}$	35.2
d_{z^2}	s	$\left(\dfrac{3}{2}\dfrac{\gamma^2}{4\kappa^2} - \dfrac{1}{2}\right)^2 \dfrac{9\kappa}{\gamma^2} e^{-(\gamma - 2\kappa)z}$	12.9
d_{z^2}	p_z	$\left(\dfrac{\gamma}{2\kappa}\right)^2 \left(\dfrac{3}{2}\dfrac{\gamma^2}{4\kappa^2} - \dfrac{1}{2}\right)^2 \dfrac{9\kappa}{\gamma^2} e^{-(\gamma - 2\kappa)z}$	35.2
d_{z^2}	d_{z^2}	$\left(\dfrac{3}{2}\dfrac{\gamma^2}{4\kappa^2} - \dfrac{1}{2}\right)^4 \dfrac{9\kappa}{\gamma^2} e^{-(\gamma - 2\kappa)z}$	166

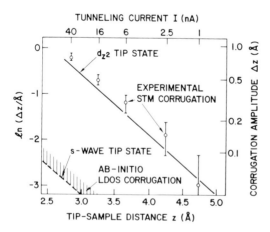

Fig. 7.13. Corrugation amplitudes of a hexagonal close-packed surface. Solid curve, theoretical corrugation amplitude for an s and a d_{z^2} tip state, on a close-packed metal surface with $a = 2.88$ Å and $\phi = 3.5$ eV. The orbital on each metal atom on the sample is assumed to be 1s-type. Measured STM corrugation amplitudes are from the data of *Wintterlin* et al. [7.6]. The corrugation amplitude for a s-wave tip state is more than one order of magnitude smaller than the experimental corrugation [7.26]

d-electron contribution [7.38], which can be more than one order of magnitude greater than the corrugation of simple metals. If on the tip apex there is also a d_{z^2} state, according to Table 7.6, the corrugation amplitude can be more than 100 times greater than the corrugation amplitudes predicted by the *Tersoff–Hamann* model [7.21]. This was indeed observed experimentally [7.7]. On the Pt(111) surface, with properly treated tips, the largest observed corrugation is about 150 times greater than the predictions from the *Tersoff–Hamann* model without considering the d states on the tip and the sample [7.21].

7.5 Effect of Atomic Forces in STM Imaging

7.5.1 Stability of STM at Short Distances

Because of the existence of force, the STM gap becomes unstable under certain circumstances. This problem has been studied extensively by *Pethica* and *Oliver* [7.39] within the classical theory of continuum elasticity. In spite of its simplicity, the theory reproduces the basic features of a large number of observed phenomena.

As proposed by *Soler* et al. [7.10], the force on the entire range of tip-sample distance under normal STM operating conditions can be described by the Morse curve

$$F = -2\kappa\varepsilon_0 (e^{-\kappa(s-s_0)} - e^{-2\kappa(s-s_0)}) , \tag{7.60}$$

where ε_0 is the binding energy of chemisorption, and s_0 is the equilibrium distance, i.e., the distance at which the net force is zero. In [7.28] we have shown that the constant κ in the expression of the Morse curve is equal to that of the sample wave function, based both on experimental evidence and theoretical arguments. The curve fitting of *Soler* et al. [7.10] for graphite images also demonstrates the accuracy of this equality.

Following *Pethica* and *Oliver* [7.39], the mechanical loop of STM responses to the force and exhibits an elastic deformation. By formally introducing an elastic constant α, the deformation is

$$\delta z = \alpha F . \tag{7.61}$$

For a well-designed, rigid STM, the deformation takes place predominately near the end of the tip. In this case, the elastic constant α is

$$\alpha = \frac{1}{a_0} \left(\frac{1}{E_T^*} + \frac{1}{E_S^*} \right) , \tag{7.62}$$

where E_S^* is the Young module of the sample surface, which may have to be corrected by a factor very close to 1 [7.39], and E_T^* is that for the tip. For many materials, the value of E^* is of the order of 10^{11} Pa or 10^{12} dyn/cm^2. The length scale a_0 is a characteristic radius of the tip end.

The observed z-piezo displacement is then the sum of $\delta z = \alpha F$ and the true tip–sample displacement δs. Using (7.60, 61), we find

$$z = s + 2\alpha\kappa\varepsilon_0 (e^{-\kappa(s-s_0)} - e^{-2\kappa(s-s_0)}) . \tag{7.63}$$

Taking the derivative with respect to s on both sides, it becomes:

$$\frac{\mathrm{d}z}{\mathrm{d}s} = 1 - \beta(e^{-\kappa(s-s_0)} - 2e^{-2\kappa(s-s_0)}) \ . \tag{7.64}$$

The dimensionless quantity $\beta = 2\alpha\kappa^2\varepsilon_0$ is a measure of the relative stiffness of the STM system with respect to the force in the gap.

In order to have a stable system, the quantity $\mathrm{d}z/\mathrm{d}s$ should be a positive number. After a short algebra, it can be shown that the condition of stability is

$$\beta \leq 8 \ . \tag{7.65}$$

For most materials, condition (7.65) is satisfied. For soft materials such as graphite. β could be much larger than 8.

7.5.2 Effect of Force in Tunneling Barrier Measurements

It is well known that in STM images of graphite surfaces, the corrugation is substantially amplified by the deformation of the sample surface [7.10], and the condition of stability is not satisfied. Observation of hysteresis was reported [7.10]. For most metals and semiconductors, the stability condition, (7.65), is satisfied. In addition, in most cases the STM images are taken under conditions when a strong attractive force is present. The dramatic amplification effect does not occur. However, the measurement of the apparent barrier height, using an ac method, based on (7.16)

$$\phi_{\mathrm{ap}} \equiv 0.952 \left(\frac{\mathrm{d} \ln I}{\mathrm{d}z} \right)^2 \ , \tag{7.66}$$

is very sensitive to a slight variation of gap distances. In fact, if the apparent barrier height *in terms of actual displacement of gap width* δs is a constant,

$$\phi_0 \equiv 0.952 \left(\frac{\mathrm{d} \ln I}{\mathrm{d}s} \right)^2 = \mathrm{const} \ , \tag{7.67}$$

but due to the force and deformation

$$\frac{\mathrm{d}z}{\mathrm{d}s} \neq 1 \ , \tag{7.68}$$

an apparent variation of the measured barrier height (7.66) should be observed. In the strong repulsive-force regime, $(\mathrm{d}z/\mathrm{d}s) \gg 1$, a very small apparent barrier height should be noted. This is a common phenomenon observed on graphite and other layered materials. In the attractive-force regime, $(\mathrm{d}z/\mathrm{d}s) < 1$, an increase of the measured value of the apparent barrier height should be observed. Actually, this was observed by *Teague* as early as 1978 [7.12] in the gold-vacuum-gold experiment. At a very short electrode distance, where the actual barrier collapses, the current becomes even higher than what was expected from an exponential dependence on the distance. A study on clean Si(111) surface with a clean W tip [7.11] reveals the entire curve of the dependence of the measured apparent barrier height, (7.66) with z-piezo displacement. The experiment was performed under the condition that a clear 7×7 pattern is observed, which indicated that both the tip (near the apex atom) and the sample are clean. By carefully moving the tip back and forth, not pressing into the sample surface too deep, the entire process is completely reversible. The experimental barrier height measurements were performed using an ac modulation method, by applying a small 0.05 Å modulation to the z-piezo at a frequency $\omega_{mod} \simeq 2$ kHz. The ac method provides a better accuracy than the dc method. The data points in Fig. 7.14 are the experimentally measured apparent barrier height as a function of tip-sample separation obtained in this manner on a clean Si(111)-(7×7) sample with a sample bias of -1 Volt with respect to the tip. As a function of tip-sample separation, the barrier height vs. distance can be separated into four distinct regimes: (1) At large separations the barrier height is approximately 3.5 eV, which is roughly equal to the average work functions of tungsten and silicon. (2) As the tip-sample separation is decreased, the barrier height first exhibits a small *increase* to about 4.8 eV. (3) Further decreasing the tip-sample separation causes the barrier height to plummet by more than a factor of ten with only a 1 Å change in tip-sample separation. (4) Pushing the tip toward the sample even further produces only a small modulation of the current dI/dz, leading to an apparent barrier height of near zero. In both cases, the observed behavior of the apparent barrier height is *continuous and reversible* if the tip is not pushed too deep.

The observed variation of the apparent barrier height can be understood quantitatively based upon (7.64). The solid curve in Fig. 7.14 is drawn with $\beta = 0.95$ and assuming the actual apparent barrier height is 3.5 eV throughout the entire region. The accurate fit indicates that the model is reasonable.

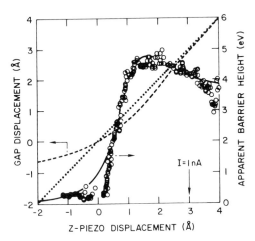

Fig. 7.14. Variation of the measured apparent barrier height with z. Circles are data points. The solid curve is derived from (7.64). The dashed curve is the actual gap displacement as a function of the measured z-piezo displacement. The dotted curve, the fictitious gap displacement in the absence of force, is included for comparison [7.11]

Using (7.61), the characteristic radius of the end of the tip can be estimated. Assuming $U_e = 5\,\text{eV}$, with the elastic constants of tungsten, $E = 34 \times 10^{11}\,\text{dyn/cm}^2$ and $v = 0.26$, it follows that $a_0 \simeq 5\,\text{Å}$. This is a reasonable value for tips which exhibit atomic resolution.

The normal tip–sample distance in STM experiments can be obtained accurately from this experiment. In Fig. 7.14, the equilibrium distance, where the net force is zero, is taken as the origin of z. As shown for the case of aluminum, because the attractive force has a longer range than the repulsive force, the absolute equilibrium distance between the apex atom and the counterpart on the sample surface is slightly less than the sum of the atomic radii of both atoms, which is about $\simeq 2\,\text{Å}$. The normal topographic images are usually taken at $I = 1\,\text{nA}$, corresponding to a distance of $\simeq 3\,\text{Å}$ from the equilibrium point, or $\simeq 5\,\text{Å}$ from nucleus to nucleus.

7.6 In-situ Characterization of Tip Electronic Structure

If the tip DOS and the sample DOS are sufficiently structured, the observed tunneling spectra mainly contain information of both tip DOS and sample DOS. For the purpose of determining the energy spectra, only the relative values of the tip DOS and the sample DOS are relevant. In this case, the accurate value of the tunneling matrix element, especially the correction factor, becomes unimportant. This situation is very similar to the original situation of the *Bardeen* tunneling theory [7.13] regarding the interpretation of *Giaever's* experiment [7.15].

The relative DOS for the sample and the tip can be defined as follows. For the case $V > 0$, by defining a parameter $u = (E - E_F)/e$, a convenient form for the relative sample DOS is:

$$\sigma(u) = \varrho_S(E)/\varrho_S(E_F) \ . \tag{7.69}$$

For the tip, we define a parameter $u = (E_F - E)/e$, and let

$$\tau(u) = \varrho_T(E)/\varrho_T(E_F) \ . \tag{7.70}$$

Obviously, because of the condition $V > 0$, the arguments u in both (7.69, 70) are always positive. (For the case $V < 0$, simply exchange the definitions of u.) Clearly, we have $\sigma(0) = 1$ and $\tau(0) = 1$. By differentiating (7.4), assuming $|M| \simeq \text{const.}$, a pair of symmetric equations are obtained [7.40]

$$\sigma(V) = g(V) + \int_0^V \tau'(V - u)\sigma(u)\,du \ , \tag{7.71}$$

$$\tau(V) = g(V) + \int_0^V \sigma'(V - u)\tau(u)\,du \ . \tag{7.72}$$

The above equations are the standard form of a well-studied class of integral equations, the *Volterra equation of the second kind*. The numerical methods for solving the Volterra integral equations are well-developed (see, for example, [7.41]). Before discussing the results of numerical calculations, we draw a few simple consequences from those equations. Using a free-electron-metal tip, i.e., if in the entire energy range of interest,

$$\tau'(V) = 0 \, , \tag{7.73}$$

then from (7.72),

$$g(V) = \sigma(V) \, . \tag{7.74}$$

In this case, the dynamic conductance equals the sample DOS up to a constant factor. Now, we measure the tunneling spectrum on the same sample using another tip of unknown DOS, and find a new dynamic conductance as a function of bias voltage, $g(V)$. By solving (7.72), we obtain the relative DOS of the unknown tip. (Similarly, if the sample is a free electron metal, i.e., $\sigma'(V) = 0$, then the dynamic conductance equals the tip DOS up to a constant factor).

As an example, we analyze the tunneling spectra published by *Pelz* [7.42], where the effect of spontaneous tip restructuring on the observed tunneling

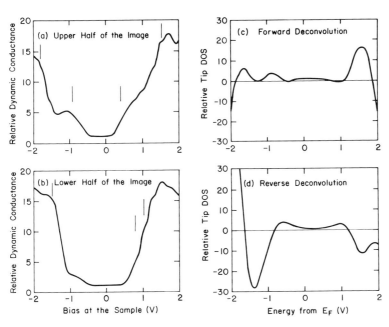

Fig. 7.15. Deconvolution of the tunneling spectra. (**a**) and (**b**), the tunneling spectra from the Si surface image in [7.42]. The observed tunneling spectra changed dramatically from the upper half to the lower half of the image. (**c**) and (**d**), results of deconvolution. It indicates that during the scan, the tip picked up a Si cluster, and the tip DOS resembles that of a Si surface [7.40]

spectra is recorded: During a single scan, the observed image suddenly changes [7.42]. Although the atomic structure on the Si(111)-(7 × 7) surface is resolved on both parts of the image, the normalized dynamic conductance changes dramatically. Using (7.70), the corresponding relative dynamic conductance is obtained. There is no a priori reason to determine which tip state is more similar to a free-electron metal. By assuming that the tip DOS before restructuring (upper half of the image) is nearly flat, the deconvolution procedure produces a tip DOS after restructuring (lower half of the image), as shown in Fig. 7.15c. A reversed deconvolution generates Fig. 7.15d. However, in Fig. 7.15d, a large portion of the obtained tip DOS is negative, which is unphysical. Therefore, the tip DOS before tip restructuring is more likely to be flat. There is still no consensus of what the accurate tunneling spectrum of the averaged Si(111)-(7 × 7) surface is. However, the general feature of Fig. 7.15a is consistent with the tunneling spectra most authors reported, whereas the general feature of Fig. 7.15b is not. As shown in Fig. 7.15c, the tip DOS after restructuring has two peaks below the Fermi level, and one peak above the Fermi level. This feature is similar to the observed tunneling spectrum of Si(111)-(7 × 7). Therefore, a likely explanation is that the tip picks up a small Si cluster, and the tip DOS becomes similar to that of silicon.

7.7 Summary

We have outlined a unified perturbation theory which can account for the tunneling phenomenon as well as the attractive force between the tip and the sample in STM. The accuracy of the new perturbation approach, the Modified Bardeen Approach (MBA), is verified by comparing with the exact solutions of the square-barrier problem and the hydrogen molecular ion problem. Explicit expressions for the tunneling matrix elements are presented. To make quantitative predictions for the STM images, two approximate methods are presented: The leading Bloch wave method, and the independent atomic orbital method. For simple periodic surfaces, both methods predict similar STM images. A measurable relation between the tunneling conductance and the attractive atomic force is established. Based on the MBA, a method for in situ characterization of tip electronic states is described.

7.8 Appendix: Modified Bardeen Integral for the Hydrogen Molecular Ion

In this Appendix, a derivation of the modification factor in the MBA treatment of the hydrogen molecular ion is presented. It is contained in a unpublished internal report, dated 1955, by *Holstein* [7.32].

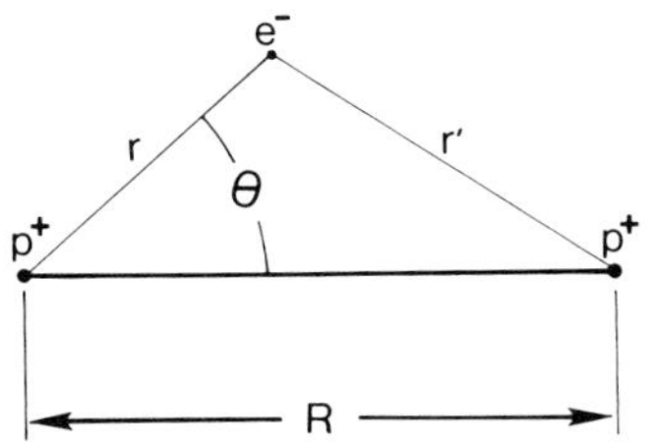

Fig. 7.16. Evaluation of the correction factor [7.32]

The exact time-independent Schrödinger equation for the electron, in atomic units, is (Fig. 7.16)

$$-\frac{1}{2}\nabla^2\psi - \left(\frac{1}{r} + \frac{1}{r'}\right)\psi = E\psi \; . \tag{7.75}$$

where r' is the distance between the electron and second proton

$$r' = \sqrt{R^2 + r^2 - 2Rr\cos\theta} \; . \tag{7.76}$$

In the absence of the second proton, (7.75) is the Schrödinger equation for the free hydrogen atom. The ground-state wave function is

$$\psi_0 = \frac{1}{\sqrt{\pi}}e^{-r} \; , \tag{7.77}$$

and the energy eigenvalue is

$$E = -\frac{1}{2} \; . \tag{7.78}$$

The presence of the second proton induces a perturbation to the wave function and the energy eigen value of the electron. For the perturbed wave function, we make the *Ansatz*

$$\psi = \psi_0 e^{-g} \; , \tag{7.79}$$

where the function $g(r)$ is to be determined. To a sufficiently accurate degree of approximation, the energy is

$$E = -\frac{1}{2} - \frac{1}{R} \; , \tag{7.80}$$

which is equivalent to taking the energy of the system equal to that of the isolated atom plus a constant potential energy term, $-1/R$. The exact energy has an additional van der Waals term, $(9/4)\,r^{-4}$, the effect of which is much smaller than the first term. By inserting (7.79, 80) into (7.75), we obtain the equation for the correction function g:

$$\nabla r \cdot \nabla g - \frac{1}{R} + \frac{1}{r'} = (\nabla g)^2 - \nabla^2 g \ . \tag{7.81}$$

To make an approximate solution, we neglect the two terms on the right-hand side of the equation. Those terms are much smaller than the terms on the left-hand side. Thus, we obtain the approximate equation for the correction function g:

$$\frac{\partial g}{\partial r} = \left(\frac{1}{R} - \frac{1}{r'} \right) \ . \tag{7.82}$$

At $r = 0$ the correction function g should be independent of θ. Also, the corrections should have an opposite sign for $+z$ and $-z$. To preserve normalization, it is accurate to set $g = 0$ at $r = 0$. Thus, (7.82) can be integrated immediately:

$$g = \int_0^r \left(\frac{1}{R} - \frac{1}{r'} \right) dr = \left(\frac{r}{R} - \log \frac{r + r' - R\cos\theta}{R(1 - \cos\theta)} \right) \ . \tag{7.83}$$

We are interested only in the values of the correction function near the medium plane, $r \simeq R/2$. From Fig. 7.16, it is clear that

$$r = r' = \frac{R}{2} \sec\theta \ . \tag{7.84}$$

for small θ, we obtain

$$g = \frac{1}{2} - \log(1 + \sec\theta) \simeq \frac{1}{2} - \log 2 \ . \tag{7.85}$$

At $r \simeq R/2$, the wave function gains a factor

$$e^{-g} = 2e^{-1/2} \simeq 1.213 \ . \tag{7.86}$$

In other words, near the separation surface between the two protons, the modified wave function of a hydrogen atom is

$$\psi = \psi_0 e^{-g} = \frac{2}{\sqrt{\pi e}} e^{-r} \ . \tag{7.87}$$

The modified Bardeen integral, (7.25), can be evaluated directly to obtain

$$M = -\frac{2}{e} R e^{-R} \ . \tag{7.88}$$

References

7.1   G. Binnig, H. Rohrer: Helv. Phys. Acta **55**, 726 (1982).
7.2   *Scanning Tunneling Microscopy* I and II, ed., H.-J. Güntherodt and R. Wiesendanger, Vols. 20 and 28, (Springer, Berlin, Heidelberg 1992)
7.3   J.A. Stroscio, R.M. Feenstra, A.P. Fein: Phys. Rev. Lett. **57**, 2579 (1986); R.M. Feenstra, J.A. Stroscio: Physica Scripta, **T19**, 55 (1987); R.M. Feenstra: Phys. Rev. B **44**, 13791 (1991)
7.4   V.M. Hallmark, S. Chiang, J.F. Rabolt, J.D. Swalen, R.J. Wilson: Phys. Rev. Lett. **59**, 2879 (1987); A. Samsavar, E.S. Hirschorn, T. Miller, F.M. Leibsle, J.A. Eades, T.C. Chiang: T.C. Phys. Rev. Lett. **65**, 1607 (1990)
7.5   R.J. Behm: Scanning tunneling microscopy: Metal surfaces, adsorption and surface reactions: In R.J. Behm, N. Garcia, H. Rohrer, *Scanning Tunneling Microscopy and Related Methods*, ed. by (Kluwer, Dordrecht 1990), pp 173–209
7.6   J. Wintterlin, J. Wiechers, H. Brune, T. Gritsch, H. Höfer , R.J. Behm: Phys. Rev. Lett. **62**, 59 (1989)
7.7   P. Zeppenfeld, C.P. Lutz, D.M. Eigler: Int'l Ultramicroscopy **42–44**, 128 (1992)
7.8   G. Binnig, H. Fuchs, Ch. Gerber, H. Rohrer, E. Stoll, E. Tosetti: Europhys. Lett. **1**, 31 (1986)
7.9   U. Dürig, J.K. Gimzewski, D.W. Pohl, D.W. Phys. Rev. Lett. **57**, 2403 (1986); U. Dürig, O. Züger, D.W. Pohl, J. Microscopy, **152**, Part 1, 259 (1988)
7.10 J.M. Soler, A.M. Baro, N. Garcia, H. Rohrer: Phys. Rev. Lett. **57**, 444 (1986); H.J. Mamin, E. Ganz, D.W. Abraham, R.E. Thompson, J. Clarke; Phys. Rev. B **34**, 9015 (1986)
7.11 C.J. Chen, and R.J. Hamers, J. Vac. Sci. Technol. B **9**, 230 (1991)
7.12 E.C. Teague: Thesis, 1978. Reprinted on Journal of Research of the National Bureau of Standards, **91**, 171 (1986)
7.13 J. Bardeen: Phys. Rev. Lett. **6**, 57 (1961)
7.14 J. Bardeen, L.N. Cooper, J.R. Schrieffer: Phys. Rev. **108**, 1175 (1957)
7.15 I. Giaever: Phys. Rev. Lett. **5**, 147, 464 (1960)
7.16 For example, L.D. Landau, L.M. Lifshitz: *Quantum Mechanics* (Pergamon, London 1977)
7.17 C.B. Duke: *Tunneling in Solids* (Academic, New York 1969)
7.18 J.R. Oppenheimer: Phys. Rev. **13**, 66 (1928)
7.19 C. Calori, R. Combescot, P. Nozieres, D. Saint-James: J. Phys. C **5**, 21 (1972) and references therein
7.20 T.E. Feuchtwang: Phys. Rev. B **20**, 430 (1979), and references therein
7.21 J. Tersoff, D.R. Hamann: Phys. Rev. Lett. **50**, 1998 (1983); Phys. Rev. B **31**, 805 (1985)
7.22 A. Baratoff: Europhysics Conf. Abstracts **7b**, 364 (1983); Physica Amsterdam) **127 B**, 143 (1984)
7.23 C.J. Chen: J. Vac. Sci. Technol. A **6**, 319 (1988)
7.24 C.J. Chen: Phys. Rev. Lett. **65**, 448 (1990)
7.25 C.J. Chen: Phys. Rev. B **42**, 8841 (1990)
7.26 C.J. Chen: J. Vac. Sci. Technol. A **9**, 44 (1991)
7.27 C.J. Chen: Mod. Phys. Lett. B **5**, 107 (1991)
7.28 C.J. Chen: J. Phys, Cond. Matter **3**, 1227 (1991)
7.29 C.J. Chen: *Introduction to Scanning Tunneling Microscopy* (Oxford Univ. Press, New York 1993)
7.30 W. Heisenberg: Z. Phys. **38**, 411 (1928)
7.31 L. Pauling: *The Nature of the Chemical Bond* (Cornell Univ. Press, Cornell 1977)
7.32 T. Holstein: Westinghouse Research Report 60-94698-3-R9 (1955)
7.33 D.R. Bates, K. Ledsman, A.L. Stewart: Phil. Trans. Roy. Soc. London, **246**, 215 (1953)
7.34 J.C. Slater: *Quantum Theory of Molecules and Solids*, Vol. 1 (McGraw-Hill, New York 1963)
7.35 C. Herring: Rev. Mod. Phys. **34**, 631 (1962)
7.36 J. Harris, A. Liebsch: Phys. Rev. Lett. **49**, 341 (1982); A. Liebsch, J. Harris, M. Weinert: Surf. Sci. **145**, 207 (1984)
7.37 F.O. Goodman: J. Chem. Phys. **65**, 1561 (1976); F.O. Goodman, H.Y. Wachman: *Dynamics of Gas–Surface Scattering* (Academic, New York 1976)

7.38 D. Drakova, G. Doyen, F.v. Trentini: Phys. Rev. B **32,** 6399 (1985)
7.39 J.B. Pethica, W.C. Oliver: Physica Scripta T **19,** 61 (1987); J.B. Pethica, A.P. Sutton, J. Vac. Sci. Technol. A **6,** 2490 (1988)
7.40 C.J. Chen: Ultramicroscopy, **42–44,** 147 (1992)
7.41 H. Brunner, P.J. van der Houwen: *The Numerical Solution of Volterra Equations,* (North-Holland, Amsterdam 1986)
7.42 J.P. Pelz: Phys. Rev. B B **43,** 6746 (1991)

8. Theory of Tip–Sample Interactions

S. Ciraci

With 10 Figures

In conventional, Scanning Tunneling Microscopy (STM) the tip–sample separation is assumed to be sufficiently large to allow only weak coupling between the electronic states. In this case the electrodes have been considered to be independent. In some operating modes of STM the tip–sample separation is purposely set small to enhance the tip–sample interaction and hence to modify the electronic and atomic structure irreversibly. Indeed, as the tip approaches the sample surface, the potential barrier is lowered, the charge density is rearranged and the ions in the vicinity of the tip are displaced to attain the minimum of the total energy at the preset tip–sample distance. Modifications of the electronic and atomic structure depending upon the tip–sample separation have led to the identification of different regimes in the operation of STM; ranging from the independent electrodes to the irreversible mechanical contact. This chapter deals with the tip–sample interaction effects. The variation of electronic structure and vacuum barrier, the character of conduction and tip force are investigated as a function of separation. Our analysis suggests that operation of the tunneling and force microscopes under a significant tip–sample interaction will bring about potential applications not only in the investigation of electronic and atomic structure but also in mesoscopic physics.

8.1 Tip–Sample Interaction

The invention of scanning tunneling microscopy [8.1] and subsequently of scanning force microscopy (SFM) [8.2] provided new techniques for studying the structure and the electronic properties of solid surfaces and molecules down to the atomic scale. The conventional view of STM usually assumes that the tip–sample separation z is large and hence the electrodes (i.e., tip and sample) are nearly independent. In this case, the tunneling can be described by the transfer Hamiltonian approach in terms of one-electron states of the bare electrodes (i.e., free tip and free sample) [8.3]. Using the transfer Hamiltonian approach and representing the tip by a single s-wave it was shown that the tunneling is approximately related to the local density of the electronic states of the sample at the center of the tip at the Fermi level, $\varrho_S(r_0, E_F)$ [8.4]. Accordingly, the tunneling current is exponentially dependent on the tip–sample separation, and STM probes the weak overlap between the wave functions of electrodes yielding a resolution down to the atomic scale.

Springer Series in Surface Sciences, Vol. 29
Scanning Tunneling Microscopy III Eds.: R. Wiesendanger · H.-J. Güntherodt
© Springer-Verlag Berlin Heidelberg 1993

Depending on the bias polarity, tunneling occurs either from filled or towards empty states of the sample and can also reveal the spectrum of the sample surface in the range of energy $E_F - eV \le \varepsilon_S \le E_F + eV$ (E_F and V being Fermi energy and bias voltage, respectively) [8.5, 6]. New techniques, such as ballistic electron emission microscopy [8.7] (to probe interfaces and to provide spatial and energy resolution of the scattering process) and field emission of electrons from an atomically sharp tip [8.8] (to obtain a stable and well collimated e-beam) are also derived from STM. Even though the coupling between the electronic states is negligible, the correlated motion of the electrons in separated electrodes gives rise to a long-range interaction. The van der Waals (VdW) forces derived from this long-range interaction may be significant depending on the shape of the tip, but are expected to be essentially uncorrugated on the atomic scale. All these modes of operation and long-range interaction in STM occur at a large separation with nearly independent electrodes being identified as the *conventional tunneling regime*.

In some typical operating modes of STM [8.9], the observation of force variation of the order of 10^{-9} N (or several eV/Å) indicated significant overlap and rearrangements between electronic charge densities of sample and tip, at least if just a few atoms are responsible for these variations. In several other studies, the tip–sample separation was purposely set small to modify atomic and electronic structure of electrodes and hence to enhance the current. STM operating at small z (corresponding to very small bias) yielded a corrugation which is much larger than that deduced from $\varrho_S(r_0, E_F)$, e.g. of the free graphite surface [8.10]. Initially the observed giant corrugation was attributed to the elastic deformation indicating a significant tip–sample interaction owing to the close proximity of the tunneling tip. Self-Consistent Field (SCF) pseudopotential calculations had shown, however, that as z decreases the potential barrier Φ_B gradually collapses, and the surface charge density of graphite is disturbed locally [8.11]. The electronic states, which are relevant for tunneling evolve into a kind of resonance states having relatively higher weights in the vicinity of the tip [8.12]. Tip-induced local disturbances of the electronic states may change into a chemical bond between the outermost tip and sample atoms [8.11]. STM observations of individual "atoms" with the periodicity of nominally flat (111) surfaces of close-packed noble [8.13] and simple [8.14] fcc metals with a corrugation much larger than one would deduce from $\varrho_S(r_0, E_F)$ suggest site-dependent tip–surface interaction effects [8.15]. Whether force variations and induced deformations along STM scans and/or changes in electronic structure (together with the contribution of d-states) must be invoked to explain such observations is understandably an important issue. The interaction energy $E_i(z)$ (or adhesion energy) between tip and sample, and the short-range force derived thereof are important outcomes of the tip–sample interaction. The tip force is of relevance for SFM when it shows significant variation with the tip position [8.16]. This occurs at relatively small z, in which short-range forces dominate the tip force. As z is decreased, the interaction energy becomes increasingly negative until the separation $z = z_e$ corresponding to maximum adhesion. The perpendicular tip force $F_{s\perp}(z) = -\partial E_i(z)/\partial z$ becomes first increasingly attractive, passes through a minimum and then decreases to become repulsive for $z < z_e$. For an atomically sharp tip it is expected that significantly strong lateral forces can also arise when the tip is positioned off high symmetry positions. If the lateral force gradient exceeds the restoring spring constant of the cantilever, the tip starts to perform a stick-slip motion on the sample surface [8.17]. These lateral forces, which are essentially conservative, can thus induce hysteresis and losses via energy transfer to shear modes, resulting in an average microscopic friction force of nonconservative nature.

It becomes clear from the above discussion that the evolution of the electronic and transport properties and the variation of the tip force with the tip–sample separation distinguish new regimes of operation in STM, which are different from the conventional tunneling between nearly independent electrodes [8.12, 18, 19]. While the conventional tunneling has developed into a powerful technique with real-space imaging capability and atomic resolution, new regimes in the presence of significant tip–sample interaction may open new horizons in the application of STM and SFM [8.18]. As the distance between the tip and sample is decreased, the overlap of the wave functions of the electrodes increases and several interrelated atomic scale interaction effects then come into play, as suggested by investigations of the transition from tunneling to electrical and mechanical contact [8.20]. The potential barrier between tip and sample is gradually lowered, which causes significant rearrangements of the charge density. This, in turn, induces an attractive (binding) interaction or adhesion energy leading to short-range attractive forces. Responding to the latter, the ions of the tip and sample are displaced even before the irreversible (plastic) deformations set in. A few ångstrom before mechanical contact, reversible local electronic and structural modifications are expected. The potential barrier collapses before the point of maximum attraction on the apex of the tip. This regime at intermediate separations is characterized by significant electronic interaction and is identified as the *electronic contact regime*. In spite of the fact that the electronic states are modified, the transport of current takes place via tunneling.

Upon a further approach of the tip, a mechanical contact is eventually formed through strong bonds with sample atoms [8.11]. If the cross section of the contact is sufficiently large the *constriction effect* due to the confinement of current carrying states becomes negligible, hence the transport occurs in the absence of any barrier. This is the ballistic conduction and involves quantum effects since the size of the contact is comparable with the Fermi wavelength λ_F. The operation of STM in this range of very small separation reveals a different regime, in which the character of transport can undergo a qualitative change [8.20–24]. Irreversible deformations are then also expected in the vicinity of the tip [8.25, 26]. We identify this regime as that of *mechanical contact*.

In Fig. 8.1 these new regimes (i.e., *electrical and mechanical contact regimes*), which occur beyond that of the conventional tunneling and involve significant

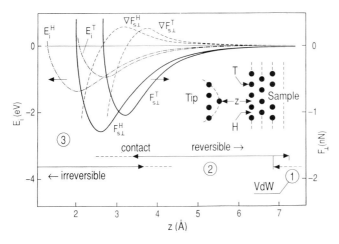

Fig. 8.1. Interaction energy $E_i(z)$, short-range perpendicular force $F_{s\perp}$ and force gradient $\nabla F_{s\perp}(z)$ (in arbitrary units) calculated for the Al(001) sample and tip positioned at the hollow (H) and top (T) sites described by the inset. Ranges of separation z for various regimes of operation in STM are schematically indicated

tip-sample interactions, are schematically represented with reference to physical quantities such as interaction energy $E_i(z)$, the perpendicular component of the short-range tip force $F_{s\perp}$, and its gradient $\nabla F_{s\perp}$ [8.27]. Of course, the extent of these regimes varies depending on the electronic and structural properties of the tip and sample. Besides, the transition between adjacent regions is not sharp.

In this chapter we present a detailed analysis of tip–sample interaction effects on the tip force and transport properties. This is achieved by examining the evolution of electronic states, the potential barrier intervening between two electrodes, and interaction energies as a function of separation. Since the theory relevant to the conventional tunneling has been treated in a number of publications [8.4, 5, 28], we consider here only the effects of long-range interactions. The focus of the chapter is, however, the electrical and mechanical contact regimes. Since the tip force which is relevant for SFM operating at small z has a short-range character, the force microscope is also included in the scope of this chapter. Based on the results of ab-initio calculations we examine the modifications of electronic states and the potential barrier leading to tip-induced localized states. The atom transfer between tip and sample attracts interest since it underlies the structural modification of the sample at the atomic scale. Here we also outline results of recent ab-initio calculations, which are relevant for the atom transfer.

Another objective of this chapter is to reveal important features of STM which pertain to mesoscopic physics. It has now become clear from recent works that the atomically stable sharp tips can be fabricated [8.8]; the cross section relevant for electron transmission is in the range of atomic dimensions or the Fermi wavelength λ_F. Therefore, we discuss the formalism of transport through a quasi one-dimensional (1D) constriction which was developed earlier [8.29]. Recently, starting from such a formalism, a method was developed to treat the transport between tip and sample surface [8.24]. Since this theory has validity in a range covering *tunneling* as well as *ballistic* conduction, physical events involving these separately or concomitantly, and the transition between them, can successfully be addressed. The effective potential barrier, which is generated due to the lateral confinement of states between tip and sample, can be easily visualized within the framework of this method, and will be shown to have important implications. Interesting features of $\log I(z)$ curve, such as saturation at the first plateau and large period "quantum" oscillations observed at the mechanical contact are also covered. Finally, we touch upon an interesting issue, namely exploring a qualitative relation between the short-range tip force and tunneling conductance to unify certain concepts in STM and SFM.

8.2 Long-Range (van der Waals) Forces

The origin of the van der Waals (VdW) force can be better described in terms of surface plasmons of the electrodes, which are coupled through their electric field [8.30]. Therefore, the VdW force is long ranged and can occur even if the electronic states of separated electrodes are decoupled and hence the exponentially decaying short-range forces are negligible. The importance of the VdW force was recognized earlier, and it was argued that, depending on the shape of the tip, the outermost atoms experience strong repulsion, even irreversible deformation, due to the body forces (or VdW forces) at the back of the apex [8.18, 19]. The role of the VdW forces in SFM has been treated by various authors [8.9, 31–33]. Existing studies rely on the integration of the basic power law over the volume of the probing tip.

Although the continuum description of electrodes leaves out the discrete nature of the tip, the generalized *Lifshitz* approach [8.34] is, nevertheless, convenient for metal electrodes. In the case of flat parallel electrodes, the electric field of a plasmon mode with a wave vector $k_{\parallel}$ varies with $\exp(-k_{\parallel}z)$ and, hence, the VdW interaction is dominated by the long wavelength modes at separation $z \gtrsim 14$ a.u. [8.30]. This is the limit leading to the *Lifshitz* formula [8.34] which, in turn, diverges as $z \to 0$, so it cannot be adequate to represent the VdW interaction for small z. With the same reason the shape of the tip becomes crucial at small z. The correct description of the VdW then requires a proper account of the collective behavior of the electrons with nonlocal microscopic dielectric theory. In SFM not only the VdW force alone, but its gradient may be important [8.32]. At large z (where the short-range force is negligible), the force gradient of the VdW force becomes relevant if it exceeds the force constant of the cantilever. This causes an inelastic instability in which the tip jumps to the range $z \simeq z_e$. At small z, the force gradient of the short range force may be added to that of the VdW force.

Integrating the Lifshitz formula with appropriate *Hamaker* constant [8.35]

$$E_W = - \frac{A}{6\pi} \int_{V_{tip}} \frac{dr}{[z(r)]^3} \ , \tag{8.1}$$

for semi-infinite tips (with conical and hemispherical ends and cylindrical shank) the VdW interaction energy can be calculated [8.33]. In (8.1) $z(r)$ is the height of the differential volume element located at r on the tip, and A is the *Hamaker* constant having the value 3.6×10^{-19} J [8.32, 35]. By varying the cone angle α, the diameter of the shank and z independently, a wide range of tip geometry can be taken into account to investigate the long-range interaction [8.33]. The VdW force, F_W is calculated by differentiating (8.1). Results relevant for the present discussion are summarized in Fig. 8.2. These calculations have indicated that for the hemispherical geometry the VdW force and its gradient are large, and their values increase with increasing diameter. For large values of the force gradient $(\nabla F_W \gg \sim 1 \text{ N/m})$ the probing tip jumps to contact and creates instabilities. On the other hand, the VdW force is only ~ 0.1 N for a conical tip having $\alpha = 45°$ and $z \simeq 14$ a.u. For the same tip and separation, the force increases by more than one order of magnitude if α increases to $75°$, but it decreases by two orders of magnitude when α is decreased to $5°$. The force gradient for the conical tip of $\alpha = 5°$ is also small and is in the range of 0.1 N/m. In contrast, the gradient of the short-range force in the increasingly attractive range is ~ 1 N/m. It appears that apart from instabilities which can be avoided by using a stiff cantilever, the VdW interactions cannot have a significant effect on atomic resolution measurements with sharp tips in STM.

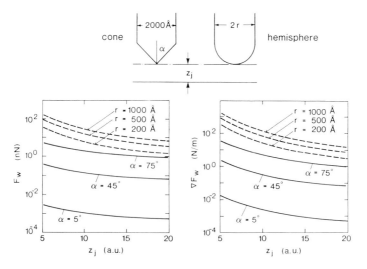

Fig. 8.2. Van der Waals force, F_W and its gradient, ∇F_W calculated as a function of tip-sample separation z_j (measured from their jellium edges) for different tip geometries

8.3 Interaction Energy: Adhesion

When two semi-infinite slabs are close to each other the electrostatic repulsion is overcome by the rearrangement of electronic states yielding net attraction (electrostatic plus exchange-correlation energy) at the expense of positive kinetic energy. The interaction energy $E_i(z)$ is obtained by subtracting the total energies of the tip E_t and sample E_S alone from the total energy of tip and sample together but separated by z [8.19, 33]

$$E_i(z) = E_{t+s}(z) - E_t - E_S \ . \tag{8.2}$$

By definition $E_i(z) < 0$ indicates an attractive (binding) interaction. The minimum of $E_i(z)$ at $z = z_e$ is identified as the binding energy E_b of the electrodes (in which the VdW interaction is not taken into account). It is customary to define the adhesive energy (per atom) for flat electrodes $E_{ad}(z) = E_i(z)/2$ which is the negative of the amount of work necessary to separate two semi-infinite slabs from z to ∞. Note that the surface energy is the negative of $E_i(z = z_e)/2$ calculated for two such slabs. The binding energies between two semi-infinite slabs can be calculated by using the jellium approach [8.36], whereas, the binding energy between an atom (or group of atoms) representing the tip and sample surface can be obtained by using supercell geometry allowing the periodic boundary conditions [8.19]. Earlier, the binding energy of a single Al atom on a graphite surface at the top and hollow site were calculated by using the pseudopotential method within the Local Density Approach (LDA). The binding energies were found rather small, $E_b = -0.33$ eV/atom for the top (T) site and -0.61 eV/atom for the hollow (H) site [8.19]. On the other hand, the binding energy between two Al(001) slabs is significantly larger, i.e., -0.92 and -1.37 eV/atom for the top and hollow sites, respectively [8.33].

The interaction energy between two thin Al(001) slabs is presented in Fig. 8.3 for the T- and H-site [8.33]. An important observation is that even for a simple metal the site dependence is significant. E_b is larger at the H-site since the resulting stacking corresponds to the natural one of the Al(001) layers in bulk Al. However, beyond the separation of maximum attraction at the T-site the $E_i(z)$ curve is lowered below that of the H-site. For $z \gtrsim 10$ a.u. both curves are merged into a single curve and decay exponentially. The interaction energy is short ranged since it is determined by the overlap of charge densities which decay exponentially. It is therefore reasonable to expect that for $z > z_e$, $E_i(z)$ can be approximated by an exponential function. Earlier, *Rose* et al. [8.37] proposed a simple universal relation in terms of the Rydberg function,

$$E_i^*(a^*) = -(1 + a^*)\exp(-a^*) \ , \tag{8.3}$$

to scale interaction energies of flat interfaces between pairs of metals. Here $E_i^* = E_i/E_b$ and $a^* = (z - z_e)/\lambda$, and λ either is taken to be the Thomas–Fermi screening length or is considered as a fitting parameter. This scaling of the

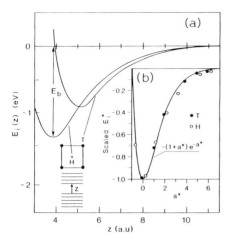

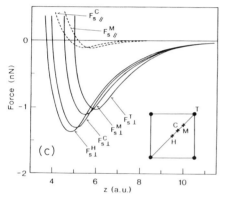

Fig. 8.3. (a) Interaction energy E_i versus separation z between two rigid Al(001) slabs at the hollow (H) and top (T) sites. The z-axis is perpendicular to the (001)-plane. E_b is the minimum value corresponding to the binding energy. (b) Scaled energy E_i^* versus scaled separation a^* according to the Rydberg function of Rose et al. [8.37a]. (c) Perpendicular $F_{s\perp}$ and lateral $F_{s\parallel}$ short range forces (in nano Newtons) on the single-layer "tip slab" versus separation z. Forces are calculated for H-, C-, M-, T-sites shown in the inset

adhesive energy has been exploited by *Dürig* et al. [8.9] to fit the variation of the force gradient with a separation in a combined atomic force and tunneling microscopy experiment. By merely taking $\lambda = 1$ a.u. for Al slabs, the scaled interaction energies E_i^* in Fig. 8.3b fit very well to the Rydberg function at both the T- and H-site.

8.4 Short-Range Forces

The force on an individual atom j of the tip can be calculated either from the derivative of the calculated interaction energy, $F_s(r) = -\nabla_j E_i(r)$, or more conveniently from $\langle \nabla_j H_{\mathrm{LDA}} \rangle$ [8.38]. Indeed, once self-consistency has been achieved, changes in the wave function due to the displacement of nuclei do not contribute to the force, since the eigenfunctions are obtained variationally [8.39]. As a consequence, the force F_s can be expressed as the sum of the ion–electron attraction (in which the electron density is calculated from the self-consistent wave functions),

$$\sim 2 \int [\varrho_S(r) + \Delta\varrho(r)] \left[\frac{\partial}{\partial \tau_j} \left(\frac{Z_j}{|\tau_j - r|} \right) \right] dr \ , \tag{8.4}$$

and the ion–ion repulsion

$$- \sum_S \frac{\partial}{\partial \tau_j} \left(\frac{Z_j Z_S}{|\tau_j - \tau_S|} \right) \ , \tag{8.5}$$

which essentially compensate each other almost completely at large separations. In the above equations, τ_S is the position vector of a sample atom, $\varrho_S(r)$ is the charge density of the bare sample, $\Delta\varrho(r)$ is the modification of charge density due to the tip–sample interaction. In Fig. 8.3c, the variation of the perpendicular and parallel components of the total short-range force F_s is shown for different lateral positions relative to the sample [8.33]. In the force calculations the sample consists of 5 layers Al(001) slab, and the tip is represented by a single Al(001) layer. The strongest attraction occurs at the smallest z_e at the H-site. As the tip slab is moved from the H-, towards the T-site the minimum of the perpendicular force curves gradually shifts to larger z and concomitantly the strength of the attraction decreases. The calculated force curves indicate a corrugation $\Delta z \simeq 1.2$ a.u. at a constant loading force $F_{s\perp}$ in the range of ± 1 nN/atom.

The force variation depicted in Fig. 8.3c is also common to the outermost tip atom. At small z the ion–ion repulsion in (8.5) is much larger than the ion–electron attraction in (8.4) and yields a repulsive force. As z increases, $F_{s\perp}$ changes sign where $E_i(z)$ for a single atom attains its minimum value leading to a net attraction. This is mainly caused by the bonding charge density $\Delta\varrho(r)$ in (8.4) which is accumulated between the tip and sample. In the strong repulsive regime at small z, ion–ion repulsion considerably exceeds the magnitude of the ion–electron part at the T-site and is also larger than the ion–ion repulsion at the H-site. Consequently, $F_\perp^T > F_\perp^H$. As shown in Fig. 8.3c ion–ion repulsion continues to determine the force variation in the range $z \lesssim 6$ a.u. For $z \gtrsim 6$ a.u. the attractive force at the T-site exceeds that at the H-site, since attraction due to the change in the charge density $\Delta\varrho(r)$ dominates the ion–ion repulsion which becomes weaker at large z. Accordingly, the outermost tip atom images the ions of the sample surface at the repulsive and at the strongly attractive force range. At the weak attractive force range the charge density of the sample is imaged. This picture of force variation at the apex atom becomes complicated if the short- and long-range forces at the back of the apex are included. For example, the outermost atom may feel strong repulsion while the total (measured) tip force is still attractive.

Short-range lateral forces which produce energy losses through the energy transfer to shear modes, are the primary cause of friction. Recent measurements by *Mate* et al. [8.17] on the lateral forces as a function of perpendicular loading force and scan velocity have revealed important microscopic features of the friction phenomena. In particular, the dissipation of energy during the stick-slip motion requires further study on the atomic scale. *Zhong* and *Tománek* [8.40] have provided a theoretical estimate of the friction coefficient μ from the atomic scale calculations on the slip of a commensurate Pd monolayer against graphite. They assumed that, in the limit of slow tracking velocity, the gain of conservative energy in going from the H-site to the T-site is fully dissipated in the opposite sequence. This, however, yields a rather unrealistic variation of energy dissipation with tip position. The experimental data [8.17] showing an average nonconservative force superposed on a conservative force, modulated with the lateral periodicity of the sample seems to be at variance with the theoretical results [8.40].

The thorough analysis of friction on the microscopic scale requires a detailed analysis of the energy dissipation. The scope of our present discussion is, however, limited to the analysis of lateral forces which are precursor to the friction. In Fig. 8.3c the variation of the lateral force $F_{s||}(z)$, which is calculated self-consistently for the two Al(001) slabs is presented. They are directed along the diagonal HT in the surface unit cell, and by symmetry they vanish at the H-, and T-sites. For the present system they are one order of magnitude smaller than the perpendicular forces in a wide range of z, since attractive contributions from all neighboring sample atoms tend to add up in $F_{s\perp}$, but they tend to cancel out in $F_{s||}$. Furthermore, $F_{s||}(z)$ is not proportional to $F_{s\perp}(z)$, although it exhibits a similar variation. For $z \gtrsim z_e$, $F_{s||}$ is directed towards T, but for $z \lesssim z_e$ its direction is reversed. Nevertheless at $z \simeq z_e$, $F_{s||}$ is finite, even though $F_{s\perp} \sim 0$. This implies that the interaction-energy surface, $E_i(r)$ does not have an absolute minimum between M- and C-points.

8.5 Deformations

Forces acting on the tip induce deformations (due to the reversible or irreversible change of atomic structure). As body forces the VdW interactions by themselves do not cause any deformation, but their resultant is added to the short-range force. At large z the tip force (and its reaction to the sample) is small and leads to small reversible deformations. In the same range the energy can further be lowered due to the *avalanche* in adhesion [8.41]. We first discuss the elastic (reversible) deformations.

The elastic deformation of the tip (and also of the sample in the vicinity of the tip) occurs in the weak attractive force range; its effect on the images was shown to be marginal [8.15]. Assume that the current I is preset at the H-site at a given z_I which is larger than the separation corresponding to maximum attraction at the T-site. According to Fig. 8.3c the force on the outermost tip atom is $F_\perp^H(z = z_I)$, and tunneling still occurs in that range, so that I sensitively depends on the location of the tip. While scanning from the H-site to the T-site the tip retracts by an amount Δz_I corresponding to the corrugation of "$\varrho_S(r, E_F)$". This causes the attractive force either to decrease or to increase depending upon the variation of $F_{s\perp}(z)$ in the weak attractive range. In the former case, which occurs when the $F_\perp^H(z)$ curve lies below the $F_\perp^T(z)$ curve and hence $|F_\perp^T(z_I + \Delta z_I)| < |F_\perp^H(z_I)|$, the outermost tip atom relaxes inwards leading to a greater separation at the T-site. The current I would decrease, but the STM feedback control will move the tip towards the sample to maintain the preset current. As a result the recorded perpendicular motion of the tip holder is therefore smaller than Δz_I in the absence of deformation. Earlier, based on the SCF-pseudopotential calculations it was found that the corrugation at constant I is slightly reduced by elastic deformations [8.15]. In the latter case, where the $F_\perp^H(z)$ curve lies above the $F_\perp^T(z)$ curve and hence $|F_\perp^T(z_I + \Delta z_I)| > |F_\perp^H(z_I)|$, the apparent corrugation can be slightly enhanced if the real corrugation is smaller than the force corrugation at constant force. Nevertheless, the effect of elastic deformation on the STM corrugation is marginal.

Smith et al. [8.41] proposed an interesting effect which may be relevant to the deformation of the tip at small z. Based on the numerical calculations they showed that outermost atomic layers avalanche together when the separation of two electrodes falls below a critical distance, even if it is much larger than the equilibrium interfacial separation. An avalanche can occur regardless of the stiffness of external supports. Normally, the interaction energy would follow the curve in Fig. 8.3a, if the interlayer spacings were kept fixed. Smith et al. [8.41] showed that the interaction energy can be further lowered if the outermost layers are allowed to relax. The larger is the gain of energy the more surface layers are involved in the relaxation. The energy of the system is lowered since the energy gained by the attraction (or bond formation) of the surface layers of two electrodes exceeds the strain energy due to the increased surface and subsurface interlayer distances.

In the range of separation, yielding increasing attraction near $z = z_e$ where $\partial F_{s\perp}/\partial z > 0$ several irreversible even hysteric effects take place, eventually leading to a plastic deformation [8.25, 26, 41, 42]. Extensive computer simulations using an empirical potential have described features of various atomistic mechanisms [8.26, 41, 42, 43] in this range of separation. Important aspects revealed from these computer simulations are surface melting, nanoindentation, formation of a connective neck, wetting mechanism and hysteresis of the retracting tip. Normally, the mechanical contact is expected to occur when the separation between the outermost tip and sample atoms is near z_e. At this point the force on the outermost tip atom is zero but the total tip force is still attractive. Advancing the tip further results in a nanoindentation on the sample surface. Nanoindentation gives rise to a local plastic deformation involving massive atomic displacement, and perhaps the mixing of tip and sample atoms. When retracting, the tip does not trace the same force and energy curve it does when advancing. This is associated with the hysteresis. In retracting the tip, a connective neck forms first between the two electrodes. Finally, even if the tip is completely separated from the sample, atoms of one electrode may wet the other. Which atoms form the connective neck, and also which electrode is wetted, depend mainly on the relative binding energies of electrode atoms (i.e., tip atom on the tip E_{tt}, sample atom on the sample E_{ss}, tip atoms on the sample E_{ts} and sample atoms on the tip E_{st}). In general, the tip atom wets the sample if $|E_{ts}| > |E_{tt}|$ and/or the cohesive energy of the sample is stronger than that of the tip.

8.6 Atom Transfer

In the previous section we described the irreversible deformations which are due to the close proximity of the tip at the mechanical contact. Here we present microscopic aspects of the atom transfer which may occur already before mechanical contact, but may lead to an irreversible modification of the atomic structure [8.44–46]. Earlier, *Gomer* [8.47] discussed possible mechanisms of atom transfer in STM using schematic variation of interaction energy near the tip and sample. Pointing out atom tunneling, thermally activated desorption and field desorption, he concluded that thermal desorption could be responsible for the sudden transfer of an atom to or from the tip.

The atom at the apex of a sharp tip has a smaller coordination and thus weaker binding compared to that in the bulk. As the tip approaches the sample, the apex atom is attracted towards the sample. To visualize what can happen we can consider a single tip atom between a flat tip and a flat sample. Then, two interaction energy curves like those in Fig. 8.3a can be visualized as being attached to each electrode in opposite directions. The resulting $E_i(z)$ curve rises to very large values in the repulsive range at very small separations from the electrodes, but has two minima separated by an activation barrier Q_b if $s \gtrsim 2z_e$ (s being the distance between two flat electrodes). Figure 8.4 illustrates the evolution of the exact interaction energy versus the position z of an Al atom between two Al(001) slabs calculated as a function of their separation s. The curve has a reflection symmetry at the center since the Al atom is taken to be located between the H-sites of both surfaces. When the atom is far away from one electrode, the interaction energy with the other becomes identical to its interaction energy curve with a single electrode. As s decreases, the two surfaces interact and hence the actual energy is expected to deviate from that obtained by superposing two $E_i(z)$ curves. Therefore calculations like that presented in Fig. 8.4 become important when s is small. As expected, Q_b decreases with decreasing s and hence the rate of the apex atom flicking back and forth to exchange its position at a given temperature increases. The rate of thermal desorption is given by [8.47]

$$\nu \exp[- Q_b(s)/k_B T] \tag{8.6}$$

where ν is the attempt frequency of 10^{12} s^{-1}. In the course of the approach to the

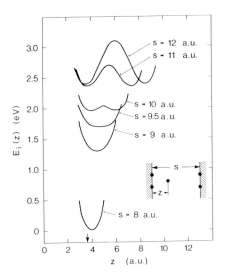

Fig. 8.4. Interaction energy of a single Al atom between two Al(001) slabs versus its distance, z from the left electrode as described in the inset. The single Al atom faces to the H-site of both Al(001) slabs

sample, the height of the apex atom from the rest of the tip gradually increases with increasing sample interaction. Eventually, the barrier collapses upon further advancing of the tip. At the instant $Q_b \approx 0$ the distance of the apex atom both from the sample surface and from the rest of the tip is larger than that corresponding to the equilibrium distances achieved with only one electrode. This behavior is reminiscent of the avalanche effect discussed in the previous section [8.41] leading to a discontinuous drop in the interaction energy. Here the tip (except the outermost atom) and sample are held rigid, so that the interlayer separations are kept fixed. The extend of the curves in Fig. 8.4 would change if the fully relaxed electrodes were taken into account.

In the phenomenon illustrated in Fig. 8.4 the strain energy generated by advancing the outermost atom from the rest of the tip is balanced by the increased attractive interaction from the sample, since the atom gets closer to the sample. In the end, the binding is achieved with a considerably larger binding energy occurring at a relatively large separation as compared to the binding with only one electrode. It means that the outermost atom is stabilized between two electrodes. This phenomenon may have several important implications in STM: (i) The probability of atom transfer towards the deeper minimum is higher for a tip-sample system yielding an asymmetric interaction energy curve. This implies that atoms of one electrode can be transferred to the one which provides stronger cohesion even before the barrier Q_b is collapsed. The asymmetry of the interaction energy curve can arise due to the fact that tip and sample are made of different materials. The tip is usually made from a hard material like W or Ir. Even if the tip and sample were made from the same material, their shapes and hence their coordination numbers would be different. The tip going away from the H-site will experience a shallower minimum. Moreover, there are special places on the sample surface (like step edges and kinks) that may provide a deeper minimum. The atom transfer as described above is expected to be crucial for the wetting. One electrode having a much deeper minimum will be unaffected by the position of the tip and will continue to attract atoms from the other side. (ii) As in field desorption, the atom transfer to one side can be enhanced. An external field of $\sim V/\text{Å}$ can do this, since it can penetrate both the tip and the sample surface [8.48]. As a result of a controlled lateral and perpendicular motion of the tip under an appropriate electric field, the atoms can be relocated to the desired positions [8.44–46]. It is also argued that atoms are transferred by electromigration. (iii) When the tip and sample are brought together in mechanical contact, the size of the single atom contact may not be sufficient to open the lowest ballistic channel with a quantum conductance $2e^2/h$, even if the atom between the electrodes may have resonance states near the Fermi level. This situation is similar to a double-barrier quantum well and gives rise to a jump in the conduction even if the first channel is not opened. The same situation was found earlier in impurity scattering in a 1D mesoscopic channel [8.49–52]. Of course, the resonance condition varies with geometrical configuration (or with s) as the tip scans above the sample surface.

8.7 Tip-Induced Modifications of Electronic Structure

In the independent electrode limit corresponding to large z, the assumption that the tip as well as the sample states are unperturbed is justifiable. However, as the tip approaches the sample surface, the overlap of the tip and sample wave functions increases and a significant electronic interaction sets in. To understand such effects let us first consider unperturbed sample and tip wave functions Ψ_s and Ψ_t with energies

$$\varepsilon_s = \langle \Psi_s | H_s | \Psi_s \rangle , \qquad \varepsilon_t = \langle \Psi_t | H_t | \Psi_t \rangle , \qquad (8.7)$$

respectively. To simplify the picture we also assume that $\langle \Psi_s | \Psi_t \rangle = 0$. For the interacting tip–sample system the total Hamiltonian H_{s+t} differs from the sum of H_s and H_t. Then, in first-order perturbation theory the hopping energy at a given z

$$U_{s,t}(z) = \langle \Psi_s | H_{s+t} | \Psi_t \rangle \qquad (8.8)$$

measures the interaction between tip and sample. When $U_{s,t}(z)$ is small the energies of the independent electrode states shift slightly without a significant mixing. In general the smaller z and $|\varepsilon_t - \varepsilon_s|$, the larger is $U_{s,t}$. If no other states engage in the interaction, these interacting electrode states form bonding, ($\Psi_+ = c_+ \psi_s + c_- \psi_t$) and antibonding ($\Psi_- = c_- \Psi_s - c_+ \Psi_t$) combinations of the unperturbed tip and sample wave functions. In terms of $\xi = [4U_{s,t}^2 + (\varepsilon_t - \varepsilon_s)^2]^{1/2}$ the coefficients are given by

$$c_+ = [1/2 + (\varepsilon_t - \varepsilon_s)/2\xi]^{1/2} , \qquad c_- = [1/2 - (\varepsilon_t - \varepsilon_s)/2\xi]^{1/2} . \qquad (8.9)$$

The energy of the bonding (Ψ_+) and antibonding (Ψ_-) states are expressed as

$$\varepsilon_{\pm} = \langle \Psi_{\pm} | H_{s+t} | \Psi_{\pm} \rangle = (\varepsilon_t + \varepsilon_s)/2 \mp \xi/2 \ . \tag{8.10}$$

The admixture $1 - c_+^2$ is a measure of the deviation from the independent electrode approximation. If $\varepsilon_t \neq \varepsilon_s$, mixing due to $U_{s,t} \neq 0$ results in a transfer of charge. Transfer of charge can occur even if z is large and STM operates in the conventional tunneling regime. The charging effect has interesting implications. For example, an additional Coulomb attraction between tip and sample is induced as a result of the charging effect.

Earlier, *Tekman* and *Ciraci* [8.12] pointed out that, owing to the local character of the perturbation, Ψ becomes increasingly localized as z is increased. For an actual tip–sample system the corresponding states become continuous, resulting in the density of ε_s and ε_t. In this situation, $\Psi_{\pm}$ become resonance states. An instructive way to look at these states is to think of the tip as creating a local perturbation in the potential near the surface of the sample. Just like a surface defect [8.53], this perturbation can lead to localized states or resonances with enhanced amplitude in the vicinity of the tip. This can give rise to an anomalous z-dependence of STM images. One important difference is that the perturbation is dragged along as the tip is scanned over the sample. The observed modulation in z (at constant tunneling current I) or in I itself (at constant mean current) therefore reflects in part changes in electronic structure due to the varying local environment of the tip. Earlier, the observed anomalous corrugation on the nominally flat (111) surfaces of simple and noble metals with the periodicity of the surface unit cell were attributed to the tip-induced states [8.15] and hence to the modifications of the width of the barrier derived thereof [8.54]. Much more recently, *Doyen* et al. [8.55] also used tip-induced resonance states [8.12] to explain the anomalous STM corrugation observed on the Al(111) surface.

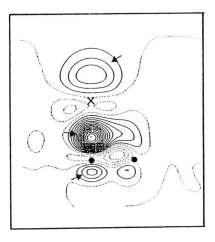

Fig. 8.5. Difference of total charge densities for a single graphite layer and the Al tip consisting of a single atom at $z = 3.8$ a.u.. Aluminium and carbon atoms are indicated by a cross and dots, respectively. Dotted contours correspond to the charge depletion. Contour spacings are 2×10^{-3} electrons/(a.u.)^3

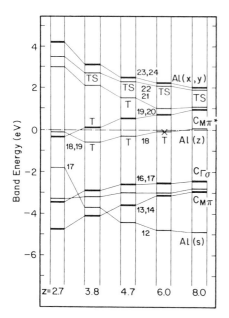

Fig. 8.6. The evolution of states of the graphite sample and the single Al atom tip as a function of separation for the H-site. C_M states originate from the M-point of the graphite Brillouin zone. Al(x, y) and T (or TS) indicate Al($p_{x, y}$) and tip-induced localized (or resonance) states, respectively

The bonding combination of Ψ_S and Ψ_t due to the tip sample interaction gives rise to charge accumulation between tip and sample surface. This can be seen in Fig. 8.5, in which the difference of total charge density [the total charge density of the combined tip–sample system minus the total charge densities of the bare electrodes, i.e. $\Delta\varrho(r) = \varrho_{S+t}(r) - \varrho_S(r) - \varrho_t(r)$], is shown. The energy difference, $\varepsilon_+ - \varepsilon_-$ increases with increasing $U_{s,t}$. As z decreases, ε_+ is lowered while ε_- raises. There might be a certain Ψ_+, which approaches to E_F at a critical z and hence participates in tunneling. For example, in Fig. 8.6 it is shown that for graphite even the states at the M-point of the Brillouin zone, which are far away from the Fermi surface, can participate in tunneling at the H-site as a result of a modification due to the close proximity of the tip. We note that in conventional tunneling the tip has to be advanced very close to the surface in order to get a significant current level from the H-site of graphite.

8.8 Calculation of Current at Small Separation

According to the basic theory of STM the tunneling current decays exponentially, $I \propto e^{-2\kappa z}$, with z and with the inverse decay length given by $\kappa = \sqrt{2m\Phi_B}/\hbar$ (or $\sqrt{\Phi_B}$ in atomic units). Because of this exponential factor the tunneling current is extremely sensitive to z. Assuming that Φ_B is independent of the lateral position of the tip, and also that the electronic states of the free sample are not disturbed by the tip, the variation of the measured tunneling current can be related to the variation of $\varrho_S(r_0, E_F)$ of the unperturbed sample. However, it becomes clear from the discussion of the previous section that at small z the electronic states are modified. The modifications have to be included in the calculation of the tunneling current. This is unfortunately very tedious [8.12].

In this section we discuss a method which starts with a realistic potential between electrodes and obtains the current I (or conductance G) by evaluating the expectation value of the current operator with respect to the current carrying states calculated by using this potential [8.24, 29, 51, 54]. Since the transfer Hamiltonian method is not used, we don't need the electronic states of bare electrodes and any induced modifications of these states explicitly. Nevertheless, these modifications are indigenous to our method since the changes in the electronic states and potential are interrelated. To formulate the transport, the tip–sample system is modeled by using two jellium electrodes separated by a vacuum potential barrier. Clearly the vacuum potential depends on the separation (the distance between two jellium edges z_j). Note that the separation z_j is related to z. That is $z = z_j + d_S + d_t$ with d_S and d_t being half of the interlayer distances in the tip and sample. SCF calculations on a sharp tip facing the metal-sample surface have predicted a potential which is approximately quadratic in the transverse (xy) plane in the vacuum gap [8.15]. Assuming cylindrical symmetry we express the potential in the vacuum gap as

$$V(z, r) = \phi(z, z') + \alpha(z, z')\eta^2 \theta(z' + d_S/2)\theta(z' - z + d_t/2) \qquad (8.11)$$

which is three dimensional. For a given tip–sample separation z, $V(z, r)$ varies with the position vector $r = r(x, y, z')$. The (xy)-plane is perpendicular to the z'-axis, and $\eta^2 = x^2 + y^2$. The maximum of $\phi(z, z')$ coincides with the saddle point of $V(z, r)$ (i.e. V_{sp}), and the barrier height is $\Phi_B = \max\{\phi(z, z')\} - E_F$. A schematic description of the potential expressed by (8.11) is presented in Fig. 8.7. This form of the potential allows a separable solution of the Schrödinger equation. Subsequent to the collapse of the barrier, the radius of the orifice with $V(z, r) < E_F$ is $\eta_m = \{[E_F - V_{sp}(d)]/\alpha\}^{1/2}$

The current carrying states are the 3D plane waves in the electrodes and the quantized states in the constriction defined by (8.11). Since ϕ and α vary with z',

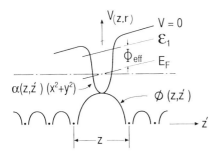

Fig. 8.7. A schematic representation of the potential expressed by (8.11)

$V(z, r)$ is divided into segments. In each segment ϕ and α are assumed to be constant. *Tekman* and *Ciraci* [8.54] have expressed the current carrying solutions $\boldsymbol{\Psi}_{k_i}$ corresponding to an incident wave k_i deep in the tip electrode as

$$\boldsymbol{\Psi}_{k_i}(z, \eta, z') = \sum [A_{nk_i}(z, z') e^{i\gamma_n(z, z')z'} + B_{nk_i}(z, z') e^{i\gamma_n(z, z')z'}] \psi_n(z, \eta, z') \quad (8.12)$$

where $\psi_n(z, \eta, z')$ is a 2D harmonic oscillator solution for a given $\alpha(z, z')$ with $j = j_x + j_y$ and $n = j + 1(j_x, j_y = 0, 1, 2, \ldots)$. The corresponding eigenenergy $\varepsilon_n(z, z') = n[2\hbar^2 \alpha(z, z')/m^*]^{1/2}$. The propagation constant is given by

$$\gamma_n(z, z') = \left\{ \frac{2m^*}{\hbar^2} [E - \phi(z, z') - \varepsilon_n(z, z')] \right\}^{1/2}. \quad (8.13)$$

The coefficients A_{nk_i} and B_{nk_i} in (8.12) are determined by using multiple boundary matching or transfer matrix methods [8.51, 56]. The total tunneling conductance in terms of the matrices of coefficients γ_n is obtained by integrating the expectation value of the current operator over the Fermi sphere. Details of derivative and final conductance expression in terms of matrices of coefficients and γ_n are given in [8.51].

For a uniform and long orifice free from any scatterer (i.e., constant ϕ and constant α) each state dipping below the Fermi level causes the conductance to jump by a quantum of conductance $2e^2/h$ [8.29, 57] and hence acts as a channel of conduction. Taking $E = E_F$, the number of eigenstates ε_n (including degeneracy u_n) which satisfy $\varepsilon_n + \Phi_B \leq E_F$ determine the number of independent conduction channels $N_c = \Sigma_n u_n$. Then the total conductance of these independent channels becomes $G = \Sigma_n 2e^2 u_n/h$. The two terminal theories [8.58] predict that the conductance is given by $G = (2e^2/h) \text{Tr}(T_t T_t^\dagger)$ in terms of the matrix of transmission amplitudes T_t. In the absence of scatterers in a 1D conductor this expression becomes identical with the one obtained by using the independent channel arguments above. Recently, *van Wees* et al. [8.59] and *Wharam* et al. [8.60] achieved the measurement of the conductance G, through a narrow constriction between two reservoirs of a 2D electron gas in a high-mobility GaAs–GaAlAs heterostructure. The constriction they made by a split gate was sufficiently narrow that its width w was comparable with the Fermi wave-length ($w \sim \lambda_F$), and also sufficiently short ($z < l_e$ electron mean free path) that electrons could move ballistically. The resulting conductance of the transport through this constriction was found to change with w (or gate voltage) in quantized steps of approximately $2e^2/h$. The sharp quantization is, however, smeared out or destroyed completely if z is not long enough and if $\phi(z, z')$ is not uniform [8.51]. In this range the character of the conduction is ballistic. Whereas for n with $\Phi_B + \varepsilon_n > E_F$, γ_n in (8.13) becomes imaginary and yields evanescent wave solutions along the z'-axis. The evanescent waves lead to the conduction by tunneling. The tunneling contribution is, however, negligible if the width of the barrier is large. According to the formalism discussed above, the conductance expression has a wide range of application and thus is appropriate for tunneling, ballistic and field emission [8.61].

8.9 Constriction Effect

The potential between the sample surface and the sharp tip displays the form of a narrow constriction, as expressed in (8.11). In the tunneling regime the maximum value of $V(z, r)$ (or saddle point value V_{sp}) at $\eta = 0$ may be higher than the energy of the highest occupied state. That is Φ_B defined in the previous section is positive. Upon decreasing the separation z, the potential barrier Φ_B decreases and eventually V_{sp} dips below E_F yielding a negative barrier. In a classical picture this situation can be viewed as a hole in the vacuum potential between electrodes leading to a ballistic transport. The contour plots of the potential for the Al tip and the Al(111) surface presented in Fig. 8.8 describe how a hole is formed in the potential barrier. To obtain the variation of the potential in the vacuum gap, the pyramidal tip was periodically repeated on the Al(111) slab resulting in a (3×3) tip array, and the charge density was calculated by using the SCF pseudopotential method. It is clearly seen that, near the plane bisecting z, $V_{sp} > E_F$ for $z = 10$ a.u., whereas for $z = 9$ a.u., V_{sp} dips into the Fermi level. In the transverse plane (or xy-plane)) the potential displays an approximately quadratic variation as in (8.11).

Clearly $V(z, r)$, having a quadratic variation in the xy-plane, is a constriction and imposes constraints in the motion of electrons in the x- and y-directions. As a result, the energy of the electron, which is confined to the radius of $[(E_F - V_{sp})/\alpha]^{1/2}$, increases. For example, the lowest eigenstate of the electron confined to the 2D quadratic potential would occur at $\varepsilon_1(z, z') + [2\hbar^2\alpha(z, z')/m^*]^{1/2}$ above V_{sp}. In the adiabatic approximation the effective

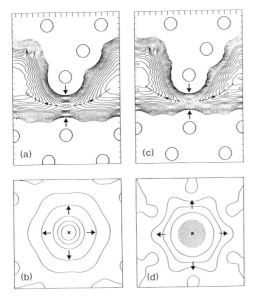

Fig. 8.8. Potential energy of a point contact which is created by a pyramidal Al tip on the Al(111) surface and calculated by the SCF pseudopotential method. (a) and (b) contour plots at $z = 10$ a.u. for (yz)- and (xy)-planes, respectively. (c) and (d) are the same as for (a) and (b) but at $z = 9$ a.u.. Solid and dotted curves correspond to $V(z, r) > E_F$ and $V(z, r) < E_F$, respectively. The potential increases in the directions indicated by arrows. The shaded area in (d) indicates the cross section of the orifice

potential governing the z-dependence of the wave function without transverse nodes can be expressed as $V_{eff} = V(z, \eta = 0, z') + \varepsilon_1(z, z')$. As long as $\Phi_{eff} = V_{eff}(z_{sp}) + [2h^2\alpha(z, z')/m^*]^{1/2} - E_F > 0$, electronic states contributing to I at small voltages experience an effective barrier of maximum height Φ_{eff} and are therefore evanescent in the constriction. This is the *constriction effect* which becomes crucial at small z, where Φ_B is either positive but very low, or negative. As a result of the constriction effect, electrons going from one electrode to the other experience an effective barrier even if $\Phi_B < 0$. Because of this effect the resistance increases.

8.10 Transition from Tunneling to Ballistic Transport

The gradual collapse of the barrier and its effect on the current have been illustrated experimentally by *Gimzewski* and *Möller* [8.20] in studies of the transition from tunneling to point contact between an Ir tip and an Ag sample. Their $\log I$ versus apparent tip-displacement, z_a plot at constant voltage, clearly shows three regimes in the operation of STM discussed at the beginning of this chapter. Initially, the current I increases first exponentially with decreasing z (or increasing z_a). This implies a tunneling behavior. The departure from linearity is due to the variation of height and width of the barrier due to the tip–sample interaction. The discontinuity observed at small z was attributed to the mechanical instability (or irreversible deformation) [8.25]. The recorded values of the conductance just after the discontinuity was only ~ 0.8 of the unit of quantum conductance $(2e^2/h)$. Upon further approach of the tip, I continues to increase and exhibits an oscillatory behavior. Other experiments with different conditions showed different behaviors. For example, in some experiments $\log I(z_a)$ did not increase, but saturated before any irreversible deformation took place. In some cases, at the point of mechanical instability $\log I(z_a)$ is even decreased.

Different behaviors in those experiments, especially the oscillatory behavior following the mechanical instability, attracted interest. Important questions to be answered were how a plateau can occur prior to the plastic deformation, and what is the origin of the observed oscillations. Initially, the oscillatory behavior following the mechanical instability in the $\log I(z_a)$ curve was interpreted as the manifestation of the quantized conductance [8.21]. *Lang* [8.22] simulated the point contact realized in the experiment [8.20] by two jellium electrodes, one of them having a Na atom attached to the jellium edge and thus representing a single atom tip. He found that the conductance G saturates at the value $\zeta 2e^2/h$ and forms a plateau when z is in the range of the distance from the Na core to the positive background edge of the tip electrode. The value of ζ was only 0.4 for Na, and was found to depend on the identity of the tip. Using the non-equilibrium Green's function method within the tight binding approximation *Ferrer* et al. [8.23] also found that G reaches $\lesssim 2e^2/h$ at the smallest z. Apparently, these studies [8.20–23] did not convey whether the transition from tunneling to ballistic regime already takes place at smallest z, or how a plateau can occur prior to the plastic deformation. *Gimzewski* and *Möller* [8.20] gave an estimate for the dimension of the contact area which lies in the range of λ_F. If this is correct, the observed transport beyond the discontinuity has to be associated with the quantum ballistic transport [8.29].

Recent calculations [8.24, 51] of the current between an atomically sharp tip and a flat sample surface performed by using the method and the model potential described in Sect. 8.8 clarified some of the features in the $\log I(z_a)$ curve. In order to link theory with experiment, α in (8.11) is obtained by using the diameter of contact given by experiment and by scaling those values calculated for the Al tip and Al sample [8.15] as a function of z. Furthermore, the electronic parameters of Ag are used for the jellium electrodes. The calculated $\log G$ (which has the same behavior as $\log I$) versus the tip displacement is illustrated in Fig. 8.9. In agreement with previous calculations [8.22, 23] the conductance associated with a uniform constriction set up by a single atom at the vertex of the tip has a value less than the quantum of conductance. Since the length of the constriction is finite and within the range of internuclear distance a_0 (i.e. the sum of atomic radii), this result implies that the energy of the first subband ε_1 is still above E_F (i.e. $\Phi_{\mathrm{eff}} > 0$) and, hence, the conductance is dominated by tunneling.

From the behavior of $\log I(z_a)$ in [8.20], it appears that ballistic transport sets in after the structural instability occurs at $z > a_0$. While z cannot be smaller than a_0, by pushing the tip further, the contact area expands due to increasing plastic deformation followed by adhesion of nearby atoms. That is, while the apparent approach of the tip towards sample continues and hence z_a increases, the separation z is stabilized at the value $\sim a_0$. This is marked by the vertical dash-dotted line in Fig. 8.9. The actual form and size of the contact after the point of mechanical instability is uncertain. Several parameters (such as the detailed atomic structure of the apex of the tip, the strengths of E_{ss}, E_{tt}, and E_{St})

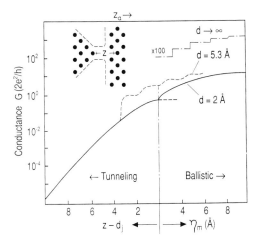

Fig. 8.9. Conductance G versus separation z (in Å) or apparent displacement z_a (in arbitrary units) of an Ag tip above an Ag sample showing the transition from tunneling to ballistic conduction. The effective length of the constriction is $d \simeq z$. The term η_m is the radius of the constriction at E_F. Quantum features are smeared out in the ballistic transport regime corresponding to $d = 2$ Å (or 3.8 a.u.) shown by solid lines beyond the tunneling regime. Quantum features become apparent at the dashed line corresponding to $d = 5.3$ Å ($\simeq \lambda_F$). The dash-dotted line shows sharp quantum steps in the constriction of infinite length

are expected to influence the contact. For example, the apex atom in the repulsive force range can be displaced in such a way that the aperture of the orifice is reduced, incidently causing G to decrease temporarily in the course of the approach. Apart from this exceptional case, we assume that the radius of the contact increases upon approach of the tip. Then, the conductance is related to the radius η_m of the orifice, which increases normally with increased plastic deformation. As the aperture (or diameter) of the contact increases, subbands due to the quantization in the constriction dip sequentially into the Fermi level, causing G to increase by the multiples of $2e^2/h$ each time. As pointed out in Sect. 8.8, in spite of these expected jumps in G, the perfect quantization with a sharp step structure can only be observed if the length of the constriction is longer than λ_F.

The variation of G in the course of plastic deformation is examined by calculating the conductance as a function of $\eta_m = \sqrt{E_F/\alpha}$, for $\phi(z, z') = 0$ at fixed effective length d (which is approximately equal to z). Results are illustrated for $d = 2$, 5.3 Å, and ∞. The first channel is opened at a radius as low as $\eta_m = 1.5$ Å followed by a rise of $\sim 2e^2/h$ in G. The pronounced oscillations (or smeared-out step structure) of $\log G(\eta_m)$ are apparent only for $d \simeq 5.3$ Å, which is comparable with λ_F. However, for $d \simeq a_0$, weak oscillations are washed out on the logarithmic scale. It is seen that the point contact between the tip and sample is not long enough to allow steps or pronounced oscillations. Therefore, the observed oscillations possibly originate from the irregular motion of the atoms as the tip is uniformly pushed towards the sample, causing irregular enlargement of the contact area. Also the atoms of a blunt tip may undergo sequential contacts, with each contact opening a new orifice and leading to abrupt changes in the current. Both cases can give rise to the variation of $\log I(z_a)$ as observed experimentally.

The behavior of $G(z_a)$ is further analyzed by using $\alpha(z)$ values corresponding to different tips. Depending on the shape and material properties of the tip, the form of $G(z_a)$ may exhibit significant changes. For example, G decreases, passing through a maximum ($\sim 2e^2/h$) if η_m is allowed to be less than the atomic radius of Ag. On the other hand, if α is small, $G(z_a)$ may reach a plateau before the point of discontinuity that results from the mechanical instability. This suggests that Φ_{eff} collapses prior to the hysteric deformation, but the value of G at the plateau may be smaller than the quantum of conductance owing to scattering by the ions in the constriction. Certain contacts may have several subbands close to E_F, each contributing to tunneling. In this case, plateaus do not occur, but $G(z_a)$ increases almost linearly. For a flat tip (with several atoms at the apex) the contact area is large and, hence, G rises above $2e^2/h$ well before tip-induced plastic deformation sets in. It should be noted that all these arguments are based on the assumption that there is neither an oxide nor a flake between the tip and sample, as this would influence the variation of $G(z_a)$ dramatically. The character of transport, and the variation of G as a function of z_a (or z) are not generic, but are strongly tip and sample specific. The plateau can appear before the point of discontinuity only under certain conditions.

8.11 Tip Force and Conductivity

In the discussion in Sects. 8.3, 4 it becomes clear that the short-range force becomes significant as long as the potential barrier between electrodes allow the wave functions to overlap. According to *Bardeen*'s theory of tunneling [8.3] the conductance is also determined by the same overlap of wave functions at the Fermi level. Therefore, it is expected that the electrostatic force and conductance are interrelated. Earlier, the reversible modifications of electronic states prior to the contact, as well as correlations between force $F_{s\perp}$ and barrier height Φ_B were pointed out [8.18, 19]. Furthermore, the shifts of the energy eigenvalues due to the proximity of the tip and hence the interaction energy of the tip–sample system were related to the hopping integral $U_{s,t}$ [8.19]. Experimentally, *Dürig* et al. [8.9] drew attention to the correlation between force gradient and tunneling conductance G in the course of the tip approach. More recently, following an idea of *Herring* [8.62], *Chen* introduced an expression which explicitly relates force with conductance [8.63]. First, in compliance with the discussion in Sect. 8.7, he obtains the interaction energy from the splitting of the coupled states through $\Sigma_{s,t}U_{s,t}(z)$. On the other hand, the hopping integral $U_{s,t}(z)$ itself is approximately equal to the tunneling matrix element $M_{s,t}(z)$ [8.64]. As a result, $E_i(z) \simeq \Sigma_{s,t}M_{s,t}(z)$, and hence the perpendicular component of the short-range force can be extracted from $\Sigma_{s,t}M_{s,t}(z)$, i.e. $F_{s\perp} \simeq -\Sigma_{s,t}\partial M_{s,t}/\partial z$. This finally leads to $F_{s\perp} = \xi\kappa\sqrt{G}$, since the tunneling conductance [8.4]

$$G \propto \sum_{s,t} |M_{s,t}(z)|^2 \delta(\varepsilon_t - E_F)\delta(\varepsilon_s - E_F) \tag{8.14}$$

can be approximated in 1D tunneling by $G \propto \exp(-2\kappa z)$ with $\kappa = \sqrt{\Phi_B}$ in atomic units. It is clear that after all these simplifying approximations one cannot expect an exact expression among $F_{s\perp}$, conductance G and κ, as *Chen* [8.63] proposed. Nevertheless, his expression can be approximately valid as long as $\Phi_B > 0$, and becomes more accurate as z increases beyond the point where Φ_B collapses, since at large z the interaction energy is better approximated by $\Sigma_{s,t}M_{s,t}(z)$. At this point, it is worth emphasizing that only the short-range force is related to the conductance. The overlap of wave functions is essential for tunneling conductance, but it is not invoked in the long-range force. Therefore, care must be taken in correlating measured tip force with conductance, since the experimentally determined force includes not only the short-range force, but also the long-range force depending upon the shape of the tip.

A possible relation between force and conductance can be sought by using calculated values. However, owing to the discretization in k-space, the accurate calculation of the conductance is tedious with the above described SCF calculations. Here we compare calculated $F_{s\perp}$ with $\kappa\exp(-\kappa z)$ (which corresponds to $\kappa\sqrt{G}$ in 1D tunneling). The difference between the maximum of the planar averaged potential at $z/2$ and E_F is taken as the barrier height Φ_B in the inverse decay length. Such a comparison indicates approximately a linear relation between the calculated short-range force $F_{s\perp}$ and $\kappa \exp(-\kappa z)$. We note when correlating $F_{s\perp}$ with G that the image potential correction may be significant for large z. This correction is not included in the planar averaged potential. Moreover, neither the constriction effect due to a sharp tip, nor the 3D character of G are taken into account. Despite the fact that several factors affecting the conductivity are not included, the linear relation between the calculated short-range force $F_{s\perp}$ and $\kappa \exp(-\kappa z)$ corresponding to 1D tunneling is interesting.

In order to go beyond the approximation $G \propto \exp(-2\kappa z)$ to calculate the conductance and to explore the relation between short-range force and conductance further, we use the jellium model following the method of *Smith* [8.36]. We first consider the electronic charge density of the semiinfinite left electrode, which is parameterized as $\varrho(z') = \varrho_+ [1 - \exp(-\beta z')/2]$ for $z' < 0$ in the positive jellium and $\varrho(z') = \varrho_+ \exp(-\beta z')/2$ for $z' > 0$ in the vacuum region. Here ϱ_+ is the positive jellium charge density corresponding to Al, and $z' = 0$ marks the jellium edge. The charge density of the right electrode can be written with the same β by replacing $-z'$ by $z' - z_j$. That the charge density is expressed in terms of a single exponent is, of course, an approximation, but is easily tractable. In the *Hohenberg* and *Kohn* theory [8.65] the ground state energy of a confined and interacting electron gas can be expressed as a functional of the electron number density $\varrho(r)$. Furthermore, this functional is minimized for the correct $\varrho(r)$. Accordingly, the exponent $\beta(z_j)$ of the charge density is determined by the minimization of the total energy (with nonlocal exchange and correlation potential Ref. 8.36) of the system consisting of two semiinfinite positive jellium at a given separation z_j. This way the modification of the electronic states due to electrode-electrode (or tip–sample) interaction can be taken into account to some extent [8.33].

Before we go further using simplified jellium approximation in calculating some transport properties, we compare its predictions with those obtained with the SCF pseudopotential method. For example, in the vacuum gap the planar averaged SCF potential is similar to that obtained by the jellium approximation, except that the barrier height in the former is slightly higher. The short-range force calculated by the jellium model compares well with the SCF pseudopotential method. For small separations the charge rearrangement cannot be described by a single $\beta(z_j)$. As a result, forces calculated by the jellium approximation become less attractive at small z_j. Also, for large z_j one expects that the decaying tail of the charge density cannot be well represented by a single decay constant. As a matter of fact, at large separation even the SCF methods are unable to give an accurate representation of the tail of the charge density in terms of the plane-wave basis set. Nevertheless, the jellium method as described above is seen to provide a reasonable representation for the adhesion energy, forces and potential between two flat simple metal electrodes in the range $5 \lesssim z_j \lesssim 12$ a.u.

The transmission probability T for electron tunneling through the vacuum barrier between two semiinfinite jellium slabs was calculated as a function of z_j by using the transfer matrix method [8.56] and by using the WKB approximation [8.33]. For the sake of comparison $\exp(-2\kappa z_j)$ and $\exp(-2\kappa_0 z_j)$ (κ_0 being the barrier height at infinite separation) are calculated. These quantities are relevant for tunneling through a vacuum barrier of constant height. The transmission probability T calculated exactly becomes unity only when $z_j = 0$ (i.e. when the two electrodes merge into one). For $z_j = 0$, $\exp(-2\kappa_0 z_j)$ also becomes unity. However, the transmission probability obtained within the WKB approximation is unity when the barrier collapses for $z_j \lesssim 5$ a.u., which is also true for $\exp(-2\kappa z_j)$. This shows that the WKB approximation fails to represent the transmission for $\Phi_B < 0$. Moreover, the transmission probability calculated by the WKB approximation overestimates the exact T, but becomes proportional to it at large separations as expected. In fact, for large separations all quantities described above become proportional since one has a rather flat and wide tunneling barrier.

Finally we explore how the (short-range) force and conductance are related in the calculations by using the jellium approximation [8.33]. To this end, the calculated force is compared with $\kappa \exp(-\kappa z_j)$ and $\kappa \sqrt{G}$ in Fig. 8.10. We note that in the jellium model the incident wave is a free electron state. Also the potential barrier is slightly underestimated as compared to the SCF results. A wide range of separation spanned in Fig. 8.10 cannot be described by the jellium aproximation, since a single $\beta(z_j)$ is not appropriate to represent the charge density for small and large separations. Moreover, for a very large separation $F_{s\perp}$ is already too small. Hence only a limited range of the curves presented in Fig. 8.10 may be relevant for the present discussion. The force versus $\kappa \exp(-\kappa z_j)$ curve in Fig. 8.10a tends to indicate a power relation in the whole range. This power relation is not a simple one, perhaps approximately linear only in a limited range of z_j, where the attractive force is weak. Since $\exp(-2\kappa z_j)$ is only an approximation to 1D conductance, force versus $\kappa \sqrt{G}$ in Fig. 8.10b (where G is the 3D conductance calculated in the jellium approximation) is more relevant for our discussion. The latter curve also depicts a power relation, but no apparent linear relation throughout the whole range. As in the Fig. 8.10a the relationship can be viewed as linear only in a limited range of z_j, especially at the weak attractive force range. Our results provide an evidence

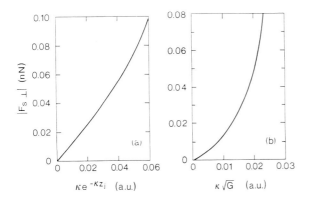

Fig. 8.10. Variation of force with $\kappa \exp(-\kappa z_j)$ (a), $\kappa \sqrt{G}$ (b); all quantities calculated by using the jellium approximation explained in the text

that the short-range force can be related to some kind of transmission property of the barrier between two metal electrodes. Keeping in mind all the approximations involved in the present model, this should be perhaps a trend rather than an exact and universal relation. The overlap between sample and tip-wave functions, which determines both conductivity and short-range force, underlies such a relationship. Experimentally, deviations from any kind of universal expression are expected, depending on the conditions. For example, the structure of the tip enters as an essential parameter because of the constriction effect and also the contribution of the VdW force. Theoretical results obtained by different methods can also deviate due to the details of representing the vacuum potential.

8.12 Summary

In this chapter we discussed the interaction effects between the tip and sample in STM and SFM by varying their separation, and also by moving one over the other along certain symmetry directions. We presented results for the interaction energy, the perpendicular and lateral forces and the vacuum potential calculated earlier by using the SCF pseudopotential method. For the sake of comparison, the same physical quantities together with transmission and conductance between two semiinfinite slabs, which were calculated within the jellium approximation, were also presented. The important aspects of the chapter are summarized as follows: (i) the interaction energy and the short-range force are site-dependent, but both can be expressed by a Rydberg function in terms of the scaled energy and distance. (ii) The van der Waals forces are important at a small tip–sample distance if the tip is flat, or has a large cross section. However, the VdW force and its gradient have negligible effects if the tip is sharp. (iii) Calculated short-range force curves of simple metal electrodes indicate a corrugation of 1.2 a.u. for constant loading force varying in the range of ± 1 nN/atom. (iv) The interaction energy curve having asymmetric double minima gives rise to the atom transfer. (v) At small separations tip-induced modifications of the electronic structure may give rise to resonance states dragged along by the tip. The tunneling current may be enhanced in the presence of these resonance states. (vi) A narrow constriction between tip and sample introduces an effective barrier; in its presence electrons tunnel, even if the potential barrier is collapsed. (vii) The ballistic transport between tip and sample is allowed if the diameter of the contact is comparable with the Fermi wavelength. However, sharp quantizations are smeared out. (viii) We presented some evidence that the tip force is related to the transmission property. This relation reflects, however, a nature of a trend rather than an exact and universal expression.

Acknowledgment. The author thanks Prof. E. Tekman for his valuable assistance and stimulating discussions during the preparation of the manuscript. This work was partially supported by the Joint Project Agreement between Bilkent University and IBM Zurich Research Laboratory.

References

8.1   G. Binnig, H. Rohrer, Ch. Gerber, E. Weibel: Phys. Rev. Lett. **49,** 57 (1982)
      G. Binnig, H. Rohrer: I.B.M. J. Res. Dev. **30,** 1 (1986)
      G. Binnig, H. Rohrer: Rev. Mod. Phys. **59,** 615 (1987)
8.2   G. Binnig, C.F. Quate, Ch. Gerber: Phys. Rev. Lett. **56,** 930 (1986)
8.3   J. Bardeen, Phys. Rev. Lett. **6,** 57 (1961)
8.4   J. Tersoff, D.R. Hamann: Phys. Rev. Lett. **50,** 1998 (1983)
8.5   A. Baratoff: Physica B + C **127,** 143 (1984)
8.6   A. Selloni, P. Carnevali, E. Tosatti, C.D. Chen: Phys. Rev. B **31,** 2602 (1984); B **34,** 7406 (1986)
8.7   L.D. Bell, M-.H. Hecht, W.J. Kaiser, L.C. Davis: Phys. Rev. Lett. **64,** 2679 (1990)
8.8   H.W. Fink: IBM J. Res. Develop **30,** 460 (1986); Phys. Scr. **38,** 160 (1988)
      H.W. Fink, W. Stocker, H. Schmidt, Phys. Rev. Lett. **65,** 1204 (1990)
8.9   U. Dürig, O. Züger, D.W. Pohl: Phys. Rev. Lett. **65,** 349 (1990)
      U. Dürig, O. Züger, Vacuum, **41,** 382 (1990)
8.10 J.M. Soler, A.M. Baro, N. Garcia, H. Rohrer: Phys. Rev. Lett. **57,** 444 (1986)
8.11 S. Ciraci, I.P. Batra: Phys. Rev. B **36,** 6194 (1987)
      I.P. Batra, S. Ciraci: J. Vac. Sci. Technol. A **6,** 313 (1988)
8.12 E. Tekman, S. Ciraci: Phys. Scr. **38,** 486 (1988); Phys. Rev. B **40,** 10286 (1989)
8.13 V.M. Hallmark, S. Chiang, J.F. Rabolt, J.D. Swalen, R.J. Wilson: Phys. Rev. Lett. **59,** 2879 (1987)
8.14 J. Wintterlin, J. Wiechers, H. Brune, T. Gritsch, H. Höfer, R.J. Behm: Phys. Rev. Lett. **62,** 59 (1989)
8.15 S. Ciraci, A. Baratoff, I.P. Batra: Phys. Rev. B **42,** 7618 (1990)
8.16 G. Binnig, Ch. Gerber, E. Stoll, T.R. Albrecht, C.F. Quate: Europhys. Lett. **3,** 1281 (1987)
      T.R. Albrecht, C.F. Quate: J. Vac. Sci. Technol. A **6,** 271 (1988)
      R. Erlandsson, G.M. McClelland, C.M. Mate, S. Chiang: J. Vac. Sci. Technol. A **6,** 266 (1988)
      H. Heinzelmann, E. Meyer, P. Grütter, H.-R. Hidber, L. Rosenthaler, H.-J. Güntherodt: J. Vac: Sci. Technol. A **6,** 275 (1988)
      O. Marti, B. Drake, S. Gould, P.K. Hansma: J. Vac. Sci. Technol. A **6,** 287 (1987)
8.17 C.M. Mate, G.M. McClelland, R. Erlandsson, S. Chiang: Phys. Rev. Lett. **59,** 1942 (1987)
8.18 S. Ciraci: Tip–surface interactions, in *Basic Concepts and Applications of Scanning Tunneling Microscopy and Related Techniques*, ed. by H. Rohrer, N. Garcia, R.J. Behm (Kluwer Academic, Dordrecht 1990) p.119
8.19 S. Ciraci, A. Baratoff, I.P. Batra: Phys. Rev. B **41,** 2763 (1990)
8.20 J.K. Gimzewski, R. Möller: Phys. Rev. B **36,** 1284 (1987)
8.21 N. Garcia (unpublished)
8.22 N.D. Lang: Phys. Rev. B **36,** 8173 (1987)
8.23 J. Ferrer, A. Martin-Rodero, F. Flores: Phys. Rev. B **38,** 10113 (1988)
8.24 S. Ciraci, E. Tekman: Phys. Rev. B **40,** 11969 (1989)
8.25 J.B. Pethica, W.C. Oliver: Phys. Scr. **19 A,** 61 (1987)
      J.B. Pethica, A.P. Sutton: J. Vac. Sci. Technol. A **6,** 2490 (1988)
8.26 J.K. Gimzewski, R. Möller, D.W. Pohl, R.R. Schlitter, Surf. Sci. **189/190,** 15 (1987)
      U. Landman, W.D. Luedtke, N.A. Burnham, R.J. Colton: Science **248,** 454 (1990)
      U. Landman, W.D. Luedtke: J. Vac. Sci. Technol. B **9,** 414 (1991)
8.27 S. Ciraci: Ultramicroscopy **42–44,** 16 (1992)
8.28 N.D. Lang: Phys. Rev. Lett. **56,** 1164 (1986); Phys. Rev. B **34,** 5947 (1986)
8.29 G. Kirczenow: Solid State Commun. **68,** 715 (1988)
      L.I. Glazman, G.B. Lesovick, D.E. Khmelnitskii, R.I. Shekhter: Pis'ma Zh. Eksp. Teor. Fiz. **48,** 238 (1988)
      A. Szafer, A.D. Stone: Phys. Rev. Lett. **60,** 300 (1989)
      E. Tekman, S. Ciraci: Phys. Rev. B **39,** 8772 (1989) and references therein
8.30 J.E. Inglesfield, E. Wikborg: J. Phys. F: Metal Phys. **5,** 1475 (1975)
      J.E. Inglesfield: J. Phys. F: Metal Phys. **6,** 687 (1976)
      J. Heinrich: Solid State Commun. **13,** 1595 (1973)

8.31 U. Hartmann: Phys. Rev. B **43**, 2404 (1991)

8.32 F.O. Goodman, N. Garcia: Phys. Rev. B **43**, 4728 (1991)

8.33 S. Ciraci, E. Tekman, A. Baratoff, I.P. Batra: Phys. Rev. B **46**, 10411 (1992); S. Ciraci, E. Tekman, M. Gökcedag, A. Baratoff, I.P. Batra: Ultramicroscopy **42**–**44**, 163 (1992)

8.34 E.M. Lifshitz: Soviet Physics, **2**, 73 (1956)

8.35 H.C. Hamaker: Physica **4**, 1058 (1937)

8.36 J.R. Smith: Phys. Rev. B **181**, 522 (1963)
J. Ferrante, J.R. Smith: Surf. Science. **38**, 77 (1973)
A. Banerjea, J.R. Smith: Phys. Rev. B **37**, 6632 (1988)

8.37 J.H. Rose, J. Ferrante, J.R. Smith: Phys. Rev. Lett. **47**, 675 (1980)
J. Ferrante, J.R. Smith Phys. Rev. B **19**, 3911 (1979)

8.38 H. Hellmann: *Einführung in Quanten Theorie* (Deutsch, Leipzig 1937)
R.P. Feynman: Phys. Rev. **56**, 340 (1939)

8.39 J.C. Slater: J. Chem. Phys. **57**, 2389 (1972)

8.40 W. Zhong, D. Tománek: Phys. Rev. Lett. **64**, 3054 (1990) Europhys. Lett. **15**, 887 (1991)
J.B. Sokoloff: Phys. Rev. Lett. **66** (1991) 965

8.41 J.R. Smith, G.B. Bozzolo, A. Banerjea, J. Ferrante: Phys. Rev. Lett. **63**, 1269 (1989)
A.P. Sutton, J.B. Pethica: J. Phys. Cond. Matt. **2**, 5317 (1990)

8.42 F. Abraham, I.P. Batra, S. Ciraci: Phys. Rev. Lett. **60**, 1314 (1988)
F. Abraham, I.P. Batra: Surf. Sci. **209**, L125 (1989)

8.43 E. Tosatti: "The Physics of Tip–surface Approaching: Speculations and open Issues" in *Highlights of the Eighties and Future Prospects in Condensed Matter Physics* ed. by L. Esaki (Plenum, New York 1990) p. 631

8.44 I.W. Lyo, P. Avouris: Science **245**, 1369 (1989)

8.45 H.J. Mamin, P.H. Guethner, D. Rugar: Phys. Rev. Lett. **65**, 2418 (1990)

8.46 D.M. Eigler, E.K. Schweizer: Nature, **344**, 524 (1990)
D.M. Eigler, C.P. Lutz, W.E. Rudge: Nature, **352**, 600 (1991)

8.47 R. Gomer: IBM J. Res. Develop. **30**, 426 (1986)

8.48 H.J. Kreuzer: Surf. Sci. **246**, 336 (1991)

8.49 C.G. Smith, M. Pepper, H. Ahmed, J.E.F. Frost, D.G. Hasko, D.C. Peacock, D.A. Ritchie, G.A.C. Jones: J. Phys. C **21**, L893 (1988)

8.50 E. Tekman, S. Ciraci: Phys. Rev. B **40**, 8559 (1990); Phys. Rev. B **42**, 9098 (1990)

8.51 E. Tekman, S. Ciraci: Phys. Rev. B **43**, 7145 (1991)

8.52 J. Faist, P. Gueret, H. Rothuizen: Phys. Rev. B **42**, 3217 (1990)

8.53 H.A. Mizes, W.A. Harrison: J. Vac. Sci. Technol. A **6**, 300 (1988)

8.54 E. Tekman, S. Ciraci: Phys. Rev. B **42**, 1860 (1990)

8.55 G. Doyen, E. Koetter, J.P. Vigneron, M. Scheffler: App. Phys. A **51**, 281 (1991)

8.56 P.E.O. Kane: In *Tunneling Phenomena in Solids*, ed. by E. Burstein, S. Lundqvist (Plenum, New York 1969)

8.57 R. Landauer, IBM J. Res. Dev. **1**, 223 (1957); Z. Phys. B **68**, 27 (1988); J. Phys. Cond. Matt. **1**, 8099 (1989)

8.58 E.N. Economou, C.M. Soukoulis: Phys. Rev. Lett. **46**, 618 (1981)
D.S. Fisher, P.A. Lee: Phys. Rev. B **23**, 6851 (1981); G. Imry: In *Directions in Condensed Matter Physics*, ed. by G. Grinstein, G. Mazenko (World Scientific, Singapore 1986) p.101. For most general treatment see: M. Büttiker: Phys. Rev. Lett. **57**, 1761 (1986); IBM J. Res. Develop. **32**, 317 (1988)

8.59 B.J. van Wees, H. van Houten, C.W. Beenakker, J.G. Williams, L.P. Kouwenhowen, D. van der Marel, C.T. Foxon: Phys. Rev. Lett. **60**, 848 (1988)

8.60 D.A. Wharam, T.J. Thornton, R. Newbury, M. Pepper, H. Ritchie, G.A.C. Jones: J. Phys. C **21**, L209 (1988)

8.61 E. Tekman, S. Ciraci, A. Baratoff: Phys. Rev. B **42**, 9221 (1990)

8.62 C. Herring: Rev. Mod. Phys. **34**, 631 (1962)

8.63 C.J. Chen: J. Phys. Cond. Matt. **3**, 1227 (1991)

8.64 F. Flores, A.M. Rodero, E.C. Goldberg, J.C. Duran: Nuovo Cimento **10**, 303 (1988)

8.65 P. Hohenberg, W. Kohn: Phys. Rev. **136**, B864 (1964)

9. Consequences of Tip–Sample Interactions

U. Landman and *W.D. Luedtke*

With 29 Figures

Understanding the atomistic mechanisms, energetics, structure and dynamics underlying the interactions and physical processes that occur when two materials are brought together (or separated) is fundamentally important to basic and applied problems such as adhesion [9.1–8], capillarity [9.9, 10], contact formation [9.3–9, 11–27], surface deformations [9.7, 8, 19–27], materials elastic and plastic response characteristics [9.3, 7, 8, 20–27], material hardness [9.28–30], micro- and nano-indentation [9.6, 13, 24–32], friction, lubrication, and wear [9.9a, 22, 33–35], fracture [9.36, 37] and atomic-scale probing, modifications and manipulations, of materials surfaces. These considerations have for over a century motivated [9.1, 3, 20–23, 38] extensive theoretical and experimental research into the above phenomena and their technological consequences. Most theoretical approaches to these problems, with a few exceptions [9.7, 8, 17–19], have been anchored in continuum elasticity and contact mechanics [9.20–28]. Similarly, until quite recently [9.39–43] experimental observations and measurements of surface forces and the consequent materials response to such interactions have been macroscopic in nature.

The quest to understand and observe natural phenomena on refined microscopic scales has led to the development of conceptual and technological devices allowing the interrogation of materials with increasing resolution. On the experimental front, prior to the inception of the Scanning Tunneling [9.44] and Atomic Force Microscopies [9.41] (STM and AFM, respectively) characterization of surface topography had been achieved using stylus profilometers [9.45] (with 1000 Å and 10 Å lateral and vertical resolutions, respectively) and the scanning capacitive microscope [9.46] (with lateral and vertical resolutions of 5000 Å, and 2 Å, respectively). Furthermore, the importance of investigating single asperity contact in order to study the fundamental micromechanical response of solids has been long recognized. Such conditions are usually associated (i.e., assumed to be valid) for tip on flat configurations, with a tip radius of 1–2 µm or less [9.13, 47–49]. This may very well be the case for clean-metal contacts [9.50–52]. Indeed, evidence for continuous contact over an entire tip of several thousand Angstroms radius was given first by *Pollock* et al. [9.12].

Our perspectives and ability to probe surface morphology and crystallography, electronic structure, the nature of interatomic forces and interactions, dynamical processes, and microscopic mechanisms of response and deformation at surfaces and interfaces, as well as our ability to manipulate materials on the atomic scale, have been revolutionized by the recent emergence and prolifer-

Springer Series in Surface Sciences, Vol. 29
Scanning Tunneling Microscopy III   Eds.: R. Wiesendanger · H.-J. Güntherodt
© Springer-Verlag Berlin Heidelberg 1993

ation of high-resolution tip-based proximal probes [9.42, 43] (while we limit our discussion to tip-based microscopies, we note other innovative experimental techniques of non-tip nature, principally the Surface-Force Apparatus (SFA) [9.9a, 11, 40], developed and used for investigations of structure, interactions, rheology and triboligical phenomena of interfacial systems, with high spatial and force resolutions). Atomic-Force Microscopy (AFM), which probes the interatomic forces between a tip and a surface and their spatial variation, was invented in 1986 [9.39] to overcome the limitations of its predecessor, STM, which can only image conducting materials. Since then the concept was generalized to include measurement and imaging of magnetic forces [9.53], new detection schemes have been introduced [9.42e, 43, 54], and AFM has been utilized in investigations of a broad spectrum of systems and phenomena including [see reviews in Refs. 9.42, 43, 55]: surface imaging of insulators, ionics, metals, amino acids, other organic molecules, and polymers; imaging and force measurements at interphase-interfaces such as liquid–solid, and liquid–vapor; studies of atomic-scale tribological and thin-film lubrication phenomena [9.7, 8, 35, 56], mechanical and rheological characteristics [9.57], and wear processes [9.58]; observations of real time biological processes, such as polymerization of the human blood protein fibrinogen on a mica surface [9.59]; and purposeful structural modifications on the nanometer scale [9.60].

By its very mode of operation, force microscopy is sensitive to the nature of interactions between the tip and surface atoms. The magnitude, range, and character of these interactions depend on the tip and substrate materials (Sect. 9.1) and can be influenced by the method of preparation of the sample and the experimental ambient conditions (e.g., surface cleanliness, relative humidity, etc.). Furthermore, under certain circumstances, the tip and/or substrate may undergo structural and/or electronic modifications induced by the interactions between them. In this context we note that considerations of the consequences of tip–sample interactions are not limited to AFM. Indeed, while the principles underlying the STM are electronic in nature, it has been demonstrated that both electronic spectrum and surface atom geometrical structure, and in particular structural deformations resulting from tip–sample interactions, are convolved in STM images [9.61]. Additionally, it has been shown [9.15b, 62] that the apparent barrier height, as measured by STM for a tungsten tip and silicon sample under ultra-high vacuum conditions, is lowered by a surface deformation induced by the adhesive interaction between the tip and the substrate.

In AFM and STM studies aimed at determination of the structural properties of the surface under investigation, structural deformations (and consequent electronic structure modifications) are often an undesirable tip-induced complication. On the other hand the ability to induce such atomic- or nano-scale structural modifications via tip–sample interactions may be regarded as a new avenue for exploration of the atomistic mechanisms of surface deformation, generation and properties of inter-materials junctions, nano-mechanical surface properties, nano-indentation, and atomic-scale mechanisms of capillary processes, wetting, thin-film drainage, lubrication, nanotribological processes and wear [9.7, 8]. Additionally, such tip-induced processes can be used for atomic-scale manipulations and nano-structural design of surfaces.

On the theoretical front, recent advances in the formulation and evaluation of the energetics and interatomic interactions in materials [9.7, 8, 63], coupled with the development and implementation of computational methods and simulation techniques [9.7, 8, 64], open new avenues for investigations of the microscopic origins of complex materials phenomena. In particular, large-scale molecular dynamics computer simulations, which are in a sense computer experiments, where the evolution of a system of interacting particles is simulated with high spatial and temporal resolution by means of direct integration of the particles' equations of motion, have greatly enhanced our understanding of a broad range of materials phenomena [9.19, 63, 64].

Although our knowledge of the interfacial processes occurring when two material bodies are brought together has significantly progressed since the original presentation by *Heinrich Hertz* before the Berlin Physical Society in January 1881 of his theory of the contact of elastic bodies [9.20], full microscopic understanding of these processes is still lacking. Moreover, it has been recognized that continuum mechanics is not fully applicable as the scale of the material bodies and the characteristic dimension of the contact between them are reduced [9.27, 47]. Furthermore, it had been observed [9.22, 30] that the mechanical properties of materials exhibit a strong dependence on the size of the sample (small specimens appear to be stronger than larger ones). Since the functions between contacting solids can be small, their mechanical properties may be drastically different from those of the same materials in their bulk form. Consequently, the application of the newly developed theoretical and experimental techniques to these problems promises to provide significant insights concerning the microscopic mechanisms and the role of surface forces in the formation of microcontacts and to enhance our understanding of fundamental issues pertaining to interfacial adherence, microindentation, structural deformations, and the transition from elastic to elastoplastic or fully developed plastic response of materials. Additionally, studies such as those described in this paper allow critical assessment of the range of validity of continuum-based theories of these phenomena and could inspire improved analytical formulations. Finally, knowledge of the interactions and atomic-scale processes occurring between small tips and materials surfaces, and their consequences, is of crucial importance to optimize, control, interpret, and design experiments employing the novel tip-based microscopies [9.6, 7, 16, 17, 39–43, 61, 65–70].

As mentioned above, tip–sample interactions can induce structural modifications of the surface and/or tip and, depending on the tip and sample materials and the separation between them, may result in bonding. In addition these interactions may induce local perturbations of the electronic spectrum, modify the tip and surface densities of states and the transmission function, as well as affect the electronic parallel wave vector conservation, thus modifying the tunneling current and its dependence on tip-to-sample distance measured via STM [9.62]. Moreover, under certain circumstances which depend on the nature of the materials and mode of operation, the structural and electronic modifications due to tip–substrate interactions may be correlated. In this chapter we focus on studies of adhesive tip–substrate interactions and their consequences, pertaining mainly to force measurements via AFM.

Following a brief discussion in Sect. 9.1 of the methodology of molecular dynamics simulations we present in Sect. 9.2 results of our investigations for several materials starting with metallic systems, including our recent results for a nickel tip interacting with an alkane film adsorbed on a gold surface, followed by studies of an ionic tip and surface systems (CaF_2) and ending with our results for a covalently-bonded system (Si).

9.1 Methodology

Prior to presentation of our results we provide in this section pertinent details of our simulation studies.

Molecular Dynamics (MD) simulations consist of integration of the equations of motion of a system of particles interacting via prescribed interaction potentials [9.64]. The interatomic interactions that govern the energetics and dynamics of the system are characteristic to the system under investigation. Thus, for an ionic material (e.g., CaF_2) the energy may be described rather adequately as a sum of pair-wise interactions between the ions, with the potential $V_{\alpha\beta}(r)$ between ions of types α and β taken to be

$$V_{\alpha\beta}(r) = z_\alpha z_\beta e^2/r + A_{\alpha\beta}e^{-r/\varrho_{\alpha\beta}} - C_{\alpha\beta}/r^6 \ , \tag{9.1}$$

where the first two terms correspond to the Coulomb and overlap-repulsion contributions, and the last term represents the van der Waals dispersion interaction; the charge on ions of type α, in units of the electron charge e, is denoted by z_α. In our simulations of CaF_2 [9.8, 71] we have employed a parameterization of the potential in (9.1) determined partly by fitting to selected experimental data for the low temperature bulk crystal (such as structure, lattice parameters, cohesive energy, elastic constants and defect formation energies) and partly by appeal to quantum-mechanical calculations [9.72]. To describe covalently bonded materials (e.g., silicon) requires, in addition to pair-interactions, three-body terms due to the directional nature of the bonds [9.73].

The nature of cohesion in metals is rather different. Here, the dominant contribution to the total cohesive energy of the system (E_{coh}) is due to the electronic distribution interacting with the metal ions embedded in it. Based on the philosophy of density-functional theory [9.74], a description of metallic systems which is amenable to molecular dynamics simulations is provided by the Embedded Atom Method (EAM) [9.75, 76]. The basic feature of this method is that the effect of the surroundings on each atom in the system can be described in terms of the average electron density which other atoms in the system provide around the atom in question. The electronic structure problem is then converted to that of embedding an atom in a homogeneous electron gas, which can be described in terms of a universal density-dependent energy function. Thus the density-dependent term gives rise to many-body interactions.

In the EAM the cohesive energy E_{coh} of the metal is written as

$$E_{coh} = \sum_i \left\{ F_i \left[\sum_{j \neq i} \varrho_j^a(R_{ij}) \right] + \frac{1}{2} \sum_{j \neq i} \phi_{ij}(R_{ij}) \right\} , \qquad (9.2)$$

where ϱ^a is the spherically averaged atomic electron density and R_{ij} is the distance between atoms i and j. In EAM the embedding function, F, and the pair-repulsion between the partially screened ions, ϕ, are determined by choosing functional forms which meet certain general requirements, and fitting parameters [9.77] in these functions to a number of bulk equilibrium properties of the solid, such as lattice constant, heat of sublimation, elastic constants, vacancy-formation energy, heat of solution (for alloys), etc. The EAM has been used with significant success in studies of metallic systems in various thermodynamic states and degrees of aggregation [9.75–78].

In our simulations of tip–sample systems the surface part of the system is modeled by a slab containing n_d layers of dynamic atoms, with n atoms per layer, exposing an (hkl) surface plane, and interacting with n_s layers of the same material and crystallographical orientation. The surface atoms interact with a dynamic crystalline tip arranged initially in a pyramidal (tapered) geometry with the bottom layer (closest to the sample surface) consisting of n_1 atom, the next layer consisting of $n_2 > n_1$ atoms and so on. In addition the tip interacts with a static holder, made of the same material as the tip, consisting of n_h atoms located in n_{hl} layers. This system is periodically replicated in the two directions parallel to the surface plane, and no boundary conditions are imposed in the direction normal to the surface.

The simulations were performed at 300 K with temperature control imposed only on the deepest layer of the dynamic substrate, (i.e., the one closest to the static substrate). No significant variations in temperature were observed during the simulations. The equations of motion were integrated using a fifth-order predictor–corrector algorithm with a time step Δt ($\Delta t = 3 \times 10^{-15}$ fs for the metallic systems and 1×10^{-15} fs for the ionic ones).

Following equilibration of the system at 300 K with the tip outside the range of interaction, the tip was lowered slowly toward the surface. Motion of the tip occurs by changing the position of the tip-holder assembly in increments of 0.25 Å over 500 Δt. After each increment the system is fully relaxed, that is, dynamically evolved, until no discernable variations in system properties are observed beyond natural fluctuations.

Analysis of the phase-space trajectories generated during the simulations allows determination of energetic, structural and dynamical properties. For example, the kinetic temperature (T) of a set of N particles is defined via the relation $(3N/2) k_B T = \sum_{i=1}^{N} m_i v_i^2 / 2$, where v_i is the velocity vector of a particle with mass m_i and k_B is the Boltzmann constant. Another quantity, of particular interest in studies of the mechanical properties of materials, is the stress tensor and individual atomic contributions to it can be derived most generally from the Lagrangian of the system [9.79, 80]. For the particular case of pair interactions the matrix of the stress tensor $\underset{\approx}{\sigma}$ is given in dyadic tensor notation by

$$\ddot{\underset{\approx}{\sigma}} \equiv \sum_i \ddot{\sigma}_i = \sum_i \left[m_i v_i v_i + \frac{1}{2} \sum_{j \neq i} \chi(r_{ij})(r_i - r_j)(r_i - r_j) \right] \Omega_i^{-1} , \tag{9.3}$$

where Ω_i is the volume per particle i, v_i is the velocity vector of the particle, r_i is it's position vector and $\chi(r) = -r^{-1}(dV(r)/dr)$, where $V(r)$ is the pair-potential. The expression for potentials beyond pair-interactions are somewhat more complicated [9.73b]. From the atomic stresses, invariants of the stress tensor can be calculated, in particular the second invariant of the stress deviator, J_2, which is proportional to the stored strain energy and is related to the Von Mises shear strain-energy criterion for the onset of plastic yielding [9.21, 24], is given by

$$J_2 = \tfrac{1}{2} \mathrm{Tr} \{ \underset{\approx}{\Gamma} \cdot \underset{\approx}{\Gamma}^{\mathrm{T}} \} , \tag{9.4a}$$

where Tr denotes the trace of the matrix product in square brackets, $\underset{\approx}{\Gamma}^{\mathrm{T}}$ is the transpose of the matrix $\underset{\approx}{\Gamma}$ defined as

$$\underset{\approx}{\Gamma} = \underset{\approx}{\sigma} - p \underset{\approx}{1} , \tag{9.4b}$$

where the hydrostatic pressure $p = \tfrac{1}{3} \mathrm{Tr} \{ \underset{\approx}{\sigma} \}$, and $\underset{\approx}{1}$ is the unit matrix.

9.2 Case Studies

In many respects, scientific issues central to the field of tribology (based upon the Greek word *tribos*, meaning rubbing), dealing with physical and chemical phenomena occurring when two interacting surfaces are in relative motion in the form of sliding, rolling, or normal approach or separation of surfaces, are of similar nature to those pertaining to the consequences of tip–sample interactions in tip-based force and scanning microscopies. Furthermore, as indicated in the introductory section to this chapter, tip-based force microscopies, as well as the surface force apparatus [9.40], open new avenues for studies of atomic and nano-scale mechanisms of tribological, lubrication, and wear processes. Therefore, it is instructive to briefly review first the evolution of ideas in tribology and some of the fundamental concepts pertaining to contact mechanics of interfacial systems.

The study of tribological (or frictional) phenomena has a long and interesting history [9.38]. Leaping over centuries of empirical observations we start with the classical friction law presented by G. Amontons in 1699 and extended by C.A. Coulomb in 1781 (although it was actually known to L. da Vinci in the fifteenth century) which asserts that relative sliding of two bodies in contact will occur when the net tangential force reaches a critical value proportional to the net force pressing the two bodies together. Furthermore, this proportionality factor, the friction coefficient, is independent of the apparent contact area, and depends only weakly on the surface roughness.

The first to introduce the notion of cohesive forces between material bodies in contact, and their contribution to the overall frictional resistance experienced by sliding bodies, was J.T. Desagulier's whose ideas on friction are contained in a book published in 1734, entitled "A Course of Experimental Philosophy". His observations which introduced adhesion as a factor in the friction, additional to the idea of interlocking asperities favoured in France, were conceived in the context of the role of surface finish where he writes " . . . the flat surfaces of metals or other Bodies may be so far polish'd as to increase Friction and this is a mechanical Paradox: but the reason will appear when we consider that the Attraction of Cohesion becomes sensible as we bring the Surfaces of Bodies nearer and nearer to Contact".

While it was recognized for many years that the *Amontons–Coulomb* laws of friction are applicable only to the description of friction between effectively rigid bodies and gross sliding of one body relative to another, the concepts of stress and the elastostatics came only later, in the writings of A.L. Cauchy, C.L.M.H. Navier and others, and the formulation by H. Hertz in 1881 of elastic contact mechanics [9.20]. Extensions of Hertz's theory to the contact of two elastic bodies including the influence of friction of the contact interface were made first by *Cattaneo* [9.81] and independently later by *Mindlin* [9.82].

Not attempting to give here a complete historical account [9.38], we note the growing realization since the beginning of the century of the role of adhesive interactions, plastic deformation and yield in determining the mechanical response and friction between bodies in contact. In particular, we mention the notion that the yield point of a ductile metal is governed by shear stress [9.24]; either the absolute maximum (Tresca criterion) or the octahedral shear stress (von Mises criterion). The relationship between the interfacial adhesive formation and shearing of intermetallic junctions and friction, was succinctly summarized by *Tabor* and *Bowden* as follows [9.33]: "Friction is the force required to shear intermetallic junctions plus the force required to plow the surface of the softer metal by the asperities on the harder surface."

The first successful theory of the contact mechanics of adhesive contact was formulated [9.25] by *Johnson, Kendall* and *Roberts* (JKR) who observed, during an investigation into the friction of automobile windscreen wipers, that in a situation of pressing together two bodies, when one or both surfaces is very compliant, the radius of the contact circle exceeded the value predicted by Hertz; moreover, when the load was removed a measurable contact area remained, and it was necessary to apply a tensile force to separate the surfaces. The JKR theory considers the adhesion between the two interfacing bodies as simply a change in surface energy only where they are in contact (i.e., infinitely short-range attractive forces). An alternative formulation [9.26] by *Derjaguin, Muller* and *Toporov*

[DMT] on the other hand asserts that the attractive force between the solids must have a finite range, and in their original formulation they assume this interaction to act in a region just outside the contact zone. (In addition the DMT theory assumes that the deformed shape of the surfaces is Hertzian, i.e., unaffected by the surface forces). More complete formulations, which allow solid–solid interactions to be a prescribed function of the local separation between the surfaces have also been suggested [9.83, 84]. Furthermore, *Maugis* and *Pollock* [9.14] have investigated, in the context of metal microcontacts, the development of plastic deformation and adherence under zero-applied load by considering the influence of surface forces, and have derived conditions for ductile or brittle modes of separation after an elasto-plastic or full plastic contact.

In an attempt to address the above issues on an atomistic level, we have embarked on a series of investigations [9.7, 8, 19] of the energetics, mechanisms and consequences of interactions between material tips and sample surfaces, using molecular dynamics simulations. Since material phenomena and processes are governed by the nature and magnitude of bonding and interatomic interactions, as well as by the characteristics of other materials (such as thermodynamic state, structure, and degree of compositional and structural perfection) a comprehensive study of any class of phenomena (interfacial processes in particular) requires systematic investigations for a range of material-dependent parameters.

In this section results of several investigations are summarized. We start with simulations of the approach to adhesive contact followed by retraction (pull-off) of a clean nickel tip interacting with a clean gold sample (Ni/Au). Simulations for the same system but when the tip is lowered beyond the point of contact formation are also described (nano-indentation). Subsequently the materials in the tip and sample are interchanged, i.e., a clean gold tip is lowered toward a clean nickel sample (Au/Ni). We also give results for a modified nickel tip, (i.e., the one obtained at the end of the Ni/Au nano-indentation, where the bottom of the tip is coated by an epitaxial gold layer) interacting with an undamaged gold sample. Following the discussion of inter-metallic interactions between a tip and a clean surface we present our recent results for a nickel tip interacting with an alkane (*n*-hexadecane) liquid film adsorbed on a gold surface. Next, consequences of the interactions between a CaF_2 surface and a tip made of the same material are described for both vertical and tangential displacements of the tip relative to the surface. Finally studies of an atomic-scale stick–slip phenomenon occurring upon sliding a silicon tip on a silicon surface are presented.

9.2.1 Clean Nickel Tip/Gold Surface

Simulated force versus distance curves for the system are shown in Fig. 9.1 as well as the calculated potential energy versus distance (Fig. 9.1c). Results for tip-to-sample approach followd by separation are shown, for adhesive contact (Fig. 9.1a) and indentation (Figs. 9.1b, c) studies [9.7a]. In these simulations the sample consists of $n_s = 3$, $n_d = 8$, $n = 450$ atoms/layer exposing the (001) face.

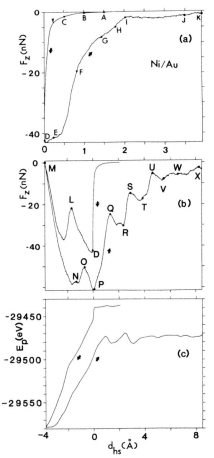

Fig. 9.1. Calculated force on the tip atoms, F_z, versus tip-to-sample distance d_{hs}, between a Ni tip and a Au sample for; (a) approach and jump-to-contact followed by separation; (b) approach, jump-to-contact, indentation, and subsequent separation; d_{hs} denotes the distance between the rigid tip-holder assembly and the static substrate of the Au surface ($d_{hs} = 0$ at the jump-to-contact point marked D). The capital letters on the curves denote the actual distances, d_{ts}, between the bottom part of the Ni tip and the Au surface; in (a): A = 5.7 Å, B = 5.2 Å, C = 4.7 Å, D = 3.8 Å, E = 4.4 Å, F = 4.85 Å, G = 5.5 Å, H = 5.9 Å, I = 6.2 Å, J = 7.5 Å, and K = 8.0 Å; in (b); D = 3.8 Å, L = 2.4 Å, M = 0.8 Å, N = 2.6 Å, O = 3.0 Å, P = 3.8 Å, Q = 5.4 Å, R = 6.4 Å, S = 7.0 Å, T = 7.7 Å, U = 9.1 Å, V = 9.6 Å, W = 10.5 Å, and X = 12.8 Å. (c) Potential energy of the system for a complete cycle of the tip approach, jump-to-contact, indentation, and subsequent separation. Forces in units of nano-newtons, energy in electron volts and distances in ångstroms

The tip consists of a bottom layer of 72 atoms exposing a (001) facet, the next layer consists of 128 atoms and the remaining six layers contain 200 dynamic atoms each. The static holder consists of 1176 atoms arranged in three (001) layers. This gives the tip an effective radius of curvature of ~ 30 Å.

The simulations correspond to a case of a rigid cantilever and therefore the recorded properties of the system as the tip-holder assembly approaches or retracts from the sample portray directly consequences of the interatomic interactions between the tip and the sample. The distance scale that we have chosen in presenting the calculated results is the separation (denoted as d_{hs}) between the rigid (static) holder of the tip and the static gold lattice underlying the dynamic substrate. The origin of the distance scale is chosen such that $d_{hs} = 0$ after jump-to-contact occurs ($d_{hs} \geq 0$ when the system is not advanced beyond the JC point and $d_{hs} < 0$ corresponds to indentation. Since the dynamic Ni tip and Au substrate atoms displace in response to the interaction between them, the distance d_{hs} does not give directly the actual separation between regions in the dynamic tip and substrate material. The *actual relative distances* d_{ts} between the bottom part of the tip (averaged z-position of atoms in the *bottom* layer of the tip) and the surface (averaged z-position of the *top layer* of the Au surface, calculated for atoms in the first layer *away from the perturbed region* in the vicinity of the tip) are given by the letter symbols in Fig. 9.1a, b. Note that the distance between the bottom of the tip and the gold atoms in the region *immediately underneath* it may differ from d_{ts}. Thus, for example, when $d_{hs} = 0$ (point D in Figs. 9.1a, b) the tip to *unperturbed* gold distance d_{ts} is 3.8 Å, while the average distance between the bottom layer of the tip and the adherent gold layer in *immediate contact* with it is 2.1 Å.

Tip–Sample Approach. Following an initial slow variation of the force between the Au sample and the Ni tip we observe in the simulations the onset of an instability, signified by a sharp increase in the attraction between the two (see Fig. 9.1a as well as Fig. 9.1b, c where the segments corresponding to lowering of the tip up to the point D describe the same stage as that shown in segment AD in Fig. 9.1a) which is accompanied by a marked decrease in the potential energy of the system (see sudden drop of E_p in Fig. 9.1c as d_{hs} approaches zero from the right). We note the rather sudden onset of the instability which occurs only for separations d_{hs} smaller than 0.25 Å (marked by an arrow on the curve in Fig. 9.1a). Our simulations reveal that in response to the imbalance between the forces on atoms in each of the materials and those due to intermetallic interactions a Jump-to-Contact (JC) phenomenon occurs via a fast process where Au atoms in the region of the surface under the Ni tip displace by approximately 2 Å toward the tip in a short time span of ~ 1 ps. The atomic configuration of the tip–surface system after the JC has occurred is shown in Fig. 9.2. After the occurrence of the jump-to-contact the distance between the bottom layer of the Ni tip and the layer of adherent Au atoms in the region immediately underneath it decreased to 2.1 Å from a value of 4.2 Å. In addition to the adhesive contact formation between the two surfaces, an adhesion-induced partial wetting of the edges of the Ni tip by Au atoms is observed.

The jump-to-contact phenomenon in metallic systems is driven by the marked tendency of the atoms at the interfacial regions of the tip and sample materials to optimize their embedding energies (which are density dependent, deriving from the tails of the atomic electronic-charge densities) while maintaining their individual material cohesive binding (in the Ni and Au) albeit strained due to the deformation caused by the atomic displacements during the JC process. In this context we note the difference between the surface energies of the two metals, with the one for Ni markedly larger than that of Au, and the differences in their mechanical properties, such as elastic moduli, yield, hardness, and strength parameters (for example, the elastic moduli are 21×10^{10} and 8.2×10^{10} N/m^2 for Ni and Au, respectively [9.85]).

Further insight into the JC process is provided by the local hydrostatic pressure in the materials (evaluated as the trace of the atomic stress tensors

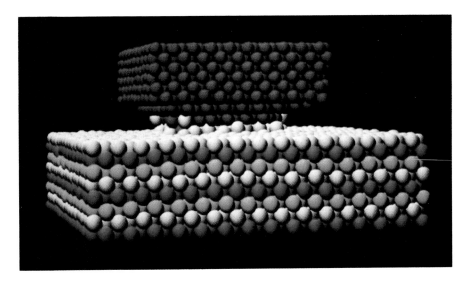

Fig. 9.2. Atomic configuration after jump-to-contact (point D in Fig. 9.1a) between a nickel tip (*red balls*) and a gold (001) surface (top atomic layer *in yellow*, second layer *in blue*, third layer *in green*, etc.). Note the gold atoms which adhere to the nickel tip resulting in a deformation of the surface in the vicinity of the contact area

[9.79]) shown in Fig. 9.3a after contact formation (i.e., point D in Fig. 9.1a). The pressure contours reveal that atoms at the periphery of the contact zone (at $X = \pm 0.19$ and $Z = 0.27$) are under extreme tensile stress (-10^5 atm $= -10^{10}$ N/m$^2 = -10$ GPa). In fact we observe that the tip as well as an extended region of the substrate in the vicinity of the contact zone are under tension. Both the structural deformation profile of the system and the pressure distribution which we find in our atomistic MD simulations are in general terms similar to those described by certain modern contact mechanics theories [9.21–24] where the influence of adhesive interactions is included.

Tip–Sample Separation After Contact. Starting from contact, the force versus distance (F_z vs. d_{hs}) curve exhibits a marked hysteresis seen both experimentally and theoretically (Fig. 9.1a) as the surfaces are separated [9.7a]. We remark that, in the simulation and the AFM measurements [9.7a], separating the surfaces *prior to contact* results in *no* hysteresis. The hysteresis is a consequence of the adhesive bonding between the two materials and, as demonstrated by the simulation, separation is accompanied by inelastic processes in which the topmost layer of the Au sample adheres to the Ni tip. The mechanism of the process is demonstrated by the pressure contours during lift-off of the tip shown in Fig. 9.3b, recorded for the configuration marked G ($d_{ts} = 5.5$ Å in Fig. 9.1a). As seen, the maximum tensile stress is located near the edges of the adhesive contact. We further observe that the diameter of the contact area *decreases*

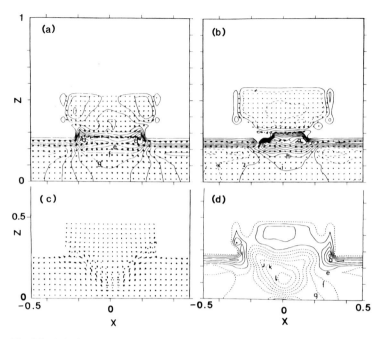

Fig. 9.3. Calculated pressure contours and atomic configurations viewed along the [010] direction, in slices through the system. The Ni tip occupies the top eight atomic layers. Short-time atomic trajectories appear as dots. Distance along the X and Z directions in units of $X = 1$ and $Z = 1$ corresponding to 61.2 Å each. Solid contours correspond to tensile stress (i.e., negative pressure) and dotted ones to compressive stress. **(a)** After jump-to-contact (point D in Fig. 9.1a). The maximum magnitude of the tensile (i.e., negative pressure, 10 GPa, is at the periphery of the contact, $(X, Z) = (\pm 0.19, 0.27)$. The contours are spaced with an increment, Δ, of 1 GPa. Thus the contours marked e, f and g correspond to -6, -5 and -4 GPa, respectively. **(b)** During separation following contact, (point G in Fig. 9.1a). The maximum tensile pressure (marked a), ~ -9 GPa, is at the periphery of the contact at (X, Z) equal to $(0.1, 0.25)$ and $(-0.04, 0.25)$. $\Delta = 0.9$ GPa. The marked contours h, i, j and k correspond to -2.5, -1.6, -0.66 and 0.27 GPa, respectively. **(c)** Short-time particle trajectories at the final stage of relaxation of the system, corresponding to point M in Fig. 9.1b, (i.e., $F = 0$). Note slip along the [111] planes in the sample. **(d)** Pressure contours corresponding to the final configuration shown in (c). Note the development of compressive pressure in the sample which maximizes in the region of the contour marked l (8.2 GPa). The increment between contours $\Delta = 1.4$ GPa. The contours marked a and e correspond to -6.4 GPa and -1.1 GPa, respectively, and those marked f and g to 0.2 and 1.6 GPa

during lifting of the tip, resulting in the formation of a thin "adhesive neck" due to ductile extension, which stretches as the process continues, ultimately breaking at a distance d_{ts} of ~ 9–10 Å. An atomic configuration of the system during the lift-off process, illustrating the formation of atomic-size contact between the tip and the sample, is shown in Fig. 9.4. The evolution of the adhesion and tear processes which we observe can be classified as mode-I fracture [9.36], reemphasizing the importance of forces operating across the crack in modeling crack propagation [9.36, 37].

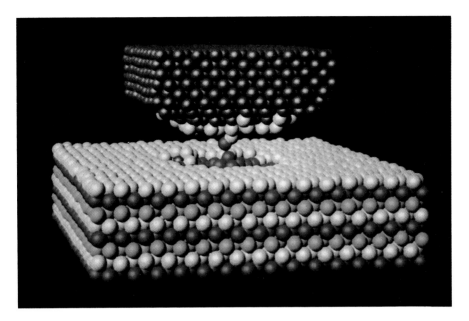

Fig. 9.4. Atomic configuration of the system during the tip–sample separation process (corresponding to point J in Fig. 9.1a), after contact, illustrating the adherence of the top Au layer to the Ni tip and the formation of an atomically thin connective neck

Indentation. We turn now to theoretical results recorded when the tip is allowed to advance past the jump-to-contact point, i.e., indentation (Figs. 9.1b, c, and 9.5). As evident from Fig. 9.1b, decreasing the separation between the tip and the substrate causes first a decrease in the magnitude of the force on the tip (i.e., less attraction, see segment DL) and an increase in the binding energy (i.e., larger magnitude of the potential energy, shown in Fig. 9.1c). However, upon reaching the point marked L in Fig.(9.1b), a sharp increase in the attraction occurs, followed by a monotonic decrease in the magnitude of the force till $F_z = 0$ (point M in Fig. 9.1b) at $d_{ts} = 0.8$ Å. The variations of the force (in the segment DLM) are correlated with large deformations of the Au sample. In particular, the nonmonotonic feature (near point L) results from tip-induced flow of gold atoms which relieve the increasing stress via wetting of the sides of the tip. Indeed the atomic configurations display a "piling-up" around the edges of the indenter due to atomic flow driven by the deformation of the Au substrate and the adhesive interactions between the Au and Ni atoms. Further indentation is accompanied by slip of Au layers [along (111) planes] and the generation of interstitial defects. In addition, the calculations predict that during the indentation process a small number of Ni atoms diffuse into the surrounding Au, occupying substitutional sites. Furthermore, the calculated pressure contours at this stage of indentation, shown in Fig. 9.3d, demonstrate that the sample surface zone in the vicinity of the edges of the tip is under tensile stress, while the

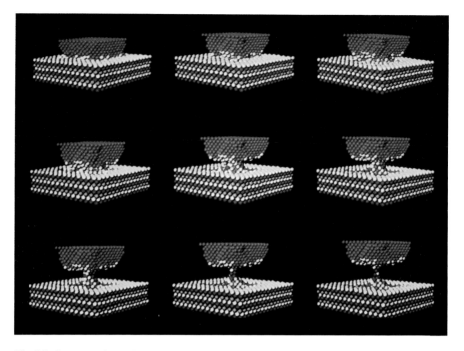

Fig. 9.5. Sequence of atomic configuration starting from a Ni tip indented in a Au (001) sample (*top left*) and during the process of retraction of the tip (*from left to right*) accompanied by formation of a connective neck

deformed region under the tip is compressed with the maximum pressure (8.2 GPa) occurring at about the fifth Au layer below the center of the Ni tip-indenter. The general characteristics of the pressure (and stress) distributions obtained in our indentation simulations correspond to those associated with the onset and development of plastic deformation in the sample 9.14, 21, 25.

Experimentally, advancing the sample past the contact point is noted by the change in slope of the force as the increasing repulsive forces push the tip and cantilever back towards their rest position. We remark that the calculated pressures from the simulations compare favorably with the *average* contact pressure of ~3 GPa determined experimentally [9.7a] by dividing the measured attractive force by the estimated circular contact area of radius 20 nm.

Tip–Sample Separation After Indentation. Reversal of the direction of the tip motion relative to the sample from the point of zero force (point M in Fig. 9.1b) results in the force- and potential energy-versus distance curves shown in Fig. 9.1b, c. The force curve exhibits first a sharp monotonic increase in the magnitude of the attractive force (segment MN in Fig. 9.1b) with a corresponding increase in the potential energy (Fig. 9.1c). During this stage the response of the system is mostly elastic accompanied by the generation of a small number of vacancies and substitutional defects in the substrate. Past this stage the force and energy curves versus tip-to-sample separation exhibit a nonmonotonic behavior which is associated mainly with process of elongation of the connective neck which forms between the sample and the retracting tip.

To illustrate the neck formation and elongation process we show in Fig. 9.6 a sequence of atomic configurations corresponding to the maxima in the force curve (Fig. 9.1b, points marked O, Q, S, U, W and X). As evident, upon increased separation between the tip–holder and the sample a connective neck forms consisting mainly of gold atoms (see atomic configurations shown in Fig. 9.5). The mechanism of elongation of the neck involves atomic structural transformations whereby in each elongation stage atoms in adjacent layers in the neck disorder and then rearrange to form an added layer, i.e., a more extended neck of a smaller cross-sectional area. Throughout the process the neck maintains a layered crystalline structure (Fig. 9.6) except for the rather short structural transformation periods, corresponding to the sharp variations in the force curve, (segments PQ, RS, TU and VW in Fig. 9.1b) and the associated features in the calculated potential energy shown in Fig. 9.1c where the minima correspond to ordered-layered structures after the structural rearrangements. We note that beyond the initial formation stage, the number of atoms in the connective neck region remains roughly constant throughout the elongation process.

Further insight into the microscopic mechanism of elongation of the connective neck can be gained via consideration of the variation of the second

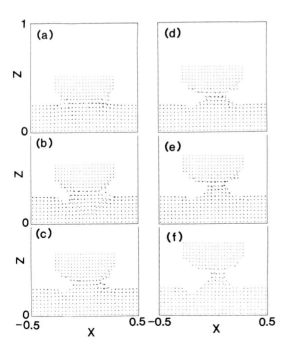

Fig. 9.6. Atomic configurations in slices through the system illustrating the formation of a connective neck between the Ni tip and the Au sample during separation following indentation. The Ni tip occupies the top eight layers. The configurations (**a**–**f**) correspond to the stages marked O, Q, S, U, W, and X in Fig. 9.1b. Note the crystalline structure of the neck. Successive elongations of the neck, upon increased separation between the tip-holder assembly and the sample, occur via structural transformation resulting in successive addition of layers in the neck accompanied by narrowing (i.e., reduction in cross-sectional area of the neck). Distance in units of X and Z, with $X = 1$ and $Z = 1$ corresponding to 61.2 Å

invariant of the stress deviator, J_2, which is related to the von Mises shear strain-energy criterion for the onset of plastic yielding [9.21, 24, 28]. Returning to the force and potential energy curves shown in Fig. 9.1b, c, we have observed that between each of the elongation events (i.e., layer additions, points marked Q, S, U, W and X) the initial response of the system to the strain induced by the increased separation between the tip–holder and the sample is mainly elastic (segments OP, QR, ST, UV in Fig. 9.1b, and correspondingly the variations in Fig. 9.1c), accompanied by a gradual increase of $\sqrt{J_2}$, and thus the stored strain energy. The onsets of the stages of structural rearrangements are found to be correlated with a critical maximum value of $\sqrt{J_2}$ of about 3 GPa (occurring for states at the end of the intervals marked OP, QR, ST and UV in Fig. 9.1b) localized in the neck in the region of the ensuing structural transformation. After each of the elongation events the maximum value of $\sqrt{J_2}$ (for the states marked Q, S, U, W and X in Fig. 9.1b) drops to approximately 2 GPa.

In this context, it is interesting to remark that the value of the normal component of the force per unit area in the narrowest region of the neck remains roughly constant (~ 1 GPa) throughout the elongation process, increasing by about 20% prior to each of the aforementioned structural rearrangements. This value has been estimated both by using the data given in Fig. 9.1b and the cross-sectional areas from atomic configuration plots (such as given in Fig. 9.6), and via a calculation of the average axial component (zz element) of the atomic stress tensors in the narrow region of the neck. We note that the above observations constitute atomic-scale realizations of basic concepts which underlie macroscopic theories of materials behavior under load [9.20–26].

A typical distribution of the stress, $\sqrt{J_2}$, prior to a structural transformation is shown in Fig. 9.7 (shown for the state corresponding to the point marked T in Fig. 9.1b). As seen, the maximum of $\sqrt{J_2}$ is localized about a narrow region

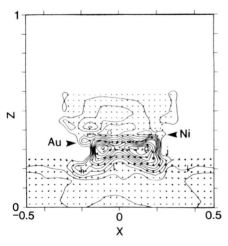

Fig. 9.7. Von Mises' shear stress ($\sqrt{J_2}$) corresponding to the configuration marked T in Fig. 9.1b [that is, just before the structural transformation resulting in the configuration (**d**) in Fig. 9.6]. The proximal interfacial layers of Ni and Au are marked by arrows. The maximum contours (2.9 GPa, marked a) occur on the periphery of the neck $(X, Z) = (\pm 0.1, 0.3)$. The increment between contours is 0.2 GPa. The contours marked h, i, j, and k correspond to 1.1, 0.9, 0.7, and 0.5 GPa, respectively. Distance along X and Z in units of $X = 1$ and $Z = 1$ corresponding to 61.2 Å

around the periphery in the strained neck. Comparison between the atomic configuration at this stage (Fig. 9.7, or the very similar configuration shown in Fig. 9.6c) and the configuration after the structural transformation has occurred (Fig. 9.6d, corresponding to the point marked U in Fig. 9.1b) illustrates the elongation of the neck by the addition of a layer and accompanying reduction in areal cross section. We note that, as the height of the connective neck increases, the magnitude of the variations in the force and potential energy during the elongation stages diminishes. The behavior of the system past the state shown in Fig. 9.6f (corresponding to the point marked X in the force curve shown in Fig. 9.1b) is similar to that observed at the final stages of separation after jump-to-contact (Fig. 9.1b), characterized by strain-induced disordering and thinning in a narrow region of the neck near the gold covered bottom of the tip and eventual fracture of the neck (occurring for a tip-to-substrate distance $d_{ts} \simeq 18$ Å), resulting in a Ni tip whose base is covered by an adherent Au layer.

The theoretically predicted increased hysteresis upon tip-substrate separation following indentation, relative to that found after contact compare Figs. 1a, b, is also observed experimentally [9.7a, b]. In both theory and experiment the maximum attractive force after indentation is roughly 50% greater than when contact is first made. While in the original experiments [9.7a] the non-monotonic features found in the simulations (Fig. 9.1b) were not discernible in the force vs. distance data, they have been observed in recent experiments performed under ultra-high vacuum conditions [9.86].

9.2.2 Gold-Covered Nickel Tip/Gold Surface

Having described in the previous section the processes occurring as a result of the interaction between a clean nickel tip with a gold surface, we turn next to a system where a nickel tip "wetted" by a gold monolayer is lowered and subsequently retracted from a clean initially undamaged Au(001) surface [9.8].

As mentioned already, the retraction of the Ni tip from the gold surface, after formation of an adhesive contact between the two, is accompanied by wetting of the tip by gold atoms. The gold-coated Ni tip used in the present simulations was obtained following a slight indentation (Sect. 9.2.1).

The force vs. distance curve obtained for the system, along with the one corresponding to the clean Ni tip (see also Fig. 9.1a) shown in Fig. 9.8 and inspection of atomic configurations reveal that, while the adhesive interaction is reduced for the coated tip, jump-to-contact instability, formation of an adhesive contact and hysteresis during subsequent retraction occur in both cases. However we should note that in the present case while a connective neck, made solely of substrate gold atoms, is formed during retraction of the tip (of dimensions similar to that formed upon indentation, see Fig. 9.5) an insignificant transfer of atoms from the surface to the tip occurs upon complete separation (i.e., while gold wets by adhering to a bare-nickel tip, no wetting occurs for a gold-covered tip).

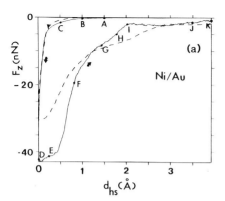

Fig. 9.8. Calculated force on the tip atoms, F_z, versus tip-to-sample distance, d_{hs}, for the case of a gold-coated Ni tip approaching and then being retracted from a clean Au (001) surface (*dashed line*), as well as for the case of a clean Ni tip (Fig. 9.1a)

9.2.3 Clean Gold Tip/Nickel Surface

In the studies discussed in Sect. 9.2.1 the tip material (i.e., Ni) was the harder one, characterized additionally by a larger surface energy. This section studies where the materials composing the tip and surface are interchanged, i.e., a Au tip interacts with a Ni (001) surface [9.7b]. In these investigations the substrate consists again of $n_s = 3$ and $n_d = 8$ layers, exposing Ni(001) surface, but with a larger number of atoms per layer ($n = 800$) than in the previous studies to allow for spreading of the gold tip (see below). The gold tip, exposing a Au(001) facet, is prepared in a similar manner to that described in Sect. 9.2.1.

Contact Formation and Separation. The simulated force on the tip and potential energy versus distance curve are shown in Fig. 9.9, and atomic configurations of the system (i.e., short-time atomic trajectories, shown for a slice of 11 Å in width through the system) recorded at selected stages during the tip-to-sample approach and subsequent separation are shown in Fig. 9.10.

Following an initial slow variation of the force between the Ni sample and the Au tip we observe the onset of an instability, signified by a sharp increase in the attraction between the two (segment AB in Fig. 9.9a) which is accompanied by a marked decrease in the potential energy of the system (see sudden drop of E_p in Fig. 9.9b as d_{hs} approaches zero). We note the rather sudden onset of the instability which occurs only for separations d_{hs} smaller than 0.25 Å (marked by an arrow on the curve in Fig. 9.9a and corresponding to $d_{ts} = 4.2$ Å). Our simulations reveal that in response to the imbalance between the forces on atoms in each of the materials, and those due to intermetallic interactions, a JC phenomenon occurs via a fast process where Au tip atoms displace by approximately 2 Å toward the surface in a short time span of ~ 1 ps (Fig. 9.10a, b where the atomic configurations before and after the JC are depicted). After the JC occurs, the distance between the bottom layer of the Au tip and the top layer of the Ni surface decreases to 2.1 Å from a value of 4.2 Å. The response in this system should be contrasted with that observed by us in simulations of a Ni tip

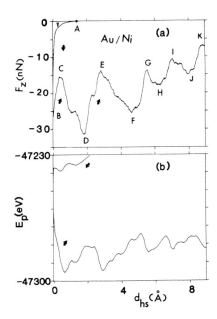

Fig. 9.9. Calculated force F_z (in **a**) and potential energy of the system E_p (in **b**), versus tip-to-sample distance, d_{hs}, between an Au tip and a Ni sample for approach and jump-to-contact followed by separation. d_{hs} denotes the distance between the rigid tip-holder assembly and the static substrate of the Ni surface ($d_{hs} = 0$ at the jump-to-contact point, marked B). The distance between the bottom layer of the tip and the top layer of the sample is 4.2 Å at the onset of the jump-to-contact i.e., last stable point upon tip lowering, [marked by an arrowhead in **a**]. That distance is 2.1 Å after contact formation (point marked B). The points marked C, E, G, I and K correspond to ordered configurations of the tip each containing an additional layer. Force in units of nN, energy in eV and distance in Å

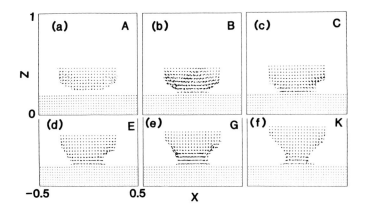

Fig. 9.10. Atomic configurations in a slice, in the XZ plane (i.e., containing the [100] and [001] directions), of width 11 Å through the system (in the [010] direction) illustrating the atomic arrangements in the system before and after jump-to-contact (in **a** and **b**, respectively) and for the ordered configurations during the tip elongation processes which occur upon retraction of the tip from the point of adhesive contact (in **c–f**). The capital letters identify the corresponding points on the F_z vs. distance curve given in Fig. 9.9a. Note the crystalline structure of the neck. Successive elongations of the neck, upon increased separation between the tip-holder assembly and the sample, occur by way of a structural transformation resulting in the successive addition of layers in the neck, accompanied by narrowing (that is, reduction in the cross-sectional area of the neck). Distance in units of X and Z, with $X = 1$ and $Z = 1$ corresponding to 70.4 Å

approaching a Au surface, (Sect. 9.2.1) where the JC phenomenon involved a bulging of the Au surface underneath the tip.

Contours of the local pressure (evaluated as the trace of the atomic stress tensors) and the square root of J_2 (second invariant of the stress deviator), after JC had occurred, are shown in Fig. 9.11a, b, respectively. We observe that the pressure on the periphery of the tip is large and negative (tensile), achieving values up to -1×10^5 atm $= -10^{10}$ N/m $= 10$ GPa, while the middle core of the tip is under a small compressive pressure. The $\sqrt{J_2}$ maximizes in the interfacial contact region achieving a value of 2.0×10^4 atm $= 2$ GPa.

As a consequence of the formation of the adhesive contact between the two materials, the interfacial region of the gold tip exhibits large structural rearrangements, both in the normal and lateral directions. The atomic configuration in the three layers of the tip closest to the Ni surface are shown in Fig. 9.12a. The structure of the proximal Au layer [large dots in Fig. 9.12a, superimposed on the atomic structure of the underlying Ni (001) surface] exhibits a marked tendency towards a (111) reconstruction [in this context we remark that only a small tendency towards a (111) structure was observed by us for a patch of a gold monolayer deposited on a Ni (001) surface]. Furthermore, the (111) reconstruction extends 3–4 layers from the interface into the Au tip. Accompanying the epitaxial surface structural rearrangement in the gold, which is partially driven by the lattice constant mismatch between gold and nickel, an increase in interlayer spacing occurs [$d_{12} = 2.44$ Å, $d_{23} = d_{34} = d_{45} = 2.53$ Å, $d_{56} = d_{67} = 2.2$ Å and $d_{78} = 2.1$ Å, compared to the interlayer spacing between (001) layers in the bulk gold of 2.04 Å where d_{nn+1} is the spacing between layers n and $n + 1$, and layer number 1 corresponds to the Au layer proximal to the Ni topmost surface layer]. The (111) reconstruction and expanded interlayer spacings in the interfacial region of the Au tip persist throughout the separation process.

Starting from contact, the force versus distance (F_z versus d_{hs}) curve exhibits a marked hysteresis as the tip is retracted from the substrate (Fig. 9.9a). We remark that separating the surfaces prior to jump-to-contact results in no hysteresis. The hysteresis is a consequence of the adhesive bonding between the two materials and, as demonstrated by the simulation, separation is accompanied by inelastic processes and the formation of an extended gold connective neck (Fig. 9.10c–f).

The hysteresis in the force vs. distance curve exhibits marked variations (Fig. 9.9a) which portray the atomistic processes of gold neck formation and elongation. In this context we remark that our earlier simulations (and accompanying AFM experiments) of a nickel tip interacting with a gold surface have also shown a marked hysteresis upon separation from adhesive contact. However, in that case, the variation of the force F_z vs. distance curve was monotonic, and the extension of the connective neck was rather limited. Variations similar to those shown in Fig. 9.9a were observed by us before (Sect. 9.2.1) only upon tip–sample separation following a slight indentation (Fig. 9.1b) of the Au surface by the Ni tip.

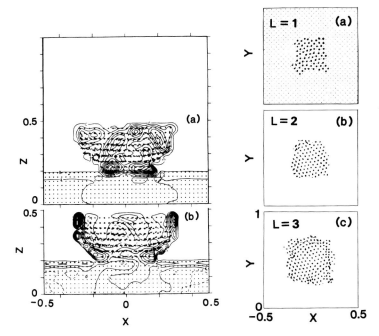

Fig. 9.11. Contours of the von Mises' shear stress ($\sqrt{J_2}$, in a) and pressure (in b) after the formation of adhesive contact between the Au tip and the Ni surface (corresponding to point B in Fig. 9.9a). Note that $\sqrt{J_2}$ maximizes at the periphery of the contact achieving a value of 2 GPa. The increment between the contours in a is 0.17 GPa. The hydrostatic pressure (in b) is tensile and large on the periphery of the tip and achieves a value of ~ -5 GPa at the peripheral region of the contact. The increment between contours is 1 GPa. Distance along the X and Z directions in units of $X = 1$ and $Z = 1$ corresponding to 70.4 Å

Fig. 9.12. Atomic configurations in layers of the Au tip closest to the Ni sample after the formation of the adhesive contact, (point B in Fig. 9.9a). L = 1 corresponds to the proximal layer (*large dots* in a) of the Au tip in contact with the Ni (001) substrate (*small dots* in a). Note the (111) reconstruction of the gold layers. L = 2 and L = 3 are the two layer above the proximal layer. Distance along the X ([100]) and Y ([001]) directions in units of 70.4 Å

The maxima in Fig. 9.9a (points marked C, E, G, I and K) are associated with ordered structures of the elongated connective neck, each corresponding to a neck consisting of one more layer than the previous one. Thus the number of Au layers at point C is 9 (one more than the original 8-layer tip), 10 layers at point E, etc. (Fig. 9.10).

The mechanism of elongation of the neck is similar to that discussed before (Sect. 9.2.1) involving atomic structural transformations whereby in each elongation stage, atoms in adjacent layers in the neck disorder and then rearrange to form an added layer, that is, a more extended neck of a smaller cross-sectional area. Throughout the process the neck maintains a layered crystalline structure except for the rather short structural transformation periods, corresponding to the sharp variations in the force curve, (segments BC, DE, FG, HI and JK in Fig. 9.9a and the associated features in the calculated potential energy shown in Fig. 9.9b where the minima correspond to ordered layered structures after the structural rearrangements.

As we mentioned before, associated with the formation of the adhesive neck, and throughout the elongation stage, the interfacial region of the neck is structurally reconstructed laterally [i.e., (111) reconstruction in layers] accompanied by expanding interlayer spacings compared to that in Au (001). The depth of the reconstructed region tends to be larger during the stretching stages, prior to the reordering which results in an additional layer. The reconstruction and increased inter-layer spacings are related to the coupled effects of lattice constant mismatch between Au and Ni and the fact that the tensile stress in the normal direction acts on interfacial layers whose areas are smaller than those in the region of the tip close to the static holder, resulting in a larger strain in the former region.

Tip Compression. Having discussed the processes of contact formation and pulloff, we show in Fig. 9.13 the force and potential energy curves obtained when the motion of the Au tip towards the Ni sample is continued past contact formation (point marked B, $d_{hs} = 0$, in Fig. 9.13. In this Figure $d_{hs} < 0$ corresponds to continued motion towards the surface). The force curve exhibits nonmonotonic variations and reverses sign, signifying the onset of a repulsive interaction.

We observe that a continued compression of the tip results in an increase in the interfacial contact area between the tip and the sample, and a "flattening" of the tip. To illustrate the process we show in Figs. 9.14 and 15 atomic configurations and pressure contours of the system, corresponding to the points marked E, F, G and H on the force curve (Fig. 9.13a). As can be seen, the number of layers in the Au tip corresponding to points E and G (Figs. 9.14a and 9.15a, respectively) decreases from 8 to 7 (contact areas ~1100 Å^2 and 1750 Å^2, respectively). The pressure contours corresponding to these configurations show tensile (negative) pressure on the sides of the tip and a concentration of compressive pressure in the interfacial sample region.

The evolution of the structural transformation, induced by continued compression, between the ordered structures, involves the generation of interstitial-layer partial dislocations in the central region (core) of the tip and subsequent transformation, as may be seen from Figs. 9.14c and 9.15c (the latter one rearranges to a 6-layer tip upon continued compression-point marked I in Fig. 9.13a- with a contact surface area of ~2400 Å^2). In the course of the lowering of the tip (between ordered configurations), the outer regions of the tip rearrange first to reduce the number of crystalline layers, leaving an interstitial-layer-defect in the core of the tip which is characterized as a high compressive pressure region (Figs. 9.14d and 9.15d). Continued lowering of the tip results in a "dissolution" (or annealing) of the interstitial-layer defect which is achieved by fast correlated atomic motions along preferred (110) directions (Fig. 9.16, show-

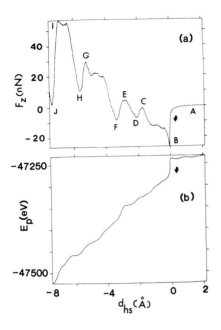

Fig. 9.13. Calculated force F_z (in **a**) and potential energy of the system E_p (in **b**), versus tip-to-sample distance d_{hs} for a Au tip lowered towards a Ni (001) surface. The point marked B ($d_{hs} = 0$) corresponds to the adhesive contact. $d_{hs} < 0$ corresponds to continued motion of the tip toward the surface past the contact point. The points marked E, G and I correspond to ordered configurations of the compressed tip, each containing one atomic layer less than the previous one (starting with 8 layer at point E). Force in units of nN, energy in eV and distance in Å

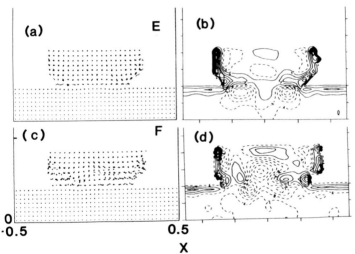

Fig. 9.14. Atomic configurations and pressure contours in an 11 Å slice through the system, corresponding to the points marked E and F in Fig. 9.13a. Note that in (**a**) the tip consists of 8 Au layers, while in (**b**), the peripheral region rearranges to form 7 layers, while the core region consists still of 8 layers, thus containing an interstitial defect layer. In (**b**), corresponding to point E, the compressive pressure (dashed contours) concentrates and maximizes just below the Ni surface, achieving a value of 6 GPa. The increment between contours is 1.2 GPa. Solid contours correspond to tensile pressure. The pressure contours during the intermediate state between ordered configurations (point F in Fig. 9.13a) exhibit concentration of compressive pressure in the core defect region (achieving a maximum value of 4.8 GPa inside the tip) and just below the surface. The increment between contours is 1 GPa

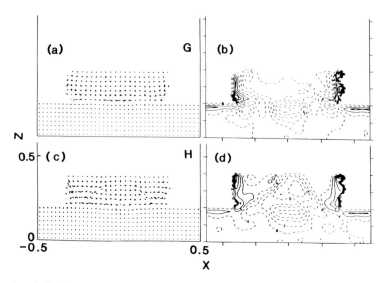

Fig. 9.15. Same as Fig. 9.14 for the configurations marked G and H in Fig. 9.13a. The ordered configuration marked G contains 7 layers in the Au tip and evolves from configuration F (Fig. 9.14c). Note the increase in contact area. Configuration H contains 6 layers at the outside region of the tip and an interstitial defect layer at the tip core region. As in Fig. 9.14 the compressive pressure (*dashed contours*) maximizes just below the surface region for the ordered configuration (in **b**) achieving a value of 7.6 GPa). During the intermediate stage (in **d**) compressive pressure concentrates near the surface region (where it achieves a maximum value of 7.7 GPa) and in the core of the tip (where a maximum value of 6.2 GPa is achieved). The increment between contours is 1.5 GPa

ing short-time trajectories in layers 1–4 of the tip, recorded in the segment FG of Fig. 9.13a). Accompanying the annealing of the defect the compressive pressure transfers from the tip to the interfacial sample region. The initial atomic rearrangement of the peripheral region of the tip, and the eventual expulsion of the remaining interstitial layer atoms from the core, contribute to the increase of the contact area between the tip and the sample.

9.2.4 Nickel Tip/Hexadecane Film/Gold Surface

Our discussion up to this point was confined to the interaction between material tips and bare crystalline samples. Motivated by the fundamental and practical importance of understanding the properties of adsorbed molecularly thin films and phenomena occurring when films are confined between two solid surfaces, pertaining to diverse fields [9.9, 87–89] such as fluid-film lubrication, prevention of degradation and wear, wetting, spreading and drainage, we have very recently initiated investigations of such systems [9.8, 90]. Among the issues which we attempt to address are the structure, dynamics, and response of confined complex films, their rheological properties, and modifications which they may

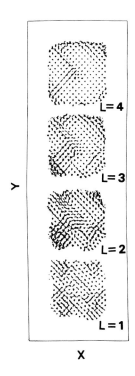

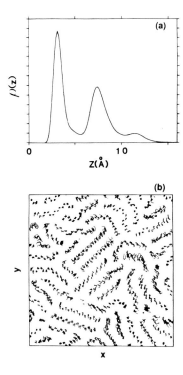

Fig. 9.16. Atomic trajectories in the interfacial layers of the compressed Au tip, recorded during part of the segment FG in Fig. 9.13a, i.e., between configurations starting from one which contains a core interstitial defect layer and ending in an ordered configuration (compare Figs. 9.14c and 9.15a). L = 1 corresponds to the proximal gold layer. The trajectories illustrate the atomic mechanism by which the core interstitial layer defect is expelled

Fig. 9.17. Density profile (in **a**) and top view of the first molecular layer adsorbed on the surface (in **b**) for a thin film of n-hexadecane adsorbed on a Au (001) surface at 300 K. Distance in units of Å. The dimension along the x and y axis in (*b*) is 61.2 Å. The origin of the z axis (normal to the surface) is at the average position of the centers of the gold atoms in the top-most surface layer

cause to adhesive and tribological phenomena, such as inhibition of jump-to-contact instabilities and prevention of contact junction formation. Furthermore, these studies are of importance in the light of recent AFM experiments on adsorbed lubricant films [9.56], aimed at elucidating basic issues pertaining to the tribology of a recording head and slider moving in close proximity and at high speeds relative to a magnetic recording media, as well as other recent studies [9.42e, 55, 57] of the structure, force characteristics, and mechanical response and relaxation mechanisms of adsorbed organic thin films (such as Langmuir–Blodgett and self-assembled films).

The molecular film which we studied, n-hexadecane ($C_{16}H_{34}$) is modeled by interaction potentials developed by *Ryckaert* and *Bellemans* [9.91], which have been employed before in investigations of the thermodynamic, structural and rheological properties of bulk liquid *n*-alkanes [9.92, 93], and adsorbed hexa-decane films of variable thickness [9.90b]. In this model the CH_2 and CH_3 groups are represented by pseudo-atoms of mass 2.41×10^{-23} g, and the inter-molecular bond lengths are fixed at 1.53 A and the bond-angles at $109°\,28'$. A 6–12 Lennard–Jones (LJ) potential describes the intermolecular interaction between sites (pseudo-atoms) in different molecules, and the intramolecular interactions between sites more than three apart. The LJ potential well-depth parameter is $\varepsilon_2 = 6.2 \times 10^{-3}$ eV, and the distance parameter is $\sigma_2 = 3.923$ Å. The range of the LJ interaction is cut-off at 9.8075 Å. An angle-dependent dihedral potential is used to model the effect of missing hydrogen atoms on the molecular conformation. The sample [Au(001)] and tip (Ni) which we use are described using the EAM potentials, as in our aforementioned studies of Ni/Au (001), (Sect. 9.2.1). Finally, the interactions between the *n*-hexadecane molecules and the metallic tip and sample are modeled using LJ potentials, determined by fitting to experimentally estimated adsorption energies [9.90b, 94], with the parameters $\varepsilon_3 = 3\varepsilon_2 = 18.6 \times 10^{-3}$ eV and $\sigma_3 = 3.0715$ Å and 3.28 Å for interactions with nickel and gold atoms, respectively. The cut-off distance of the molecule–surface interaction is $2.5\,\sigma_3$. All details of the simulation pertaining to the metallic tip and sample are as those given in Sect. 9.2.1.

In our first study [9.8], the hexadecane film was composed of 73 alkane molecules (1168 pseudo-atoms) equilibrated initially on the Au (001) surface at a temperature of 300 K. The constrained equations of motion for the molecules were solved using a recently proposed method [9.95], employing the Gear 5th-order predictor corrector algorithm.

The equilibrated adsorbed molecular film is layered prior to interaction with the Ni tip (Fig. 9.17a) with the interfacial layer [the one closest to the Au (001) sample, [Fig. 9.17b] exhibiting a high degree of orientational order. The molecules in this layer tend to be oriented parallel to the surface plane.

The lowering of the (001) faceted Ni tip to within the range of interaction causes the first adherence of some of the alkane molecules to the tip, resulting in partial "swelling" of the film [9.56, 96] and a small attractive force on the tip. Continued approach of the tip toward the sample causes "flattening" of the molecular film, accompanied by partial wetting of the sides of the tip, and reduced mobility of the molecules directly underneath it [see short time trajectories shown in Fig. 9.18a, corresponding to a distance $d_{ts} = 9.5$ Å between the bottom layer of the tip and the topmost layer of the Au (001) surface]. The arrangement of molecules in the interfacial layer of the film is shown in Fig. 9.19a. At this stage, the tip experiences a repulsive force $F_z = 2$ nN.

A continued lowering of the tip induces drainage of the second-layer molecules from under the tip, increased wetting of the sides of the tip, and "pinning" of the hexadecane molecules under it. Side views for several tip-lowering stages are shown in Fig. 9.18b–d, corresponding to $d_{ts} = 6.5$ Å (in b), 5.1 Å (in c), and 4.0 Å (in d and e). (The corresponding recorded forces on the tip, after relaxation, for these values of the tip-to-surface separations are 0 nN, 25 nN, and -5 nN, respectively.) Note that for $d_{ts} = 5.1$ Å and $d_{ts} = 4.0$ Å the region of the surface of the gold sample directly under the tip is deformed and the above d_{ts} values represent averages over the whole surface area (in this context we mention that for $d_{ts} = 5.1$ Å the average pressure in the contact area between the tip and the sample is ≈ 2 GPa). We also remark that we have observed that during the later stages of the tip-lowering process, drainage of entangled, or "stapled", molecules from under the tip is assisted by transient local inward deformations of the substrate which apparently lower the barriers for the relaxation of such unfavorable conformations of the confined alkane molecules. The arrangement of molecules in the first adsorbed alkane layer and in the region above it for $d_{ts} = 6.5$ Å is shown in Fig. 9.19b, c. The molecules are oriented preferentially

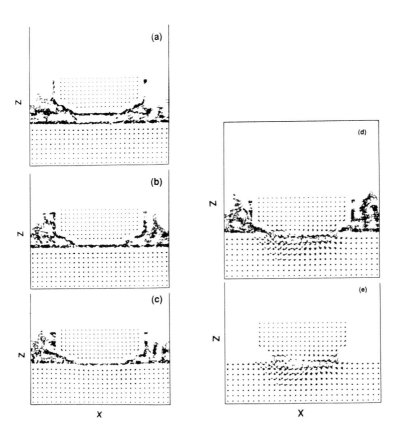

Fig. 9.18. Side views of short-time trajectories of the Ni tip – hexadecane film – Au (001) surface system at four stages of the tip lowering process: **(a)** $d_{ts} = 9.5$ Å, **(b)** $d_{ts} = 6.5$ Å, **(c)** $d_{ts} = 5.1$ Å, **(d, e)** $d_{ts} = 4.0$ Å, where in **(d)** both the metal atoms and alkane molecules are displayed, and in **(e)** only the metal atoms are shown. The values of d_{ts} are average distances between atoms in the bottom layer of the nickel tip and those in the topmost layer of the gold substrate

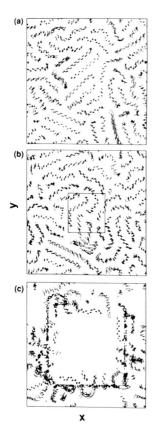

Fig. 9.19. (a) Arrangement of molecules in the first interfacial layer adsorbed on the Au (001) substrate for $d_{ts} = 9.5$ Å and in the first molecular layer **(b)** and the region above it **(c)**, at $d_{ts} = 6.5$ Å. Note that the molecules above the first layer drained from under the tip. The inner marked square in **(b)** denotes the projected area of the bottom layer of the tip

parallel to the surface, particularly in the region under the tip (exhibiting in addition a reduced mobility).

Comparison of the response of the system with that described in Sect. 9.2.1 for the bare gold surface (Fig. 9.1a) reveals that while in the latter case the force between the tip and the substrate is attractive throughout [and remains attractive even for a slight indentation of the surface (Fig. 9.1b)], the overall force on the tip in the presence of the adsorbed alkane film is repulsive for relative tip-to-sample distances for which it was attractive in the other case (except for the initial stages of the tip approach process). However, we note for the smallest tip-to-sample separation which we investigated here (average distance $d_{ts} = 4.0$ Å before relaxation), the onset of the intermetallic contact formation occurring by displacement of gold atoms towards the nickel tip accompanied by partial drainage of alkane molecules, resulting in a net attractive force on the tip of about -5 nN, (the intermetallic contribution to this force is about -20 nN and the alkane repulsive contribution is about 15 nN).

From these results we conclude that the lowering of a faceted nickel tip towards a gold surface covered by a thin adsorbed n-hexadecane film results first in small attraction between the film and the tip followed, upon further lowering of the tip, by ordering (layering) of the molecular film. During continued approach of the tip toward the surface the total interaction between the tip and the sample (metal substrate plus film) is repulsive, and the process is accompanied by molecular drainage from the region directly under the tip, wetting of the sides of the tip, and ordering of the adsorbed molecular monolayer under the tip. Further lowering of the tip is accompanied by inward deformation of the substrate and eventual formation of intermetallic contact (occurring via displacement of surface gold atoms towards the tip) which is accompanied by partial molecular drainage and results in a net attractive force on the tip. The implications of these results to the analysis of AFM measurements of the thickness and rheology of adsorbed films [9.56], and the dependence of the results on the extent of the film, and on the nature of the adsorbed molecular film and its interaction with the substrate and tip, are currently under investigation in our laboratory.

The second case study which we discuss illustrates the consequences of tip–sample interactions in investigations of capillary phenomena occurring upon approach and subsequent retraction of a blunted tip, to and from a liquid alkane film.

In this study the simulated system is composed of a large gold static substrate, a static nickel tip, and a film of alkane molecules whose number is controlled such as the approximately conserve the chemical potential of the system (i.e., a pseudo grand-canonical molecular dynamics simulation where the thickness of the adsorbed molecular liquid film in regions away from the tip is monitored throughout the wetting and growth of the capillary liquid column, and molecules are added at the edges of the computational cell to compensate for those that were pulled-up toward the tip by the adhesive and capillary forces). Since in these simulations the film is not compressed beyond the thickness of about ~ 20 Å, and in view of the much stronger interactions between metal atoms than those characteristic to the molecular liquid, the results are insensitive to the atomic dynamics of the metal surface and tip.

Four configurations of the alkane film are shown in Fig. 9.20, each corresponding to an equilibrium state of the system at a given separation between the tip and the surface. The configurations shown in Figs. 9.20a, b correspond to a distance $d_{ts} = 19.3$ Å and $d_{ts} = 28.1$ Å between the bottom layer of the Ni tip and the topmost layer of the gold surface. As seen in these stages, the liquid film in the region under the tip consists of 4 and 6 distinct layers, respectively, while the arrangement of molecules on the sides of the liquid-junction is less ordered, receding to a two-layer adsorbed film further away from the tip. At this point it is of interest to note that the degree of layering is found to exhibit a marked sensitivity to the distance d_{ts} between the tip and the substrate, maximizing when d_{ts} is commensurate with an integral number of liquid layer spacings. Upon

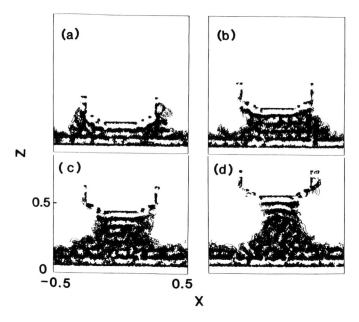

N

0.5

0

−0.5   0.5

X

Fig. 9.20. Short-time trajectories, obtained via MD simulations at 350 K, of hexadecane molecules forming a capillary column between a gold substrate and a nickel tip at four stages of the tip-lifting process (trajectories were plotted in a 23 Å wide slice through the middle of the system). The distances between the tip and the substrate for the configurations shown in (**a**–**d**) are $d_{ts} = 19.3$ Å, 28.1 Å, 36.9 Å and 45.7 Å, correspondingly. The length scale of the calculational cell, which was periodically repeated along the x and y directions is 77.5 Å

raising the tip to a distance $d_{ts} = 36.9$ Å (Fig. 9.20c) and subsequently to $d_{ts} = 45.7$ Å (Fig. 9.20d) the layered nature of the confined film (in the middle region of the capillary column) diminishes. The short-time trajectories of hexadecane molecules at the top, middle and bottom of the liquid column of height 36.9 Å, shown in Fig. 9.21, illustrate the ordered nature of the film next to the tip and surface, and reduced order at the middle of the capillary junction.

The transition from a layered to a partially layered liquid junction is reflected in the force versus distance curve shown in Fig. 9.22, portraying the energetics of the system. Clearly, the force required in order to elongate the column in the height range corresponding to layered configurations is significantly larger than that needed to separate the two interfaces when part of the liquid junction between them acquires liquid-like properties.

Finally, these studies allow a quantitative analysis of the density distribution and energetics of the capillary structures on the molecular level, as illustrated in Fig. 9.23, where contours of the segmental density for a capillary column of height $d_{ts} = 36.9$ Å, are shown. First, it is of interest to note that even for such microscopic capillary junctions the density at the core achieves that of the bulk

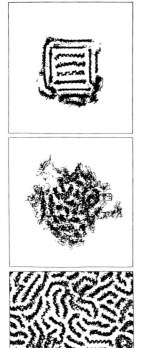

y

x

Fig. 9.21. Short-time trajectories of hexadecane molecules in three regions of the capillary column formed between a gold surface and a nickel tip separated by 36.9 Å (corresponding to the configuration shown in Fig. 9.20c). The bottom configuration corresponding to the region of the film closest to the gold surface illustrates preferential orientation of the molecules parallel to the surface and islands of intermolecular ordering. The middle panel illustrates lack of order in the middle of the liquid junction, while the molecular configuraion shown at the top, corresponding to a region closest to the bottom of the nickel tip, illustrates a high degree of preferential parallel molecular orientation and intermolecular order. Length scale as in Fig. 9.20

liquid [9.90b] at the same temperature $(T = 350 \text{ K})$. Secondly, the density profiles of the column exhibits a gradual decrease from the core outwards. The radii of curvature determined from such plots together with contours of the molecular pressure tensor distribution in the column, allow a direct microscopic evaluation of the Young–Laplace equation and disjoining pressure [9.9a, c] of the liquid, and provide detailed molecular-level guidance for the analysis of AFM data [9.56, 97].

9.2.5 CaF$_2$ Tip/CaF$_2$ Surface

In the previous sections, results pertaining to intermetallic contacts were discussed. Here we turn to results obtained in simulations of a CaF$_2$ tip interacting with a CaF$_2$ (111) surface [9.8]. As remarked in Sect. 9.1, the nature of bonding in ionic materials is different from that in metallic systems, including long-range Coulombic interactions.

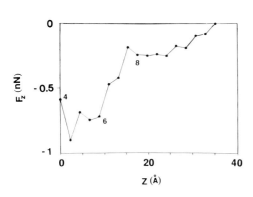

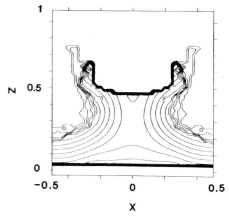

Fig. 9.22. Normal force on the nickel tip (in nN) versus the distance between the tip and the gold surface, Z (in Å) with the origin taken as that corresponding to a four-layer capillary column (Fig. 9.20a). Dots indicate distances for which the system was relaxed during the tip-lifting process. The numbers on the graph denote the number of layers in the capillary column (for the 8-layer column this number of ambiguous due to disordering in the middle part, see Fig. 9.20c)

Fig. 9.23. Segmental density contours for the 36.9 Å high capillary column (Fig. 9.20a) of hexadecane molecules, obtained via MD simulations at 350 K. The density at the middle (core) of the column is 0.033 segment/Å^3, which is the same as that of bulk hexadecane at the above temperature [9.90b]. The spacing between contour lines is $\Delta\varrho = 0.003$ segment/Å^3. The linear dimension along the x and z directions is 77.5 Å

In these simulations the sample is modeled by 3 static layers interacting with 12 layers of dynamic atoms, with 242 Ca^{+2} cations in each calcium layer and 242 F$^-$ anions in each fluorine layer exposing the (111) surface of a CaF$_2$ crystal (the stacking sequence is ABAABA ... where A and B correspond to all F$^-$ and all Ca^{+2} layers, respectively. The top surface layer is an A layer). The CaF$_2$ tip is prepared as a (111) faceted microcrystal containing nine (111) layers, with the bottom layer containing 18 F$^-$ anions, the one above it 18 Ca^{+2} cations followed by a layer of 18 F$^-$ anions. The next 3 layers contain 50 ions per layer, and the 3 layers above it contain 98 ions in each layer. The static holder of the tip is made of 3 CaF$_2$ (111) layers (242 ions total). The system is periodically replicated in the two directions parallel to the surface plane and no periodic boundary conditions are imposed in the normal, z, direction. The long-range Coulomb interactions are treated via the Ewald summation method and temperature is controlled to 300 K via scaling of the velocities of atoms in the three layers closest to the static substrate. The integration time step $\Delta t = 1.0 \times 10^{-15}$ s, and motion of the tip occurs in increments of 0.5 Å over a time span of 1 ps (10^{-12} s). As before, after each increment in the position of the tip–holder assembly the system is allowed to dynamically relax.

Curves of the average force, F_z, on the tip atoms recorded for the fully relaxed configurations, versus distance d_{ns}, are shown in Fig. 9.24a along with

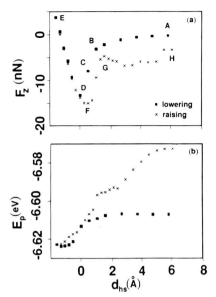

Fig. 9.24. Calculated force on the tip atoms, F_z, in (a), and potential energy per atom in the tip, E_p, versus tip-to-sample distance d_{hs}, for a CaF_2 tip approaching (*filled squares*) and subsequently retracting (*crosses*) from a CaF_2 (111) surface. The distance from the bottom layer of the tip to the top-most surface layer, d_{ts}, for the points marked by letters is: A (8.6 Å), B (3.8 Å), C (3.0 Å), D (2.3 Å), E (1.43 Å), F (2.54 Å), G (2.7 Å) and H (3.3 Å). Distance in Å, energy in eV, and force in nN

the corresponding variations in the potential energy of the tip atoms (Fig. 9.24b). From Fig. 9.24a we observe, following a gradual increase in the attraction upon approach of the tip to the surface, the onset of an instability marked by a sharp increase in attraction occurring when the bottom layer of the tip approaches a distance $d_{ts} \approx 3.75$ Å from the top layer of the surface. This stage is accompanied by an increase in the interlayer spacing in the tip material, i.e., tip elongation, and is reminiscent of the jump-to-contact phenomena which we discussed in the context of intermetallic contacts, although the elongation found in the present case (~ 0.35 Å) is much smaller than that obtained for the metallic systems.

Decreasing the distance between the tip–holder assembly and the substrate past the distance corresponding to maximum adhesive interaction (which occurs at $d_{ts} \approx 2.3$ Å) results in a decrease in the attractive interaction, which eventually turns slightly repulsive (positive value of F_z), accompanied by a slight compression of the tip material. Starting from that point ($d_{hs} = 26.5$ Å, $d_{ts} \approx 1.4$ Å) and reversing the direction of motion of the tip–holder assembly (i.e., detracting it from the surface) results in the force curve denoted by crosses in Fig. 9.24a, b.

As clearly observed from Fig. 9.24a, the force versus distance relationship upon tip-to-sample approach and subsequent separation exhibits a pronounced hysteresis. The origin of this behavior, which is also reflected in the tip potential-energy versus distance curve shown in Fig. 9.24b, is a plastic deformation of the crystalline tip, leading to eventual fracture. At the end of the lifting processes part of the tip remains bonded to the substrate. Atomic configurations corresponding to those stages marked by letters on the force curve (Fig. 9.24a) are shown in Fig. 9.25.

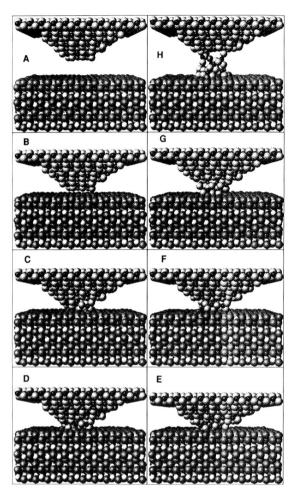

Fig. 9.25. Atomic configurations corresponding to the marked points in Fig. 9.24a. Small and large balls correspond to Ca^{+2} and F^- ions, respectively. The images were obtained for a cut in the middle of the system

Tip Sliding. Starting from the tip–sample configuration under a slight attractive load (Fig. 24a, $h_{ts} = 1.7$ Å, $F_z = -3.0$ nN) lateral motion of the tip parallel to the surface plane is initiated by translating the tip–holder assembly in the $\langle \bar{1}, 1, 0 \rangle$ direction in increments of 0.5 Å followed by a period of relaxation, while maintaining the vertical distance between the tip–holder and the sample at a constant value. This then corresponds to a constant-height scan in the language of atomic-force microscopy. We have also performed constant-load simulations which will not be discussed here.

The recorded component of the force on the tip atoms in the direction of the lateral motion, as a function of the displacement of the tip–holder assembly, is shown in Fig. 9.26a and the corresponding potential energy of the tip atoms is given in Fig. 9.26b. As seen, the force on the tip exhibits an oscillatory variation

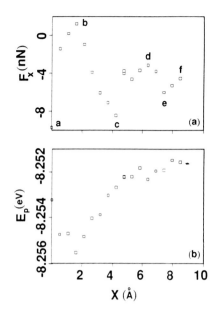

Fig. 9.26. Tangential force on the tip atoms, F_x, in the $\langle 1\bar{1}0 \rangle$ direction, and per-ion potential energy in the tip, E_p, versus distance (X, along the $\langle 1\bar{1}0 \rangle$ direction), calculated for a CaF_2 tip translated at constant height parallel to a CaF_2 (111) substrate surface in the $\langle 1\bar{1}0 \rangle$ direction. Note the oscillatory character of the force curve, portraying an atomic stick-slip process. Note also the increase in E_p with translated distance. The marked points in (**a**) correspond to minima and maxima of the F_x curve along the $\langle 1\bar{1}0 \rangle$ (X) direction. Distance in Å, energy in eV and force in nN

as a function of lateral displacement which is a characteristic of atomic-scale stick–slip behavior. Inspection of the atomic configurations along the trajectory of the system reveals that the lateral displacement results in shear-cleavage of the tip. The sequence of atomic configurations shown in Fig. 9.27 reveals that the bottom part of the tip remains bonded to the sample, and sliding occurs between that portion of the tip material and the adjacent layers. This results indicates that under the conditions of the simulation (i.e., small load), atomic layers of the tip may be transferred to the sample upon sliding resulting in tip wear. From the average value of the recorded variation in the tangential force on the tip (Fig. 9.26a), and the contact area we estimate that the critical yield stress associated with the initiation of slip in the system is ~ 9 GPa, in good correspondence with other simulations of shear deformations of perfect bulk crystalline CaF_2.

9.2.6 Silicon Tip/Silicon Surface

In our earlier studies [9.19a, c] we investigated the interaction between silicon tips and silicon surfaces (i.e., a case of reactive tip–sample system). Our simulations, in both the constant-tip height and constant-force scan modes, revealed that the local structure of the surface can be stressed and modified as a consequence of the tip–sample dynamical interaction, even at tip–sample separations which correspond to weak interaction. For large separations these perturbations anneal upon advancement of the tip while permanent damage can occur for smaller separations. For this system (employing the interatomic potentials

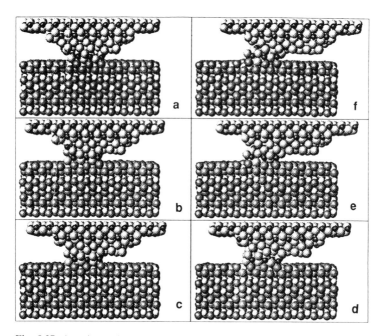

Fig. 9.27. Atomic configurations corresponding to the marked points in Fig. 9.26a. Note the interlayer slip occurring in the tip as the tip-holder assembly is translated from left to right. The bottom three layers of the tip adhere to the surface leading to an adhesive wear process of the sliding tip

constructed by *Stillinger* and *Weber* [9.73a], which include two- and three-body interactions reflecting the directional bonding character in covalent materials) we did not find long-range elastic deformations, which may occur in other circumstances (such as a graphite surface [9.61]) depending upon the elastic properties of the material and the nature of interactions. Furthermore, we found [9.19a, c] that the characteristics of the data depend upon the geometry of the scan, the degree of perfection of the sample, and the temperature.

In the following we focus on consequences of the interaction of a large dynamic tip (consisting of 102 silicon atom, arranged and equilibrated initially in four layers and exposing a 16 atoms (111) facet) scanning a sample surface consisting of 6 (111) layers of dynamic Si atoms (100 atoms per layer), at 300 K.

Lateral scans in both constant-height and constant-force modes were performed. In constant-force scanning simulations, in addition to the particle equations of motion, the center of mass of the tip–holder assembly, Z, is required to obey $M\ddot{Z} = (F(t) - F_{ext}) \cdot \hat{Z} - \gamma\dot{Z}$, where F is the total force exerted by the tip-atoms on the static holder at time t, which corresponds to the force acting on the tip atoms due to their interaction with the substrate, F_{ext} is the desired (prescribed) force for a give scan, γ is a damping factor, and M is the mass of the

holder. In these simulations the system is brought to equilibrium for a prescribed value of $F_{z,ext}$, and the scan proceeds as described above while the height
of the tip–holder assembly adjusts dynamically according to the above feedback
mechanism.

In Fig. 9.28a–d we show the results for a constant-force scan, for $F_{z,ext} =$
$-13\ \varepsilon/\sigma$ (corresponding to -2.15×10^{-8} N, i.e., negative load). Side views of
the system trajectories at the beginning and end stages of the scan are shown in
Fig. 9.28a, b, c, respectively. As seen, the tip–sample interactions induce local
modifications of the sample and tip structure, which are transient (compare the
surface structure under the tip at the beginning of the scan (Fig. 9.28a), exhibiting outward atomic displacements of the tip-layer atoms, to that at the end of
the scan (Fig. 9.28c, where that region relaxed to the unperturbed configuration).
The recorded force on the tip holder along the scan direction (x) is shown in Fig.
9.28d, exhibiting a periodic modulation, portraying the periodicity of the sample
(at the same time the normal force F_z fluctuates around the prescribed value and
no significant variations are observed in the force component normal to the scan
direction).

Most significant is the stick-slip behavior signified by the asymmetry in F_x
(observed also in the real-space atomic trajectories in Figs. 9.28a and 9.28b).
Here, the tip atoms closest to the sample attempt to remain in a favorable
bonding environment as the tip–holder assembly proceeds to scan. When the
forces on these atoms due to the other tip atoms exceed the forces from the
sample, they move rapidly by breaking their current bonds to the surface and
forming new bonds in a region translated by one unit cell along the scan
direction. The detailed energetics of the atomic-scale stick–slip phenomenon can
be elucidated from the variations in the potential and kinetic energies of the tip
atoms along the scan [9.19a]. It is found that during the stick stage, the potential
energy of the strained bonds between the tip and substrate atoms increases. The
slip stage is signified by a discontinuity in the force along the scan direction and
by a sharp decrease in the potential energy which is accompanied by a sudden
increase in the kinetic temperature of the tip atoms as a result of the disruption
of the bonds to the substrate and rapid motion of the tip atoms to the new
equilibrium positions. We note that the excess kinetic energy (local heating)
acquired by the tip during the rapid slip, dissipates effectively during the
subsequent stick stage, via the tip-to-sample interaction. In this context we
remark on the possibility that in the presence of impurities (such as adsorbed
molecules) the above mentioned transient heating may induce (activate) interfacial chemical reactions.

We note that our constant-force simulation method corresponds to the
experiments in [9.35] in the limit of a stiff wire (lever) and thus the stick–slip
phenomena which we observed are a direct consequence of the interplay between the surface forces between the tip and sample atoms and the interatomic
interactions in the tip. The F_x force which we record corresponds to the
frictional force. From the extrema in F_x (Fig. 9.28d) and the load ($F_{z,ext}$) used we

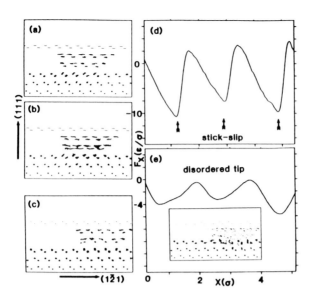

Fig. 9.28. (**a–c**) Particle trajectories in a constant-force simulation, $F_{z,\text{ext}} = -13.0$ (i.e., -2.15×10^{-8} N), direction just before (**a**) and after (**b**) a stick-slip event viewed along the $(10\bar{1})$ and towards the end of the scan (**c**), for a large, initially ordered, dynamic tip. (**d**) The rcorded F_x, exhibiting stick-slip behavior. (**e**) The F_x force in a constant-force scan ($F_{z,\text{ext}} = 1.0$) employing a glassy static tip, exhibiting the periodicity of the sample. Shown in the inset are the real-space trajectories towards the end of the scan, demonstrating the tip-induced sample local modifications. Distance in units of $\sigma = 2.095$ Å, and force in units of $\varepsilon/\sigma = 1.66 \times 10^{-9}$ N

obtain a coefficient of friction $\mu = |F_x|/|F_{z,\text{ext}}| = 0.77$, in the range of typical values obtained from tribological measurement in vacuum, although we should caution against taking this comparison rigorously.

The frictional force obtained in simulations employing a disordered rigid 102-atom tip, prepared by quenching of a molten droplet, scanning under a load $F_{z,\text{ext}} = 1.0$ are shown in Fig. 9.28e. The significance of this result lies in the periodic variation of the force, reflecting the atomic structure of the sample. This demonstrates that microscopic investigations of structural characteristics and tribological properties of crystalline samples are not limited to ordered tips.

Results for a constant-force scan at a positive load ($F_{z,\text{ext}} = 0.1$, i.e., 1.66×10^{-10} N), employing the large faceted tip, are shown in Fig. 9.29. As seen from Fig. 9.29d, the center-of-mass height of the tip and holder assembly from the surface, Z, exhibit an almost monotonic decrease during the scan, in order to keep the force on the tip atoms around the prescribed value of 0.1 (9.29e). At the same time the potential energy of the tip increases. This curious behavior corresponds to a "smearing" (wear) of the tip as revealed from the real-space trajectories shown in Fig. 9.29a–c. Comparison of the atomic configurations at the beginning (Fig. 9.29a), during the scan (Fig. 9.29b corresponding to $X \sim 1\sigma$,

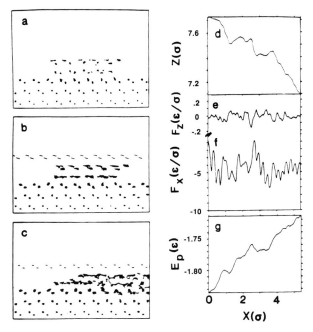

Fig. 9.29. Constant-force scan simulation at $F_{z,\text{ext}} = 0.1$ (i.e., 1.66×10^{-10} N), employing a large (102 atoms), initially ordered dynamic tip. (**a**–**c**) Real-space particle trajectories at selected times during the scan, beginning (**a**), middle (**b**), and end (**c**), respectively. Note that the bottom layer of the tip adheres to the sample surface (**c**); (**d**) Center-of-mass height of the tip holder assembly during the scan, as a function of scan distance, $\sigma = 2.095$ Å. Note the decrease in height associated with the adherence of the bottom tip atoms to the sample (**c**). (**e**), (**f**) Normal force F_z and tangential force in the direction of the scan F_x during the scan. (**g**) Potential energy of the tip atoms E_p during the scan

demonstrating a slip) and towards the end of the scan (Fig. 9.29c) shows that as a result of the interaction between the tip and sample atoms, the bottom layer of the tip adheres to the sample and thus, in order to maintain the same force on the tip holder throughout the scan (as required in the constant-force scan mode), the tip assembly must move closer to the sample. These simulations demonstrate that in reactive tip–sample systems, even under relatively small loads, rather drastic structural modifications may occur, such as wear of the tip resulting in "coating" of the sample by the tip (or vice versa).

Acknowledgement. This work was supported by the U.S. Department of Energy, the Air Force Office of Scientific Research and the National Science Foundation. Simulations were performed on the CRAY Research, Inc. computers at the National Energy Research Supercomputer Center, Livermore, CA, through a grant from DOE, and at the Pittsburgh Supercomputer Center.

References

9.1   D. Tabor: J. Coll. Interface Sci. **58**, 2 (1977)
      M.D. Pashely, D. Tabor: Vacuum **31**, 619 (1981)
9.2   N. Gane, P.F. Pfaelzer, D. Tabor: Proc. R. Soc. (London) A **340**, 395 (1974)
9.3   H.M. Pollock: Vacuum **31**, 609 (1981)
9.4   D. Maugis: Le Vide **186**, 1 (1977)
9.5   R.G. Horn, J.N. Israelachivili, F. Pribac: J. Coll. Interface Sci. **115**, 480 (1987) and references therein
9.6   N.A. Burnham, R.J. Colton: J. Vac. Sci. Technol. A **7**, 2906 (1989)
9.7   U. Landman, W.D. Luedtke, N.A. Burnham, R.J. Colton: Science **248**, 454 (1990)
      U. Landman, W.D. Luedtke: J. Vac. Sci. Technol. **9**, 414 (1991)
9.8   U. Landman, W.D. Luedtke, E.M. Ringer: Wear **153**, 3 (1992)
9.9   J.N. Israelachivili: *Intermolecular and Surface Forces*, 2nd ed. (Academic, London 1992)
      R.G. Horn: J. Am. Ceram. Soc. **73**, 1117 (1990)
      R.J. Hunter, *Foundations of Colloid Science*, Vols. 1 and 2 (Oxford University, Oxford 1987 and 1989)
      Thin Liquid Films, ed. by I.B. Ivanov (Dekker, New York 1988)
9.10 J.S. Rowlinson, B. Widom: *Molecular Theory of Capillarity* (Clarendon, Oxford 1982)
      A.W. Adamson: *Physical Chemistry of Surfaces*, (Wiley, New York 1982)
9.11 D. Tabor, R.H.S. Winterton: Proc R. Soc. (London) A **312**, 435 (1969)
9.12 H.M. Pollock, P. Shufflebottom, J. Skinner: J. Phys. D **10**, 127 (1977)
      H.M. Pollock, ibid. **11**, 39 (1978)
9.13 N. Gane, F.P. Bowden: J. Appl. Phys. **39**, 1432 (1968)
9.14 D. Maugis, H.M. Pollock: Acta Metall. **32**, 1323 (1984), and references therein
9.15 U. Dürig, J.K. Gimzewski, D.W. Pohl: Phys. Rev. Lett. **57**, 2403 (1986)
      U. Dürig, O. Züger, D.W. Pohl: J. Microsc. **152**, 259 (1988)
9.16 J.M. Gimzewski, R. Möller: Phys. Rev. B **36**, 1284 (1987)
9.17 J.B. Pethica A.P. Sutton: J. Vac. Sci. Technol. A **6**, 2494 (1988)
9.18 J.R. Smith, G. Bozzolo, A. Banerjea, J. Ferrante: Phys. Rev. Lett. **63**, 1269 (1989)
9.19 U. Landman, W.D. Luedtke, M.W. Ribarsky: J. Vac. Sci. Technol. A **7**, 2829 (1989); Mater. Res. Soc. Symp. Proc. **140**, 101 (1989) See also M.W. Ribarsky, U. Landman: Phys. Rev. B **38**, 9522 (1988)
      U. Landman, W.D. Luedtke, A. Nitzan: Surf. Sci. **210**, L177 (1989)
      U. Landman, W.D. Luedtke: Appl. Surf. Sci. **60/61**, 1 (1992)
9.20 H. Hertz, J. Reine: Angew. Math. **92**, 156 (1882); also in *Miscellaneous Papers* (Macmillan, London 1896), p. 146; see review by K.L. Johnson: Proc. Instrm. Mech. Engrs. **196**, 363 (1982)
9.21 G. Dieter: *Mechanical Metallurgy* (McGraw-Hill, New York 1967)
9.22 E. Rabinowicz: *Friction and Wear of Materials* (Wiley, New York 1965)
9.23 S.P. Timoshenko, J.N. Goodier: *Theory of Elasticity*, 3rd edn. (McGraw-Hill, New York 1970)
9.24 K.L. Johnson: *Contact Mechanics* (Cambridge Univ. Press, Cambridge 1985)
9.25 K.L. Johnson, K. Kendall, A.D. Roberts: Proc. R. Soc. (London) A **324**, 301 (1971)
9.26 B.V. Derjaguin, V.M. Muller, Yu. P. Toporov: J. Coll. Interface Sci **53**, 314 (1975)
      V.M. Muller, B.V. Derjaguin, Yu. P. Toporov: Colloids Surfaces **7**, 251 (1983)
9.27 P.A. Pashley: Colloids Surfaces **12**, 69 (1984)
9.28 D. Tabor: *The Hardness of Metals* (Clarendon, Oxford 1951)
9.29 J.B. Pettica, R. Hutchings, W.C. Oliver: Philos. Mag. A **48**, 593 (1983)
9.30 N. Gane: Proc. R. Soc. (London A **317**, 367 (1970), and references therein
9.31 P.J. Blau, B.R. Lawn (eds): *Microindentation Techniques in Materials Science and Engineering* (American Society for Testing and Materials, Philadelphia, 1985)
9.32 M.F. Doerner, W.D. Nix: J. Mater. Res. **1**, 601 (1988)
9.33 F.P. Bowden, D. Tabor: *Friction* (Anchor Press/Doubleday, Garden City, NY 1973) p. 62

9.34 See articles in Mater. Res. Soc. Symp. Proc. **140**, 101 (1989), ed. by L.E. Pope, L.L. Fehren-
      bacher, W.O. Winer (Materials Research Society, Pittsburgh, 1989)
      F.P. Bowden, D. Tabor, *Friction and Lubrication Solids* (Clarendon, Oxford 1950)
9.35 C.W. Mate, G.M. McClelland, R. Erlandsson, S. Chiang: Phys. Rev. Lett. **59**, 1942 (1987)
      S.R. Cohen, G. Neubauer, G.M. McClelland: J. Vac. Sci. Technol. A **8**, 3449 (1990)
9.36 R. Thomson: Solid State Phys. **9**, 1 (1986)
9.37 B.R. Lawn: Appl. Phys. Lett. **47**, 809 (1985)
9.38 D. Dowson: *History of Tribology* (Longman, London 1979)
9.39 G. Binnig, C.F. Quate, Ch. Gerber: Phys. Rev. Lett. **56**, 930 (1986)
9.40 J.N. Israelachvili: Acc. Chem. Res. **20**, 415 (1987): Proc. Nat. Acad. Sci. U.S.A. **84**, 4722 (1987)
      J.N. Israelachvili, P.M. McGuggan, A.M. Homola: Science **240**, 189 (1988)
9.41 G. Binnig, H. Rohrer, Ch. Gerber, E. Weibel: Phys. Rev. Lett. **50**, 120 (1983)
9.42 See reviews by P.K. Hansma, J. Tersoff: J. Appl. Phys. **61**, R1 (1986)
      R.J. Colton, J.S. Murday: Naval Res. Rev. **40**, 2 (1988)
      J.S. Murday, R.J. Colton: Mater. Sci. Eng. B, in press J.S. Murday,
      R.J. Colton: in *Chemistry and Physics of Solid Surfaces, VIII*, ed. by R. Vanselow, R. Howe,
      Springer Ser. Surf. Sci., Vol. 22(Springer, Berlin, Heidelberg 1990)
      N.A. Burnham, R.J. Colton: in *Scanning Tunneling Microscopy: Theory and Applications*, ed. by
      D. Bonnell (VCH Publishers, 1992)
9.43 D. Sarid: *Scanning Force Microscopy* (Oxford Univ. Press, New York 1991)
9.44 G. Binnig, H. Rohrer: Helvetica Physica Acta **55**, 726 (1982)
9.45 R. Young, J. Ward, F. Scire: Rev. Sci. Instrum. **43**, 999 (1972)
9.46 E.C. Teague, F.E. Scire, S.M. Baker, S.W. Jensen: Thin Solid Films (1982)
9.47 M.D. Pashley, J.B. Pethica, D. Tabor: Wear **100**, 7 (1984)
9.48 J. Skinner, N. Gane: J. Appl. Phys. **5**, 2087 (1972)
9.49 D. Maugis, G. Desatos-Andarelli, A. Heurtel, R. Courtel: ASLE Trans. **21**, 1 (1976)
9.50 J.B. Pethica, W.C. Oliver: Physica Scripta T **19**, 61 (1987)
9.51 J.B. Pethica: Phys. Rev. Lett. **57**, 323 (1986)
9.52 Q. Guo, J.D.J. Ross, H.M. Pollock: Mater. Res. Soc. Proc. **140**, 51 (1989)
9.53 Y. Martin, H.K. Wickramasinghe: Appl. Phys. Lett. **50**, 1455 (1987)
      P.J. Grütter, M.H. Heinzelmann, L. Rosenthaler, H.-R. Hidber, H.-J. Güntherodt: J. Vac. Sci.
      Technol. A **6**, 279 (1988)
      D. Rugar, H.J. Mamin, P. Guethner, S.E. Lambert, J.E. Stern, I. McFadyen, T. Yogi: J. Appl.
      Phys. **68**, 1169 (1990)
9.54 S.A. Joyce, J.E. Houston: Rev. Sci. Instrum. **62**, 710 (1991)
9.55 J. Frommer: Angewandte Chem. (1992)
9.56 C.M. Mate, M.R. Lorenz, V.J. Novotny: J. Chem. Phys. **90**, 7550 (1989)
      C.M. Mate, V.J. Novotny: J. Chem. Phys. **94**, 8420 (1991)
9.57 S.A. Joyce, R.C. Thomas, J.E. Houston, T.A. Michalske, R.M. Crooks: Phys. Rev. Lett. **68**, 2790
      (1992)
9.58 Y. Kim, J.-L. Huang, C.M. Lieber: Appl. Phys. Lett. **59**, 3404 (1991)
9.59 B. Drake, C.B. Prater, A.L. Weisenhorn, S.A.C. Gould, T.R. Albrecht, C.F. Quate, D.S.
      Cannell, H.G. Hansma, P.K. Hansma: Science **243**, 1586 (1989)
9.60 N.A. Weisenhorn, J.E. MacDougal, S.A.C. Gould, S.D. Cox, W.S. Wise, J. Massie, P. Maivald,
      V.B. Elings, G.D. Stucky, P.K. Hansma: Science **247**, 1330 (1990);
      For surface manipulations using STM tips see: G.M. Shedd, P.E. Russell: Nanotechnology **1**,
      67 (1990)
      R.S. Becker, J.A. Golvchenko, B.S. Swartzentruber: Nature **325**, 419 (1987)
      D.M. Eigler, E.K. Swizer, Nature **344**, 524 (1990)
      L.J. Whitman, J.A. Stroscio, R.A. Dragoset, R.J. Celotta: Science **251**, 1206 (1991)
      H.J. Mamin, P.H. Guethner, D. Rugar: Phys. Lett. **65**, 2418 (1990)
      J.S. Foster, J.E. Frommer, P.C. Arnett: Nature **331**, 324 (1988)
      R. Emch, J. Nagami, M.M. Dovek, C.A. Lang, C.F. Quate: J. Microsc. **152**, 129 (1988)

Y.Z. Li, R. Vazquez, R.P. Andres, R. Reifenberger: Appl. Phys. Lett. **54**, 1424 (1989)

I.W. Lyo, P. Avouris: Science **245**, 1369 (1989)

H.J. Mamin, S. Chiang, H. Birk, P.H. Guethner, D. Rugar: J. Vac. Sci. Technol. B. **9**, 1398 (1991)

See also T.T. Tsong Phys. Rev. B **44**, 13703 (1991)

9.61 J.M. Soler, A.M. Baro, N. Garcia, H. Rohrer: ibid. **57**, 444 (1986); see comment by J.B. Pethica, ibid., p. 3235

9.62 See discussion by S. Ciraci: In *Scanning Tunneling Microscopy and Related Methods*, ed. by R.J. Behm, N. Garcia, H. Rohrer (Kluwer, Dordrecht 1989)

9.63 See articles in *Atomistic Simulations of Materials, Beyond Pair Potentials*, ed. by V. Vitek, D.J. Srolovitz (Plenum, New York 1989)

Many-Body Interactions in Solids, ed. by R.M. Nieminen, M.J. Puska, M.J. Manninen, Springer Proc. Phys., Vol. 48 (Springer, Berlin, Heidelberg 1990)

9.64 See reviews by F.F. Abraham: Adv. Phys. **35**, 1 (1986); J. Vac. Sci. Technol. B **2**, 534 (1984)

U. Landman: In *Computer Simulation Studies* in *Condensed Matter Physics: Recent Developments*, ed. by D.P. Landau, K.K. Mon, H.B. Schuttler (Springer, Berlin, Heidelberg 1988), p. 108

D.W. Heermann: *Computer Simulation Methods*, 2nd. edn. (Springer, Berlin Heidelberg 1990)

9.65 F.F. Abraham. I.P. Batra: S. Ciraci: Phys. Rev. Lett. **60**, 1314 (1988)

9.66 R.J. Colton, S.M. Baker, R.J. Driscoll, M.G. Youngquist, J.D. Baldeschwieler, W.J. Kaiser: J. Vac. Sci. Technol. A **6**, 349 (1988)

9.67 D. Tománek, C. Overney, H. Miyazaki, S.D. Mahanti, H.J. Güntherodt: Phys. Rev. Lett. **63**, 876 (1989)

9.68 N.A. Burnham, D.D. Dominguez, R.L. Mowery, R.J. Colton: Phys. Rev. Lett. **64**, 1931 (1990)

9.69 W. Zhong, D. Tomanek: Phys. Rev. Lett. **64**, 3054 (1990)

9.70 N.A. Burnham, R.J. Colton, H.M. Pollock: J. Vac. Sci. Technol A **9**, 2548 (1991)

9.71 E. Ringer, U. Landman (to be published)

9.72 C.R.A. Catlow, M. Dixon, W.C. Mackrodt: in *Computer Simulations of Solids*, Lecture Notes in Physics, Vol. 166 (Springer, Berlin, Heidelberg 1982) p. 130

see also M. Gillan: in *Ionic Solids at High Temperatures*, ed. by A.M. Stoneham (World Scientific, Singapore 1989) p. 57

9.73 F.H. Stillinger, T.A. Weber: Phys. Rev. B **31**, 5262 (1985)

U. Landman, W.D. Luedtke, M.W. Ribarsky, R.N. Barnett, C.L. Cleveland: Phys. Rev. B **37**, 4637 (1988)

9.74 P. Hohenberg, W. Kohn: Phys. Rev. B **136**, 864 (1964)

9.75 See review by M. Baskes, M. Daw, B. Dodson, S. Foiles: Mater. Res. Soc. Bull. **13**, 28 (1988)

9.76 S.M. Foiles, M.I. Baskes, M.S. Daw: Phys. Rev. B **33**, 7983 (1986)

9.77 The parameterization used in our calculations is due to J.B. Adams, S.M. Foiles, W.G. Wolfer: J. Mater. Res. Soc. **4**, 102 (1989)

9.78 E.T. Chen, R.N. Barnett, U. Landman: Phys. Rev. B **40**, 924 (1989); ibid. **41**, 439 (1990)

C.L. Cleveland, U. Landman, J. Chem. Phys. **94**, 7376 (1991)

R.N. Barnett, U. Landman: Phys. Rev. B **44**, 3226 (1991)

W.D. Luedtke, U. Landman, Phys. Rev. B **44**, 5970 (1991)

9.79 T. Egami, D. Srolovitz: J. Phys. **12**, 2141 (1982)

9.80 M. Parrinello, A. Rahman: Phys. Rev. Lett. **45**, 1196 (1980)

9.81 C. Cattaneo: Rend. Accad. Naz. dei Lincei, Ser. 6, fol. 27 (1938); Part I, pp. 342–348, Part II, pp. 434–436; Part III, pp. 474–478

9.82 R.D. Mindlin: J. Appl. Mech. **16**, 259 (1949); see also J.L. Lubkin: in *Handbook of Engineering Mechanics*, ed. by W. Flugge (McGraw-Hill, New York 1962)

9.83 V.M. Muller, V.S. Yushchenko, B.V. Derjauin: J. Coll. Interface Sci. **77**, 91 (1980); ibid. **92**, 92 (1983)

9.84 B.D. Hughes, L.R. White: Quat. J. Mech. Appl. Math. **32**, 445 (1979)

A. Burgess, B.D. Hughes, L.R. White (1990) – unpublished results

9.85 Mater. Eng. **90**, C120 (1979)

9.86 C.M. Mate (private communication)
9.87 D.Y. Chan, R.G. Horn, J. Chem. Phys. **83**, 5311 (1985)
9.88 J.N. Israelachvili, P.M. McGuiggan, A.M. Homola, Science **240**, 189 (1988)
     A.M. Homola, J.N. Israelachvili, P.M. McGuiggan, M.L. Gee, Wear **136**, 65 (1990)
9.89 J. Van Alsten, S. Granick: Phys. Rev. Lett. **61**, 2570 (1988)
     H.-W. Hu, G.A. Carson, S. Granick: Phys. Rev. Lett. **66**, 2758 (1991)
     S. Granick: Science **252**, 1374 (1991)
9.90 For simulations of the structural and dynamical properties of thin alkane films confined
     between two solid boundaries and the dynamics of film collapse upon application of load see
     M.W. Ribarsky, U. Landman: J. Chem. Phys. **97**, 1937 (1992)
     For simulations of alkane films adsorbed on metal surfaces see T.K. Xia, J. Ouyang, M.W.
     Ribarsky, U. Landman: Phys. Rev. Lett. **69**, 1967 (1992); see also [9.8]
9.91 J.P. Ryckaert, A. Belmans: Discuss. Faraday Soc. **66**, 96 (1978)
9.92 J.H.R. Clark, D. Brown: J. Chem. Phys. **86**, 1542 (1987)
9.93 R. Edberg, G.P. Morriss, D.J. Evans: J. Chem. Phys. **86**, 4555 (1987)
9.94 Q. Dai, A.J. Gellman: Surf. Sci. (in press)
     R. Zhang, A.J. Gellman: J. Phys. Chem. **95**, 7433 (1991)
     A.V. Hamza, R.J. Madix: Surf. Sci. **179**, 25 (1987)
9.95 R. Edberg, D.J. Evans, G.P. Morriss: J. Chem. Phys. **84**, 6933 (1986)
9.96 M.L. Forcada, M.M. Jakas, A. Gras-Marti: J. Chem. Phys. **95**, 706 (1991)
9.97 W.D. Luedtke, U. Landman: Comput. Mater. Sci. **1**, 1 (1992)

10. Theory of Contact Force Microscopy on Elastic Media

G. Overney

With 13 Figures

The Scanning Force Microscope (SFM) was invented by *Binnig* et al. in 1986 [10.1]. Since the first presentation of the SFM, the field of surface imaging has experienced fast development. In the overview by *Meyer* and *Heinzelmann* [10.2], the experimental aspects of force microscopy have been described. In this chapter, I describe a possible theoretical description of the elastic interaction between the SFM tip and a solid surface.

10.1 Description of a Scanning Force Microscope

The power of the SFM lies in the ability to resolve isolated atomic defect structures such as steps or impurity atoms on both conducting and insulating surfaces. The SFM uses an "atomically" sharp tip which scans the sample surface at a sample-to-tip separation of a few Ångstroms. During a horizontal scan of the surface along the x-direction, the SFM measures the equilibrium tip height z_t for a constant preset external force (load) F_{ext}. Because the probing tip is attached to a cantilever-type spring (lever), the external force is controlled by keeping the deflection of the lever constant. This mode of operation is called the *equiforce mode*. Of course, there are other possibilities of imaging modes (e.g., variable deflection mode, constant gradient mode, and the spectroscopic modes), but the equiforce mode is the one most widely used. With the SFM, several groups have achieved atomic resolution on a variety of systems. NaCl [10.3], LiF [10.4], PbS [10.5], and AgBr [10.6] are examples for ionic crystals that have been measured with the SFM on an atomic scale. For almost five years, it has been possible to achieve atomic resolution on layered materials such as graphite [10.7–10], boron nitride [10.11], mica, $MnPS_3$, MoS_2, $TaS(e)_2$ [10.12]. These experiments prove that the SFM is indeed a practical tool to "see" atoms. Because these measurements are done in real space, no Fourier transform is necessary to get information about the position of the atoms. The SFM can "locally" probe the surface to get information that has previously been invisible for spectroscopical measurements. Nevertheless, the interpretation of SFM images can be quite difficult. It is important to understand the so-called "contrast mechanisms" that are responsible for the imaging of a surface using the SFM. In the case of layered materials, dragging of flakes or shearing of layers have been used to explain the relative ease of resolving these materials on an

Springer Series in Surface Sciences, Vol. 29
Scanning Tunneling Microscopy III Eds.: R. Wiesendanger · H.-J. Güntherodt
© Springer-Verlag Berlin Heidelberg 1993

atomic scale. In Chap. 11 the important influence of frictional forces on SFM images will be discussed. Although it has become evident that frictional forces can dramatically affect the imaging of surfaces using the SFM-lever, I will not consider these dissipative forces in this chapter.

10.2 Elastic Properties of Surfaces

A microscopic description of the tip–sample interaction is very difficult. There are a lot of different approaches, but none of them is so universal that it can describe all the phenomena that occur by the interaction between a SFM tip and a sample surface. The different forces between the atoms of the SFM tip and the surface atoms are listed in [10.2]. We could classify these interactions in their range (short-range or long-range forces) or by their origin (e.g., metallic-adhesive forces, magnetic forces, van der Waals forces, etc.). Any of these forces has a microscopic origin that should be described by quantum mechanics. But a complete quantum mechanical treatment is very far from the possibilities of today's fastest computer systems. Therefore we have to make assumptions. One possible assumption that has been used in present theories [10.13] is to assume that the surface is rigid and the tip is replaced by an infinite "periodic" tip. This allows us to use periodic boundary conditions. Another possibility to describe the tip–surface interaction is to use a semi-empirical potential. These potentials are often used within a molecular dynamics simulation. With such empirical potentials, we are able to calculate systems that are close to reality. Typically these potentials use adjustable parameters. Their parameters are obtained from experiments or ab initio calculations. But sometimes the justification of their mathematical form and/or the choice of the parameters is not very convincing. In this chapter, I present a new general approach to determine long-range distortions near isolated impurities in extended systems. I first use ab initio calculations to determine the elastic response of the substrate to non-local external forces. In a second step, I use these results in the framework of the continuum elasticity theory to determine deformations near structural impurities [10.14, 15]. This kind of approach can be used to describe a variety of geometries which are computational beyond the scope of a first principles calculation. In the following, I completely neglect the scanning process. Only a "static" SFM-tip which does not move in the scan-direction (*x*-direction) is considered. Experimentally this "no-scanning mode" is used to record the so-called *force–distance curves*. By measuring force–distance curves, the movement of the tip and sample are approximately perpendicular to the sample surface. From these curves, a lot of information about the tip–surface interaction can be obtained.

10.2.1 Continuum Elasticity Theory for Layered Materials

Using the continuum elasticity theory, it is possible to obtain several analytic and universal results. I apply this formalism to graphite and determine the deformations due to intercalants (impurity atoms between the graphite layers) and/or the SFM tip. The main goal is to get valuable information about the *local* surface rigidity. I will show that under well defined conditions the SFM is a unique tool to measure such *local* changes in the elastic properties of the surface on an atomic scale, which has been predicted in [10.14] and observed on a nanometer scale by *Thibaudau* et al. [10.16]. In the work of *Thibaudau* et al., they observed tracks induced by swift Kr ions in mica. The tracks are directly associated with softer areas in the mica surface. From an experimental point of view, they concluded that SFM is indeed a powerful tool to study local changes in elastic properties. Other systems that are very suitable to study *local elasticity* are the Graphite Intercalation Compounds (GICs). Recent experiments show that the SFM is able to achieve atomic resolution on these materials [10.17].

To describe the tip/surface interactions within a continuum elasticity theory, I have to model the graphite surface and the SFM tip. I assume a semi-infinite system of graphite layers. Each graphite layer is considered as a two-dimensional elastic continuous medium, or a thin elastic plate. This model for graphite layers was first used by *Komatsu* and *Nagamiya* [10.18]. The SFM tip is characterized by a spatial distribution of forces acting on the substrate. In the same way, I model an isolated intercalant impurity in the first gallery (between topmost and second graphite layer). This means that the intercalant is modeled by an incompressible "stick" of finite thickness which is perpendicular to the first and second graphite layer. The thickness of the tip and the intercalant is a model parameter that can be adjusted to the actual situation. To characterize the semi-infinite system of graphite layers, I need the flexural rigidity D, transverse rigidity K (proportional to C_{44}), c-axis compressibility G (proportional to C_{33}), and the interlayer spacing d. The relations between D, K, and G and the elastic tensor components are given by

$$K = C_{44}d_{C-C}^2 \ ,$$
$$G = C_{33}/d \ , \tag{10.1}$$
$$D = \frac{8}{3}C_{44}h^3\left(\frac{C_{12} + C_{44}}{C_{11}}\right) ,$$

where d_{C-C} is the in-plane C–C bond length and h is called the effective layer thickness. h is only meaningful in cases where it can be measured (e.g., macroscopic systems). Therefore, I will determine D from the dispersion relation of an appropriate spring-model rather than from the tensor components C_{nm}. In this continuum model, the vertical distortions of the graphite layers $w_n(r)$ due to a general distribution of forces (originating from the SFM tip or an intercalant impurity) are solutions of a set of partial differential equations presented in [10.19].

Deformations due to the SFM Tip. I consider a cylindrical SFM tip with radius R_0 at position $r = 0$. The total external force (load) F_{ext} is evenly distributed by a constant "hydrostatic" pressure on the substrate and acts on the first layer. The force distribution due to this tip is given by

$$F_1(r) = -\frac{F_{ext}}{\pi R_0^2}\, \theta(R_0 - r) \tag{10.2}$$

and

$$F_n(r) = 0, \quad n \geq 2 \ . \tag{10.3}$$

Here, $\theta(R_0 - r)$ is the well-known Heaviside step function

$$\theta(R_0 - r) \equiv \begin{cases} 0, & \text{if } r > R_0 \\ 1, & \text{if } r < R_0 \ . \end{cases} \tag{10.4}$$

I use the model parameter R_0 to distinguish between "sharp" and "dull" SFM tips. The net force applied to the surface through the tip is equal to the external load applied to the tip, $\int dr\, F_1(r) = -F_{ext}$. The vertical distortion $w_n(r)$ of the nth layer in real space is given by

$$w_n(r) = -\frac{F_{ext}}{\pi \varrho_0 \sqrt{GD}} \int_0^\infty dq \, \frac{J_0(qr/l_0)\, J_1(q\varrho_0)[L(q)]^{n-1}}{X(q) - 1 - L(q)}, \quad n \geq 1 \ , \tag{10.5}$$

where q is the length of the dimensionless vector in the momentum space and $\varrho_0 = R_0/l_0$. l_0 is a characteristic length given by $l_0 \equiv (D/G)^{1/4}$. J_0 and J_1 are the Bessel functions of order zero and one, respectively. The functions $X(q)$ and $L(q)$ are given by

$$X(q) \equiv q^4 + 2\delta q^2 + 2 \tag{10.6}$$

and

$$L(q) = \tfrac{1}{2}\{X(q) - [X(q)^2 - 4]^{1/2}\} \ . \tag{10.7}$$

In the limiting case of a rigid system (flexural rigidity $D \to \infty$) and a finite force, the layer distortions w_n are zero.

Equation (10.5) simplifies further in the case of a "sharp" δ-function-like tip [10.14], corresponding to $\varrho_0 \to 0$, since

$$\lim_{\varrho_0 \to 0} \frac{J_1(q\varrho_0)}{q\varrho_0} = \frac{1}{2} \ . \tag{10.8}$$

Deformations due to an Intercalant Impurity. Let us now consider an intercalant atom (such as K or Cs) sandwiched in the first gallery. As mentioned above, this intercalant is modelled by an incompressible "stick" of finite thickness which is perpendicular to the first and second graphite layer. Similar to the SFM tip, the intercalant exerts a force F_{inc} on the upper and lower layer which is evenly distributed through a constant "hydrostatic" pressure. The intercalant is placed at the origin and the intercalant-induced substrate distortions are due to a force distribution, similar to (10.2). In this case, in (10.2) F_{ext} is to be replaced by F_{inc} and R_0 by R_{inc}. R_{inc} is therefore the radius of a circle inside which the total force F_{inc} is evenly distributed. In the following, $\varrho_{inc} = R_{inc}/l_0$ and the real space integration over the force distribution acting on the first and second layer is given by $\int dr\, F_{1,2}(r) = \pm F_{inc}$.

The force distribution is constructed by imposing boundary conditions on the interlayer distance in the first gallery, or equivalently on the difference of displacements $\Delta w(r) = w_1(r) - w_2(r)$ in the presence of the intercalant. In the absence of an SFM tip, at the position of the intercalant (given by $r = 0$)

$$\Delta w(r = 0) = \frac{1}{2\pi} \int_0^\infty dq\, \frac{|\tilde{F}_{1,2}(q)|q[2X(q) - L(q) - 3]}{[X(q) - 1][X(q) - L(q)] - 1} . \tag{10.9}$$

It is reasonable to assume that $\Delta w(r = 0)$ depends only on the intercalant species (size and compressibility). A possible interaction between the tip and the intercalant (e.g., via charge transfer), which would modify this quantity, is neglected in the present discussion. In the case of an incompressible intercalant impurity, $\Delta w(r = 0) = d_{inc} - d$, where d_{inc} is the diameter of the intercalant and d is the interlayer spacing. I can then calculate F_{inc} from (10.9). Once F_{inc} is known, it is easy to obtain the displacements of the graphite layers due to an intercalant

$$w_1(r) = \frac{F_{inc}}{\pi \varrho_{inc}\sqrt{GD}} \int_0^\infty dq\, \frac{J_0(qr/l_0) J_1(q\varrho_{inc})[X(q) - L(q) - 1]}{[X(q) - 1][X(q) - L(q)] - 1}$$

$$\vdots \tag{10.10}$$

$$w_n(r) = \frac{F_{inc}}{\pi \varrho_{inc}\sqrt{GD}} \int_0^\infty dq\, \frac{J_0(qr/l_0) J_1(q\varrho_{inc})[2 - X(q)][L(q)]^{n-2}}{[X(q) - 1][X(q) - L(q)] - 1}, \quad n \geq 2 .$$

Deformations due to the SFM Tip and an Intercalant Impurity. Let us consider the SFM tip and the intercalant impurity simultaneously and define their positions (with respect to the origin) as r_1 and r_2, respectively. To simplify the problem, I assume that the total layer distortion in the presence of an SFM tip *and* an intercalant is given by the linear superposition of distortions due to each of these separate interactions. If the SFM tip is at a finite distance R from the intercalant, given by $R = |r_1 - r_2|$, the force F_{inc} has to be adjusted in order to keep $\Delta w(r = r_2)$ unchanged. From now on, the position of the intercalant is in the origin ($r_2 = 0$). The modified intercalant force $F_{inc}(R)$ can be obtained from

$$\Delta w(r = 0) = \Delta w^{SFM} + \Delta w^I . \tag{10.11}$$

Here, Δw^{SFM} can be obtained using the expression in (10.5)

$$\Delta w^{\text{SFM}} = w_1(|r| = R) - w_2(|r| = R) \tag{10.12}$$

and Δw^{I} is given by using (10.10) for the layer distortions, as

$$\Delta w^{\text{I}} = w_1(r = 0) - w_2(r = 0) . \tag{10.13}$$

Once $F_{\text{inc}}(R)$ has been determined, the deformation of each layer due to an SFM tip and an intercalant impurity can be calculated using (10.5, 10) and the superposition principle.

10.2.2 Atomic Theory

In the above section, I presented a parametrized set of equations to calculate the deformations near an SFM tip and/or an intercalant impurity. It is obvious that the electronic structure (bonding characteristic, charge transfer between graphite layers and intercalant impurities) is strongly related to the elastic parameters which I used in the previous part of this chapter. In this section, I use an ab initio theory to obtain all these parameters.

The elastic response of a solid to external forces can be obtained from first-principles total energy calculations within the Density Functional Theory [10.20, 21]. I use the Local Density Approximation (LDA) to this theory [10.21] which expresses the total energy of the system in the ground state by

$$E_{\text{tot}} = \sum_{\text{occ}} \varepsilon_{n,k} - \frac{1}{2} \int d\mathbf{r}\, V_{\text{H}}(\varrho)\varrho - \int d\mathbf{r}\, V_{\text{xc}}(\varrho)\varrho + \int d\mathbf{r}\, \varepsilon_{\text{xc}}(\varrho)\varrho + E_{\text{ion-ion}} . \tag{10.14}$$

Here, V_{H} is the Hartree potential due to the electron charge density ϱ and V_{xc} is a local exchange-correlation potential which is conventionally taken from a first-principles calculation of the electron gas. $\varepsilon_{\text{xc}}(\varrho)\varrho$ is the exchange-correlation energy density, and $E_{\text{ion-ion}}$ is the electrostatic repulsion energy between the bare ions. The correct electron density ϱ and the LDA eigenvalues $\varepsilon_{n,k}$ in the ground state are obtained by solving self-consistently the *Kohn–Sham* equations [10.21]

$$\left[-\frac{\hbar^2}{2m}\nabla^2 + V_{\text{ion}}(\mathbf{r}) + V_{\text{H}}(\varrho) + V_{\text{xc}}(\varrho) \right] \psi_{n,k}(\mathbf{r}) = \varepsilon_{n,k}\psi_{n,k}(\mathbf{r}) \tag{10.15}$$

and

$$\varrho(\mathbf{r}) = e \sum_{\text{occ}} |\psi_{n,k}(\mathbf{r})|^2 . \tag{10.16}$$

V_{ion} is the potential due to the ion cores which – in the case of graphite – is replaced by an ab initio pseudopotential generated within the scheme of

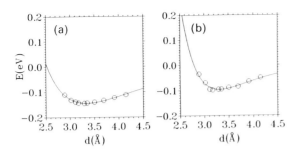

Fig. 10.1. Binding energy of graphite (with respect to isolated layers, per carbon atom) as a function of the interlayer spacing d, (**a**) for hexagonal (or AB) and (**b**) for AA stacking of layers. The solid line represents a Morse potential fit to the calculated data points [10.19]

Hamann et al. [10.22]. The exchange-correlation energy is determined using the parametrization of *Hedin* and *Lundqvist* [10.23]. I expand the wave functions in a linear combination of local Gaussian-type orbitals [10.24]. The LDA charge density and potentials are obtained by sampling the Brillouin zone with a fine mesh using the special-point scheme of *Chadi* and *Cohen* [10.25].

Using the first-principles calculation, I determined the equilibrium structure and elastic properties of graphite. The calculated in-plane C–C bond length $d_{C-C} = 1.42$ Å is in excellent agreement with the experimental results. Also the equilibrium interlayer spacing $d = 3.35$ Å, determined from the energy versus d curve shown in Fig. 10.1, is identical to the observed value [10.26].

In a next step, the frequencies of the in-plane and the out-of-plane phonon modes have been determined. As shown in Fig. 10.2a, the in-plane E_{2g_2} mode consists of in-plane displacements of the α-sublattice with respect to the β-sublattice. The out-of-plane A_{2u} mode consists of small displacements along the c-axis of the α-sublattice with respect to the β-sublattice, as shown schematically in Fig. 10.2b (for this mode the C–C bond length has been kept constant). The vibration frequency can be obtained from a frozen phonon calculation by calculating the total energy as a function of displacement.

The phonon frequencies are given by the curvature of the total energy with respect to the displacement at equilibrium. These calculated values for the phonon frequencies are $\omega(E_{2g_2}) = 1541$ cm^{-1} and $\omega(A_{2u}) = 809$ cm^{-1}. These values are in very good agreement with experimental data [10.27] $\omega(E_{2g_2}) = 1582$ cm^{-1} and $\omega(A_{2u}) = 868$ cm^{-1} and previously calculated values [10.28, 29] $\omega(E_{2g_2}) = 1598$ cm^{-1} and $\omega(A_{2u}) = 839$ cm^{-1}. In all calculations, I assumed the AA layer stacking which is appropriate for stage one intercalated graphite.

In the calculation of the flexural rigidity D [10.30], I made the simplifying assumption that the weak interlayer interaction can be safely dropped when compared to the dominating in-plane interaction. This is true especially when the out-of-plane distortions $w(r)$ are very small. Hence a system of decoupled graphite layers is considered which individually obey the equation of motion [10.18, 30]

$$\eta \frac{d^2 w(r)}{dt^2} = - D\nabla_r^4 w(r) \ . \tag{10.17}$$

(a)

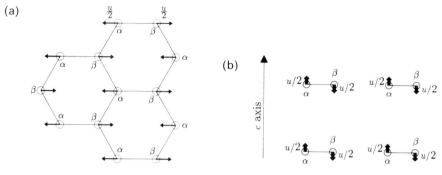

(b)

Fig. 10.2. (a) Schematic top view of the E_{2g_2} phonon mode of graphite. In-plane displacements $u/2$ of atoms in the α- and β-sublattice are indicated by arrows. **(b)** Schematic side view of the A_{2u} phonon mode of graphite. Out-of-plane displacements $u/2$ of atoms in the α- and β-sublattice are indicated by arrows; the nearest-neighbor C–C bond length is kept constant [10.19]

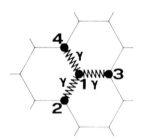

Fig. 10.3. Schematic top view of the graphite lattice and the spring model used to describe the out-of-plane mode. The harmonic spring constant is γ [10.19]

To determine the value of D, I construct a harmonic lattice spring model (Fig. 10.3). The long wavelength behaviour of this model corresponds to (10.17). Then, the total harmonic distortion energy is given by

$$E = \frac{1}{2}\gamma \sum_i \left(3u_{i1} - \sum_j{}' u_{j2}\right)^2 + \frac{1}{2}\gamma \sum_i \left(3u_{i2} - \sum_j{}' u_{j1}\right)^2 . \tag{10.18}$$

In the long wavelength limit, (10.18) leads to

$$\omega^2(k) = \frac{9}{16}\left(\frac{\gamma}{M_C}\right)k^4 d_{C-C}^4 , \tag{10.19}$$

which is also the form of the dispersion relation obtained from (10.17). Here, M_C is the mass of a carbon atom and $d_{C-C} = 1.42$ Å. This way, D is related to the microscopic force constant γ. I have determined γ from first-principles LDA calculations by creating a short wavelength A_{2u} out-of-plane distortion (discussed above and shown in Fig. 10.2b) and comparing the LDA energy with that obtained using (10.18). The value of γ is 0.2584×10^5 dyn/cm. Using this value of

γ, (10.19), and the Fourier transformed (10.17), the flexural rigidity D can be determined, and I obtain the numerical value $D = 7589$ K, which compares very favorably with the experimental value [10.30] $D = 7076 \pm 420$ K. I combined this value with experimental values [10.27] for the c-axis compressibility $G = 789$ K Å^{-4} and the transverse rigidity $K = 932$ K Å^{-2} in the continuum elasticity theory calculations described in the previous section.

10.3 Interaction Between SFM and Elastic Media

Now I apply this theory first to determine the distortion of graphite layers due to the SFM tip and an intercalant impurity and present results in Fig. 10.4. The calculated equilibrium z coordinates of carbon atoms in the three topmost graphite layers near the SFM tip (solid and dotted lines) are compared to their positions in the absence of external forces (dashed lines) in Fig. 10.4a, c. I also investigated how the equilibrium geometry is affected by local changes of the flexural rigidity D, which can occur due to charge transfer near intercalant impurities. In Fig. 10.4, I compare results for pristine graphite, obtained using $D = 7589$ K and given by solid lines, to those obtained using a reduced value $D = 3795$ K and given by dotted lines. In these calculations, a cylindrical SFM tip with a radius $R_0 = 2.75$ Å is considered. The "hydrostatic" force distribution within R_0 is given by (10.2). The calculation yields surface distortions w_n (note that $w_1 = z$ for the topmost layer [10.31]) and the healing length λ_p of pristine graphite, corresponding to the distance at which the layer distortion decreases

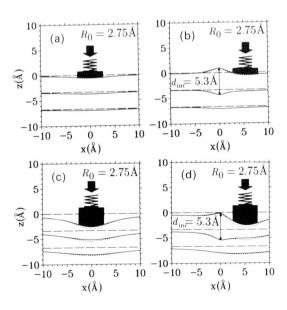

Fig. 10.4. Equilibrium structure of graphite interacting with an SFM tip, for a load 10^{-9} N (**a, b**) and 5×10^{-9} N (**c, d**). Vertical displacements of carbon atoms in the three topmost layers are shown in a plane perpendicular to the layers, for a flexural rigidity $D = 7589$ K (*solid lines*) and $D = 3795$ K (*dotted lines*). The position of unperturbed layers is given by the dashed lines. Results for pristine graphite in (**a**) and (**c**) are compared to the case of a single intercalant (modeled by an incompressible infinitely thin "stick" of length $d_{inc} = 5.3$ Å) in (**b**) and (**d**). The radius R_0 of the cylindrical SFM tip is 2.75 Å [10.19]

to half its maximum value. In the linear response theory, surface distortions are proportional to the applied load F_{ext}, but the healing length should be independent of F_{ext}. Clearly, both λ_p and w_n are expected to depend on the surface flexural rigidity D.

In the case the flexural rigidity D is reduced to half its value, our calculations for a load $F_{ext} = 10^{-9}$ N show that λ_p decreases from 5.56 Å to 4.85 Å and that the maximum layer distortion z_t (at the tip site $r = 0$) increases from 0.52 Å to 0.60 Å. Since the vertical displacement of the SFM tip $\delta z_t = z_t(F_{ext}) - z_t(F_{ext} = 0)$ due to a load depends sensitively on D, the measurement of δz_t as a function of F_{ext} should provide a unique experimental access to this elastic constant and its variations along the surface (e.g., *local* surface rigidity). While the SFM can determine force differences to a high accuracy, the calibration of the zero-load point $F_{ext} = 0$ is very difficult and uncertain. It is useful to note in this context that, within the linear elasticity theory, the value of δz_t is not affected by a miscalibration of the force by $\tilde{F}$, since $\delta z_t \approx z_t (F_{ext} + \tilde{F}) - z_t(\tilde{F})$ if harmonic response is assumed.

In the continuum model, which I presented here, the vertical position z_t of an SFM tip does not depend on its horizontal position x during a scan of the surface at a constant load. In this case, the SFM image does not provide any information about the substrate distortions near the tip, shown in Fig. 10.4a, c, and the healing length λ_p of graphite. The healing length can be easily probed by the SFM near a structural defect, such as a step or an intercalant atom in the first gallery. In Fig. 10.4b, d, the results for a model intercalant representing a K atom are presented. I use the value $d_{inc} = 5.3$ Å for the diameter of the intercalant, which is larger than the graphite interlayer spacing $d = 3.35$ Å. The force distribution of the intercalant is modeled by a pair of equal δ-function-like forces acting in opposite directions on the first and second layer. The special case of an "infinitely thin" intercalant ($R_{inc} \to 0$) can be treated in analogy to the δ-function-like tip, using (10.8). In Fig. 10.4b, d the horizontal distance between the SFM tip and the intercalant is assumed to be 5 Å. As in Fig. 10.4a, c, the total load on the tip F_{ext} is distributed evenly onto the substrate within the tip radius $R_0 = 2.75$ Å. In Fig. 10.4a, b the load is $F_{ext} = 10^{-9}$ N; in Fig. 10.4c, d $F_{ext} = 5 \times 10^{-9}$ N. Results for the topmost three layers of pristine graphite, obtained using $D = 7589$ K and given by the solid line, are compared to those obtained using a reduced value $D = 3795$ K and given by the dotted line.

The SFM image, reflecting the vertical tip position z_t as a function of the horizontal distance r_t from the intercalant, is shown in Fig. 10.5 for different SFM tip loads and tip shapes, and for a "rigid" and a "soft" surface. The "rigid" surface is given by $D = 7589$ K; and in the case of the "soft" surface, D is chosen to be 3795 K. Figure 10.5a shows the image in the case of zero load $F_{ext} = 0$, which obviously reflects the equilibrium surface structure *in the absence* of the tip. More interesting are the images obtained for a nonzero load $F_{ext} = 10^{-9}$ N applied to the surface via a δ-function SFM tip (Fig. 10.5b) and a cylindrical SFM tip with a constant force distribution inside the radius $R_0 = 2.75$ Å (Fig. 10.5c). The intercalant is modeled by an incompressible, infinitely thin "stick" in

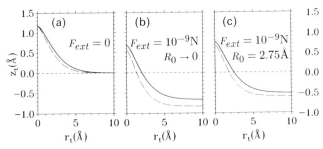

Fig. 10.5. Vertical position z_t of the SFM tip as a function of its horizontal distance r_t from a model K intercalant atom with diameter $d_{inc} = 5.3$ Å. **(a)** Results for zero load on the tip for pristine graphite with $D = 7589$ K (*solid line*) are compared to distortions for a reduced $D = 3795$ K (*dashed line*). **(b)** Corresponding results for a load $F_{ext} = 10^{-9}$ N applied via a δ-function SFM tip. **(c)** Results for a load $F_{ext} = 10^{-9}$ N and a cylindrical SFM tip with a constant force distribution inside the radius $R_0 = 2.75$ Å [10.19]

all three cases. A comparison of Fig. 10.5b, c confirms the intuitive result that differences between the "rigid" and the "soft" surface vanish gradually with increasing tip size R_0. In general, I expect the flexural rigidity D to change in the case of a charge transfer between intercalant atoms and the layers in graphite intercalation compounds, in analogy to similar observed changes of d_{C-C} [10.28, 29]. Systems with intercalants should be correctly described by a locally variable $D(r)$, which shows the strongest deviation from the pristine graphite value at the intercalant site $r = 0$. These local changes get partly smoothed out by a finite-size tip, and the observed SFM corrugation is expected to lie between the calculated curves for the "soft" and "rigid" surface.

The influence of the tip size on the layer distortions (and hence the equilibrium tip position z_t) is shown in Fig. 10.6. I considered a cylindrical SFM tip with a constant "hydrostatic" force distribution inside the variable radius R_0, given by (10.2). The distortion of the topmost graphite layer at the center of the SFM tip, $w_1(r = 0)$, is shown as a function of R_0 for two different loads ($F_{ext} = 10^{-9}$ N and 5×10^{-9} N). Results presented in Fig. 10.6 have been obtained using the flexural rigidity of pristine graphite $D = 7589$ K.

Once the equilibrium positions of the carbon atoms have been determined for a given force distribution, the total charge density ϱ of the deformed graphite can be approximated by a superposition of atomic charge densities obtained from LDA. This approach is justified by the level of agreement between the charge densities of undistorted graphite layers, which have been obtained alternatively from a first-principles LDA surface calculation and from a super-position of atomic charge densities. These charge densities differ by less than 5%, mainly due to incorrect description of the π-bonds in graphite by the superposition of atomic charge densities. This small level of inaccuracy does not affect the corrugations which are observed in the SFM. Figure 10.7 shows the calculated charge distribution which – in the framework of embedded-atom-like

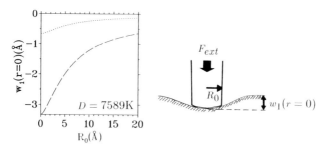

Fig. 10.6. Distortions of the topmost graphite layer $w_1(r = 0)$ at the center of the SFM tip as a function of the effective tip radius R_0. The total loads on the tip are $F_{ext} = 5 \times 10^{-9}$ N (*dashed line*) and $F_{ext} = 10^{-9}$ N (*dotted line*) [10.19]

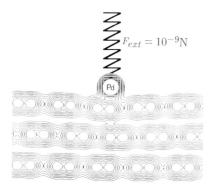

Fig. 10.7. Total charge density ϱ of a 1-atom Pd SFM tip interacting with the elastic surface of hexagonal graphite near the on-top site, for a load $F_{ext} = 10^{-9}$ N. For this load, the distance between the Pd SFM tip and the graphite surface is 2.42 Å [10.34]. Contours of constant ϱ are shown for the three topmost graphite layers in the xz plane perpendicular to the surface. The ratio of two consecutive charge density contours $\varrho(n + 1)/\varrho(n)$ is 1.5

theories – is indicative of the interatomic interactions within the crystal and between a SFM tip and the substrate. The surface deformations for the applied load $F_{ext} = 10^{-9}$ N correspond to those shown in Fig. 10.4a, b.

The predictions for apparent tip corrugations still hold in the presence of van der Waals forces between the tip and the substrate. These long-range forces do not result in additional SFM corrugations and are effectively compensated for in the experiment by an additional force applied on the tip suspension [10.32]. In order to estimate the effect of van der Waals forces on the SFM image, the SFM tip is modeled by a conus with an opening angle $\alpha = 45°$. Following *Anders* and *Heiden* [10.33], I obtained a total van der Waals force $F_{cone}(z) = A_h \times (\tan \alpha)^2/(6z)$, where A_h is the Hamaker constant for the graphite sample ($A_h = 3 \times 10^{-19}$ J). I found that the magnitude and site-dependence of van der Waals interactions between the SFM tip and the surface is negligible as compared to the closed-shell repulsion in the weak repulsive region (distance between SFM tip and surface smaller than ≈ 2.5 Å). In this distance range, I obtained for a conical tip a force ≈ 0.25 nN and 0.16 nN for a tip/sample separation of 2.0 Å and 3.0 Å, respectively, which is much smaller than the forces obtained from the LDA calculations published in [10.34]. From [10.34], forces of 5.0 nN and 4.1 nN are derived for a distance of 2.0 Å and 3.0 Å, respectively. At much larger tip/sample separations, the van der Waals forces dominate the tip–substrate interaction, but do not show measurable corrugations on the atomic scale.

10.3.1 Local Flexural Rigidity

As already mentioned, the flexural rigidity D of graphite intercalation compounds is spatially modified due to charge transfer. In the case of stage-1 alkali-metal graphite intercalation compounds (e.g., LiC_6), I show qualitatively that D has to change as a function of the distance between intercalant and the probing tip. Only in cases where the elastic parameters change locally, is the SFM suitable to detect these changes.

To get the qualitative justification of a *local* flexural rigidity $D(r)$, let us consider the LiC_n-system with $n = 6, 8$. To achieve a very high precision by calculating the charge transfer between intercalant Li-atoms and the graphite host, I again use the density functional theory within the local density approximation. Because surface effects are not important by calculating the charge transfer, it is appropriate to consider just the bulk structure. Figure 10.8 depicts the structure of LiC_2, LiC_6, and LiC_8. Only LiC_6 has been observed experimentally.

To verify my LDA calculations, I determined the change of the C–C bond length due to charge transfer. It is obvious that not only the elastic properties change, but also the geometry of the host is slightly modified by the intercalation process. For the equilibrium C–C bond length of AA-stacked graphite, I have calculated 1.415(7) Å. The experimental value is 1.421 Å [10.35]. In the following, all changes of d_{C-C} (denoted as Δd_{C-C}) will be relative to the LDA result for AA-stacked graphite [$= 1.415(7)$ Å] [10.36].

In the case of the KC_8-structure, I get for $d_{C-C} = 1.430(5)$ Å ($\Delta d_{C-C} = 0.015$ Å) to minimize the total energy. This is in reasonable agreement with the experimental result [10.35]. In the case of LiC_6, d_{C-C} is 1.4338 Å ($d_{C-C}^{exp} = 1.435$ Å from [10.37]) and $c = 3.65$ Å ($c^{exp} = 3.70$ Å from [10.37]). For LiC_8, I got $d_{C-C} = 1.4289$ Å and $c = 3.66$ Å. Finally, d_{C-C} and c for LiC_2 are 1.4837 Å and 3.45 Å, respectively. In Fig. 10.9, I have plotted the change in total energy as a function of the nearest neighbour distance d_{C-C} of the graphite layers for LiC_6 and KC_8. Figure 10.9 depicts very well that anharmonic effects do not occur by such small changes of the lattice parameters. The fitted second-order polynomial "hits" every LDA-value almost exactly.

Following the tight binding crystal orbital calculations from *Kertesz* et al. [10.38], it is qualitatively clear why d_{C-C} increases as a function of decreasing n in LiC_n. In this approach it has been shown that the nodal structure of the one-electron Bloch function at the "K" point in the Brillouin zone (Fermi level) is nonbonding at the level of first neighbour interactions but slightly antibonding for the second neighbours. By the donor GIC's, the total-electron-difference

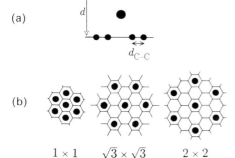

(a)

d_{C-C}

Fig. 10.8. (a) Shows the structure of graphite intercalation compounds from a side-view. In (b), the structure of MC_2 (1×1), MC_6 ($\sqrt{3} \times \sqrt{3}$), and MC_8 (2×2) are drawn. The large black dots label the intercalant atoms

(b)

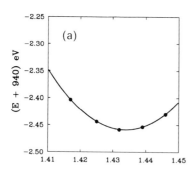
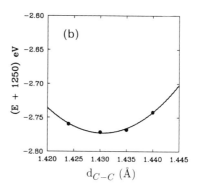

Fig. 10.9. Total energy is plotted as a function of d_{C-C} for LiC_6 (a) and KC_8 (b), respectively. The fitted solid curve is a second order polynomial

density [10.39] has its peak density located in π-like contours on each C site. For LiC_6, this has been shown by *Holzwarth* et al. [10.40]. For LiC_8 and KC_8, I achieved the same result. Figure 10.10 exhibits the total-electron-difference density of LiC_6 and LiC_8, respectively. This effect is now responsible for an enlarged d_{C-C} due to the charge transfer between intercalants and graphite.

Let us go back to the *local* flexural rigidity. Because changes of elastic properties are directly related to structural modifications that originate from charge transfer, I can estimate the spatial variation of elastic parameters. Are these changes in the flexural rigidity large enough to be detected using the SFM? To answer this question, the amount of charge that is stored in the p_z-orbitals and in the sp^2-orbitals for LiC_8 has to be considered. The LiC_6-structure cannot be used to get information about the local charge distribution inside a graphite layer because all carbon atoms are equivalent in this structure. In Fig. 10.11, I have plotted the difference charge density for LiC_6 and LiC_8 slightly above one graphite layer. It can be easily seen that in case of LiC_8 there are two in-equivalent carbon atoms. In Fig. 10.12, these two different sites in LiC_8 are

(a)

(b)

Fig. 10.10. A contour plot of the total-electron-difference charge density in a plane perpendicular to the graphite layers. The contours show lines of equal charge density. The dashed lines indicate a lower charge density for LiC_n than for graphite. (a) for LiC_6 and (b) for LiC_8, respectively. A linear contour increment of 10^{-3} el./(a.u.)^3 is used. The highest contour value is 15.46×10^{-3} el./(a.u.)^3 in (a) and 11.06×10^{-3} el./(a.u.)^3 in (b), respectively. The carbon atoms are located between the filled p_z-orbitals

(a)

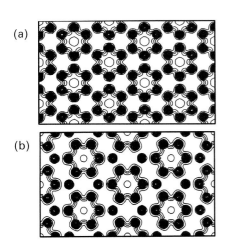

(b)

Fig. 10.11. Total-electron-difference charge density is plotted in a plane parallel to the graphite layer. (a) for LiC_6 and (b) for LiC_8, respectively. In both cases the distance between the contour-plane and the graphite layer is 0.4 Å. A linear contour increment of 10^{-3} el./(a.u.)^3 is used. The highest contour value is 15.13×10^{-3} el./(a.u.)^3 in (a) and 10.81×10^{-3} el./(a.u.)^3 in (b), respectively. The carbon atoms are sitting just below the black dots which indicate a large difference charge density $\Delta\varrho$

denoted by '1' and '2'. Now I build boxes around those two inequivalent sites which have half the size of the unit cell for a graphite monolayer (Fig. 10.12). It is important to emphasize that it is insufficient to consider only the charge located in the p_z-orbitals. Due to intercalation with alkali metal atoms, the charge in the bonding sp^2-orbitals is reduced. Therefore, there is a second mechanism that modifies the structure and elastic properties of the graphite layers. This charge depletion is also known as the formation of "bounding" graphite layers due to intercalation with alkali metals (described in [10.40] and shown in Fig. 10.13).

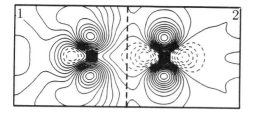

Fig. 10.12. Total-electron-difference charge density for LiC_8. The contours of constant charge density are plotted in a plane perpendicular to the graphite layers. The carbon atoms are indicated by black squares. A linear contour increment of 10^{-3} el./(a.u.)^3 is used. The highest contour value is 11.08×10^{-3} el./(a.u.)^3

Fig. 10.13. Total-electron-difference charge density for LiC_6. The contours of constant charge density are plotted inside a graphite layer. A linear contour increment of 10^{-4} el./(a.u.)^3 is used. The highest contour value is 31.21×10^{-4} el./(a.u.)^3. The dashed lines indicate less charge density in case of LiC_6 than compared to graphite

By plotting the valence-electron contribution of the difference charge density, this statement of bounding carbon layers can be easily verified [Ref. 10.40, Fig. 5c]. The ratio between the charge located in box '1' and that located in box '2', $\Delta\varrho_1/\Delta\varrho_2$, is around 5 (site '1' is closer to the intercalant atom than site '2'). Therefore, the charge at site '1' is five times larger than at site '2' for the LiC_8 structure. This localized charging of the graphite host layers due to intercalation proves that the structural and elastic properties of the carbon layers have to change *locally* as well. From Sect. 10.2.2, we know that the flexural rigidity is directly related to the in-plane coupling constant γ between neighboring carbon atoms (see the lattice model in Fig. 10.3). This coupling constant depends on the geometry as well as on the charge-distribution in the bonding (sp^2-orbitals) and anti-bonding states (π-orbitals) of the graphite structure. But, using the above result, we see that these changes are very *localized*. Because of the quite lengthy LDA calculations, I cannot present a quantitative result for $D(r)$. But I would like to refer the work of *Chan* et al. [10.28]. They have calculated changes of the E_{2g}-mode as a function of charge transfer between donor intercalant atoms (e.g., Li, K, Cs, and Rb) and the graphite layers. They have found an almost linear dependence between $\omega(E_{2g})$ and $\Delta\varrho$ for a charge donation to the graphite planes. In the case of a charge depletion from the graphite planes (acceptor compounds), Chan et al. obtained a small but nonlinear increase of $\omega(E_{2g})$ that causes a stiffening of the graphite layers. I have to mention that they did not calculate the "true" structure fully self-consistent, nor did they optimize the interlayer separation. Nevertheless their results are reasonable and can also be applied to higher-stage compounds (e.g., stage-2 compounds). In their LDA approach, they represented the intercalant by ions of charge Z, located midway in the interlayer spaces between each pair of facing graphitic hexagon centers (1×1 structure as shown in Fig. 10.8). With different values of Z, they simulated acceptor and donor compounds. It is now reasonable to assume that $\omega(A_{2u})$ changes as a function of $\Delta\varrho$ in a similar way compared to $\omega(E_{2g})$. But further LDA calculations will give quantitative results for local changes of elastic properties.

10.4 Conclusions and Outlook

Recent experiments proved that the scanning force microscope is quite successful in measuring *local* effects at surfaces (e.g., steps and local elastic properties). On the atomic scale, a lot of questions have still to be answered to obtain a better understanding of the tip–surface interaction. The theory described in this chapter is just one possibility to describe the tip–surface interactions in the elastic regime. But it completely neglects the scanning process which is highly non-trivial. The above treatment of the problem does also not take into account multi-tip effects.

For a better understanding of the contrast mechanisms, SFM measurements in ultra-high vacuum (UHV) have to be done. UHV–SFM will provide experiments under well-defined conditions. From such measurements, it will be possible to adjust parametrized hamiltonians to describe the tip–surface interaction within a theory based on quantum mechanical assumptions.

Acknowledgements. I thank Professors S.D. Mahanti, H. Miyazaki, and D. Tománek, as well as Mr. W. Zhong from Michigan State University for stimulating discussions. I also thank Dr. H. Heinzelmann and Dr. E. Meyer from the SXM group of the University of Basel for many good suggestions.

References

10.1   G. Binnig, C.F. Quate, C. Gerber: Phys. Rev. Lett. **56,** 930 (1986)
10.2   E. Meyer, H. Heinzelmann: In: *Scanning Tunneling Microscopy II*, ed. by R. Wiesendanger and H.-J. Güntherodt, Springer Ser. in Surf. Sci., Vol. 28 (Springer, Berlin, Heidelberg 1992)
10.3   G. Meyer, N.M. Amer: Appl. Phys. Lett. **56,** 2100 (1990)
10.4   E. Meyer, H. Heinzelmann, H. Rudin, H.-J. Güntherodt: Z. Phys. B **79,** 3 (1990)
10.5   H. Heinzelmann, E. Meyer, D. Brodbeck, G. Overney, H.-J. Güntherodt: unpublished.
10.6   E. Meyer, H.-J. Güntherodt, H. Haefke, G. Gerth, M. Krohn: Europhys. Lett. **15,** 319 (1991)
10.7   G. Binnig, C. Gerber, E. Stoll, T.R. Albrecht, C.F. Quate: Europhys. Lett. **3,** 1281 (1987)

10.8   E. Meyer, H. Heinzelmann, P. Grütter, T. Jung, T. Weisskopf, H.-R. Hidber, R. Lapka, H. Rudin, H.-J. Güntherodt: J. Microsc. **152**, 269 (1988)

10.9   O. Marti, B. Drake, S. Gould, P.K. Hansma: J. Vac. Sci. Technol. A **6**, 287 (1988)

10.10 P.J. Bryant, R.G. Miller, R. Yang: Appl. Phys. Lett. **52**, 2233 (1988)

10.11 T.R. Albrecht, C.F. Quate: J. Appl. Phys. **62**, 2599 (1987)

10.12 E. Meyer, R. Wiesendanger, D. Anselmetti, H.-J. Güntherodt, F. Lévy, H. Berger: Europhys. Lett. **9**, 695 (1989)

10.13 I.P. Batra, S. Ciraci: J. Vac. Sci. Technol. A **6**, 313 (1988)

10.14 D. Tománek, G. Overney, H. Miyazaki, S.D. Mahanti, H.-J. Güntherodt, Phys. Rev. Lett. **63**, 876 (1989); ibid. **63**, 1896(E) (1989)

10.15 S. Lee, H. Miyazaki, S.D. Mahanti, S.A. Solin: Phys. Rev. Lett. **62**, 3066 (1989)

10.16 F. Thibaudau, J. Cousty, E. Balanzat, S. Bouffard: Phys. Rev. Lett. **67**, 1582 (1991)

10.17 E. Meyer, R. Overney: private communications

10.18 K. Komatsu, T. Nagamiya: J. Phys. Soc. Jpn. **6**, 438 (1951)

10.19 G. Overney, D. Tománek, W. Zhong, Z. Sun, H. Miyazaki, S.D. Mahanti, H.-J. Güntherodt: J. Phys.: Cond. Mat. **4**, 4233 (1992)

10.20 P. Hohenberg, W. Kohn: Phys. Rev. **136**, B864 (1964)

10.21 W. Kohn, L.J. Sham: Phys. Rev. **140**, A1133 (1965)

10.22 D.R. Hamann, M. Schlüter, C. Chiang: Phys. Rev. Lett. **43**, 1494 (1979)

10.23 L. Hedin, B.J. Lundqvist: J. Phys. C **4**, 2064 (1971)

10.24 An s-orbital is given by $\exp(-\alpha r^2)$ and the three p-orbitals are given by $x\exp(-\alpha r^2)$, $y\exp(-\alpha r^2)$, and $z\exp(-\alpha r^2)$. The combination $(x^2 + y^2 + z^2)\exp(-\alpha r^2)$ describes an s-orbital

10.25 D.J. Chadi, M.L. Cohen: Phys. Rev. B **8**, 5747 (1973)

10.26 In the case $d \to \infty$, the energy $E(d)$ corresponds to isolated layers. I avoided the lengthy LDA total energy calculation for an isolated graphite layer, but rather treated it as a free parameter in the Morse fit. I obtained for the total energy of a monolayer $-155.525\,\text{eV}$ [(a), AB stacking] and $-155.550\,\text{eV}$ [(b), AA stacking]. The small disagreement of 25 meV between these monolayer energies results from an incomplete basis in the LDA calculations. This uncertainty must be considered when comparing the absolute stability of AB versus AA stacked graphite in Figs 10.1a and b

10.27 H. Zabel: In: *Graphite Intercalation Compounds I*, ed. by H. Zabel and S.A. Solin, Springer Ser. in Mat. Sci. Vol. 14 (Springer Berlin, 1990)

10.28 C.T. Chan, K.M. Ho, W.A. Kamitakahara: Phys. Rev. B **36**, 3499 (1987)

10.29 C.T. Chan, W.A. Kamitakahara, K.M. Ho, P.C. Eklund: Phys. Rev. Lett. **58**, 1528 (1987)

10.30 D is related to the bending modulus B by $D = \eta B$. The proportionality constant $\eta = 2M_C/(3\sqrt{3}d_{C-C}^2/2)$ is the area mass density (in this case, M_C is the mass of a carbon atom and $d_{C-C} = 1.42\,\text{Å}$). The experimental value for $B = (2.55 \pm 0.15) \times 10^{-5}\,\text{cm}^4/\text{sec}^2$ is listed in [10.27]

10.31 The absolute vertical position z of atoms in the nth layer (with respect to the undistorted topmost layer) is given by $z(r) = -(n - 1)d + w_n(r)$, where $d = 3.35\,\text{Å}$ is the interlayer spacing

10.32 F.O. Goodman, N. García: Phys. Rev. B **43**, 4728 (1991)

10.33 M. Anders, C. Heiden: Poster at the Fifth Int'l Conf. on Scanning Tunneling Microscopy and NANO I, Baltimore (1990)

10.34 W. Zhong, G. Overney, D. Tománek: Europhys. Lett. **15**, 49 (1991)

10.35 D.E. Nixon, G.S. Parry: J. Phys. C **2**, 1732 (1969)

10.36 In the paper of *Chan* et al. [10.28], the LDA value is 1.419 Å for AB-stacked graphite. The different stacking geometry may be responsible for this small difference

10.37 D. Guérard, A. Hérold: Carbon **13**, 337 (1975)

10.38 M. Kertesz, F. Vonderviszt, R. Hoffman: Mat. Res. Soc. Symp. Proc. **20**, 141 (1983)

10.39 To obtain the total-electron-difference density, the density of the intercalation compound is compared with that of graphite modified to the same structure

10.40 N.A.W. Holzwarth, S.G. Louie, S. Rabii: Phys. Rev. B **28**, 1013 (1983)

11. Theory of Atomic-Scale Friction

D. Tománek

With 17 Figures

Friction between two solids is not only one of the most common, but also one of the most complex and least understood processes in nature [11.1]. At rough interfaces, plastic deformations and abrasion – both associated with the re-arrangement of interatomic bonds – are responsible for energy dissipation during the relative motion of the two solids. A fundamentally different behavior is observed at "perfect", weakly-interacting interfaces. There, friction without wear corresponds to energy transfer from macroscopic degrees of freedom (describing the relative motion of the bodies in contact) to microscopic degrees of freedom (such as phonons or electronic excitations) which occur as heat. Considerable success has been achieved recently in the quantitative measure-ment of friction forces on the atomic scale [11.2] and the understanding of the underlying microscopic mechanisms in the case of sliding friction without wear [11.3]. This success has been made possible by an increasing sophistication in the characterization of interfaces, from the use of rather rough interfaces [11.4, 5] to atomically flat areas [11.6], and imaginative adaptations of the Scanning Force Microscope (SFM) [11.7] for friction measurements. On the other hand, rapid development of computational techniques and the availability of large computer resources have made quantitative predictions for the friction process possible [11.8, 9]. The success on both the experimental and theoretical side has opened up a new research field called *nanotribology*.

    In this chapter, I will discuss the new possibilities, but also the limitations of scanning force microscopy in obtaining fundamental understanding of both the *sliding* and *rolling friction* processes. I will start with a brief discussion of the irreversibility in the friction process and possible ways to model an "ideal friction machine" based on the SFM. Next, I will review existing first-principles calculations for friction associated with a single atom sliding on a substrate. Finally, I will discuss limits of nondestructive adsorbate–substrate interactions which are related to the onset of wear in SFM measurements.

11.1 Microscopic Origins of Friction

Let us consider two bodies in contact, A and B, which are in relative motion. Under normal conditions, a friction force F_f occurs in this situation along the direction of motion in addition to the reaction forces of classical mechanics

Springer Series in Surface Sciences, Vol. 29
Scanning Tunneling Microscopy III   Eds.: R. Wiesendanger · H.-J. Güntherodt
© Springer-Verlag Berlin Heidelberg 1993

[11.10]. This force is related to the applied load F_{ext} between the two bodies as

$$F_f = \mu F_{ext} . \tag{11.1}$$

The coefficient of friction, μ, ranges typically between 10^{-2} for smooth interfaces and 1 for rough interfaces. Large values of μ reflect the fact that interatomic bonds at the A–B interface are being broken or rearranged during the relative motion of the bodies. At an interface with essentially flat areas, we can consider several cases which can be classified according to relative bond strengths. This will allow us to establish the conditions under which friction without wear can be expected.

Let us first discuss the basic situations occurring during friction *with wear*. If interatomic A–A bonds are comparable in strength to B–B bonds, and both much weaker than A–B bonds, the relative motion between A and B will eliminate large asperities at the interface in a "sandpaper action". This process is typically accompained by long-ranged plastic deformations and dislocation motion away from the interface. Note that, in this case, a lubricant C is often used [11.11]. The situation of A–B bonds stronger than A–A and B–B bonds will lead to spontaneous bonding at the interface (similar to the case of strong C–C bonds which make C an adhesive). The relative motion of A and B will in this case result from a fracture inside the weaker of the solids along the interface. If B–B bonds are much weaker than A–A bonds, the relative motion of A on B will lead to a "plowing" of B by A.

Friction *without wear* can occur in the ideal case of a defect-free interface between single crystals A and B. In order to avoid plastic deformations, we also require the load F_{ext} on the interface not to exceed the elastic limit within the "softer" solid. Furthermore, the A–A bonds should be comparable in strength to B–B bonds, and at the same time be stronger than A–B bonds. It was recognized a long time ago that, in this case, the only source of friction and its modulation should be atomic-scale corrugations of the A–B interaction potential [11.12].

These modulations and the associated friction force have first been observed successfully on highly oriented pyrolytic graphite using the friction force microscope (FFM) with a tungsten tip [11.2]. As for the theory, two different approaches have been used to determine atomic-scale friction. Molecular dynamics calculations with parametrized pair potentials have been used to simulate the stick–slip motion and to determine the friction force between a Si SFM tip and a Si substrate [11.8]. An independent approach, based on an *ab initio* density functional calculation, has been used to determine the trajectory of a Pd atom moving along a graphite surface and to estimate the associated friction force [11.9].

The variety of processes which are expected to occur during sliding friction can be easily illustrated in a somewhat simple-minded model, namely two haircombs in relative motion, with their teeth (representing surface atoms) in contact. This is shown in Fig. 11.1. "Friction" occurs due to the snapping motion as the teeth slide against each other (Fig. 11.1a). In the case that the teeth are only weakly interlocked, friction energy is dissipated into the vibrations of

(a)

(b)

Fig. 11.1. Haircomb models illustrating the origin of atomic-scale friction. (**a**) Sliding friction at an "ideal" A–B interface. (**b**) Sliding friction as observed in a friction force microscope (a modified SFM capable of measuring forces in the direction of the trajectory). The teeth of the combs represent atoms at the surface of materials A and B

the teeth of the "softer" comb. In the case of extremely weak interlocking, no snapping motion of the teeth and consequently no friction is expected. No wear occurs in this process. In the case of strong interlocking, individual teeth are likely to break off, corresponding to friction with wear. Without loss of generality, this process can be investigated even if one of the combs has a single tooth left, as shown in Fig. 11.1b. This is the basic model of sliding friction between A and B being investigated by a sharp SFM tip of material A interacting with the substrate B.

The microscopic description of friction without wear must address the fact that friction is a non-conservative process. In other words, the friction force depends on the direction of motion between two bodies in contact and hence can *not* be obtained as a derivative of a potential. A closed-loop integral over such a force yields a nonzero value of the dissipated energy W_f,

$$W_f = \oint F_x(x)\,dx \neq 0 \ . \tag{11.2}$$

The origin of the dependency of the force on the direction of motion is illustrated in Fig. 11.2. The solid line with overhangs in the hatched area can be obtained as a gradient of a potential. The overhangs cause a nonunique relationship between F_x and x. For a given direction of motion between A and B (either the $+x$ or $-x$ direction), the system will follow the trajectory indicated by arrows and thereby undergo a sequence of instabilities. The locations where these

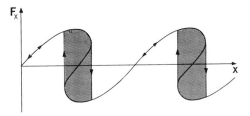

Fig. 11.2. Force F_x on the friction force microscope tip along the trajectory of the scan. Nonzero average value of $\langle F_x \rangle$, corresponding to friction, results from instabilities and non-uniqueness in the F_x versus x relationship

instabilities (or "snapping") occur, depend on the direction of motion. The resulting hysteresis reflects the microscopic irreversibility in the process. The degrees of freedom predominantly involved in the instabilities are those of A or B which are easiest to excite, giving rise to "tip-induced" or "substrate-induced" friction. The energy dissipated in friction, W_f, is related to the hatched area in the hysteresis curve. In the following, we will investigate the microscopic orgin of this hysteresis and discuss friction in a quantitative way.

11.2 Ideal Friction Machines

11.2.1 Sliding Friction

In order to describe friction in a given system better than phenomenologically, we must carefully examine the microscopic processes which lie at the origin of the hysteresis in the "force versus position" curve in Fig. 11.2. I will start the discussion with the sliding friction between an SFM tip and the substrate. After describing the real instrument which is being used to study atomic-scale friction, I will simplify this system to an idealized "Friction Force Microscope" (FFM) which gives rise to sliding friction without wear during a surface scan. I will describe two different models of the FFM which contain the essential physics leading to friction and which I will call "ideal friction machines" [11.13]. These models should be sufficiently realistic to allow a comparison with the experimental equipment described in the following.

One of the first FFM's is a modified scanning force microscope with a tungsten tip which has been used to observe atomic-scale friction without wear on graphite [11.2]. A more recent realization of the friction force microscope [11.14] is shown in Fig. 11.3. Like the SFM, the FFM consists of an "atomically sharp" tip of material A, suspended on a soft cantilever, which is brought into nondestructive contact with a well-defined substrate B. The vertical deflection of the cantilever is regulated in order to keep the applied load F_{ext} constant during the surface scan. The instrument shown in Fig. 11.3 uses the torque on the cantilever due to a horizontal force on the tip to measure the atomic-scale friction force F_f between the tip A and the substrate B. The torsion of the cantilever can be measured independently from F_{ext} by a laser beam which is

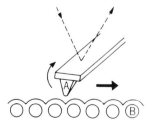

Fig. 11.3. Schematic picture of a friction force microscope. Both vertical motion and torsion of the cantilever are observed by the reflected laser beam

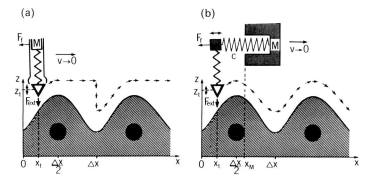

Fig. 11.4. Two models of the Friction Force Microscope (FFM). In both models, the external suspension M is guided along the horizontal surface x direction at a constant velocity $v = dx_M/dt \rightarrow 0$. The load F_{ext} on the "sharp" tip (indicated by ∇) is kept constant along the trajectory $z_t(x_t)$ (shown by arrows). (a) A "maximum friction microscope", where the tip is free to move up, but gets stuck at the maximum z_t between $\Delta x/2$ and Δx. (b) A "realistic friction microscope", where the position of the tip x_t and the suspension x_M may differ. In this case, the friction force F_f is related to the elongation $x_t - x_M$ of the horizontal spring from its equilibrium value [11.13]

reflected from the cantilever. This experiment gives direct information about F_f as a function of F_{ext}. It is interesting to note that the torque induced by the horizontal force can increase the vertical tip deflection at topographic surface features and hence enhance the contrast of the SFM image.

Two idealized models of a friction force microscope are shown in Fig. 11.4. In both models, the microscope suspension M moves quasi-statically along the surface x-direction with its position x_M as the externally-controlled parameter. The tip is assumed to be stiff in respect to excursions in the surface y-direction. We restrict our discussion to the case of tip-induced friction and assume a rigid substrate which applies for friction measurements on graphite [11.2, 9].

In the "maximum-friction microscope" [11.9], the full amount of energy needed to cross the potential energy barrier ΔV along Δx is dissipated into heat [11.15]. This process and the corresponding friction force can be observed in an imperfect scanning force microscope which is shown in Fig. 11.4a. A vertical spring connects the tip and the external microscope suspension M. The horizontal positions of the tip and the suspension are *rigidly* coupled, $x_t = x_M = x$. For $0 < x < \Delta x/2$, the load F_{ext} on the tip is kept constant by moving the suspension up or down. For $\Delta x/2 < x < \Delta x$, however, the tip gets stuck at the maximum value of z_t. At Δx, the energy ΔV stored in the spring is abruptly and completely released into internal degrees of freedom which appear as heat.

The potential energy $V(x)$ during the sliding process is shown in Fig. 11.5a [11.9]. The force on the tip in the negative x-direction, as defined in Fig. 11.4a, is given by

$$F_f(x_M) = \begin{cases} \partial V(x_M)/\partial x_M & \text{if } 0 < x_M < \Delta x/2 \\ 0 & \text{if } \Delta x/2 < x_M < \Delta x \end{cases} \qquad (11.3)$$

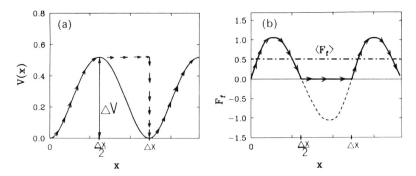

Fig. 11.5. (a) Potential energy of the tip $V(x)$, (b) the friction force $F_f(x)$ and the average friction force $\langle F_f \rangle$ in the "maximum friction" microscope. The arrows indicate the tip trajectory corresponding to a relaxed vertical tip position $z_{t, min}$ for a constant load on the tip [11.13]

and shown in Fig. 11.5b. The non-zero value of the average friction force $\langle F_f \rangle$, indicated by the dash-dotted line in Fig. 11.5b, is a consequence of the mechanism which allows the tip to get stuck. $F_f(x_M)$ is a non-conservative force since it does depend on the scan direction. In absence of the "sticking" mechanism, F_f would be given by the gradient of the potential energy *everywhere*, as indicated by the dashed line in Fig. 11.5b; it would be independent of the scan direction and hence conservative. In such a case, F_f would inhibit sliding for $0 < x_M < \Delta x/2$ and promote sliding for $\Delta x/2 < x_M < \Delta x$. The average value of this force would be zero, resulting in no friction.

A more realistic construction of the friction force microscope is shown in Fig. 11.4b. In this "realistic friction microscope", the SFM-like tip-spring assembly is *elastically* coupled to the suspension in the horizontal direction, so that the horizontal tip position x_t may differ from x_M. While the equilibrium height of the tip, $z_{t, min}(x_t)$, is independent of the scan direction in this model, a "snapping motion" leading to friction can still occur in this instrument, specifically in the case of a soft horizontal spring and a strongly corrugated tip–substrate potential. In the following, I will discuss the conditions for the onset of friction in this model more quantitatively.

For a given x_t, the SFM tip experiences a potential $V(x_t, z_t) = V_{int}(x_t, z_t) + F_{ext} z_t$ consisting of the tip–surface interaction V_{int} and the work against F_{ext}. The tip trajectory $z_{t, min}(x_t)$ during the surface scan is given by the minimum of $V(x_t, z_t)$ with respect to z_t. For this trajectory, $V(x_t) = V(x_t, z_{t, min})$ represents an effective tip–substrate potential. This potential $V(x_t)$ depends strongly on F_{ext} and is corrugated with the periodicity of the substrate due to variations of the chemical bond strength and of $z_{t, min}$, as shown in Fig. 11.6a. The corrugation of the potential $V(x_t)$ will elongate or compress the horizontal spring from its equilibrium which corresponds to $x_t = x_M$. The "instantaneous friction force" is given by

$$F_f(x_M) = -c(x_t - x_M) \ , \tag{11.4}$$

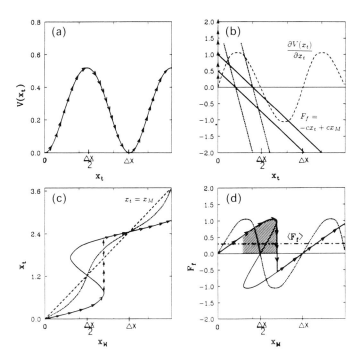

Fig. 11.6. Microscopic friction mechanism in the "realistic friction microscope". Results for a soft spring, giving nonzero friction, are compared to a zero-friction microscope with a stiff spring. (**a**) Potential energy of the tip $V(x_t)$. (**b**) Graphical solution of (11.7) yielding the equilibrium tip position at the intersection of the derivative of the potential $\partial V(x_t)/\partial x_t$ (*dashed line*) and the force due to the horizontal spring F_f. Solid lines, for different values of x_M, correspond to a soft spring and dotted lines correspond to a stiff spring. (**c**) The calculated equilibrium tip position $x_t(x_M)$. (**d**) The friction force F_f as a function of the FFM position x_M and the average friction force $\langle F_f \rangle$ for the soft spring (*dash-dotted line*) [11.13]

where c is the horizontal spring constant. The total potential energy V_{tot} of the system consists of $V(x_t)$ and the energy stored in the horizontal spring,

$$V_{tot}(x_t, x_M) = V(x_t) + \tfrac{1}{2} c(x_t - x_M)^2 \ . \tag{11.5}$$

For a given horizontal position x_M of the FFM suspension, the equilibrium position of the tip x_t is obtained by minimizing V_{tot} with respect to x_t. We obtain

$$\frac{\partial V_{tot}}{\partial x_t} = \frac{\partial V(x_t)}{\partial x_t} + c(x_t - x_M) = 0 \tag{11.6}$$

or, with (11.4),

$$F_f = -cx_t + cx_M = \frac{\partial V(x_t)}{\partial x_t} \ . \tag{11.7}$$

A graphical solution of (11.7) is shown in Fig. 11.6b and the resulting relation $x_t(x_M)$ is shown in Fig. 11.6c.

If, for a stiff spring, the force constant c exceeds the critical value $c_{crit} = -[\partial^2 V(x_t)/\partial x_t^2]_{min}$, we obtain a single solution x_t for all x_M. This situation is indicated by the dotted line in Figs. 11.6b, c. The friction force F_f is given by (11.4) and shown by the dotted line in Fig. 11.6d. Since F_f is independent of the scan direction, it is conservative, resulting in $\langle F_f \rangle = 0$. Consequently, we expect no friction to occur during a nondestructive surface scan in a standard SFM which we can consider as a limiting case of the instrument shown in Fig. 11.4b for $c \to \infty$.

A more interesting case arises if the horizontal spring is soft, $c < c_{crit}$. This situation is represented by the solid line in Figs. 11.6b, c. In this case, the solution $x_t(x_M)$ of (11.7) displays a sequence of instabilities. These instabilities lead to a stick–slip motion of the tip as x_M increases, similar to "plucking a string". The hysteresis in the $x_t(x_M)$ relation (Fig. 11.6c) results in a dependence of the force F_f on the scan direction. The friction force $F_f(x_M)$ in this case is shown by the solid line in Fig. 11.6d. It is a nonconservative/dissipative force and averages to a non-zero value of $\langle F_f \rangle$, given by the dash-dotted line. The energy released from the elongated spring into heat is represented by the shaded area in Fig. 11.6d.

The present theory predicts occurrence of friction only for very soft springs or a strongly corrugated potential $V(x_t)$. The latter fact can be verified experimentally since the corrugations $\Delta V(x_t)$ increase strongly with increasing applied load [11.9]. Consequently, for a given c, the friction force is zero unless a minimum load F_{ext} is exceeded. On the other hand, for given F_{ext}, no friction can occur if c exceeds the critical value $c_{crit}(F_{ext})$.

A similar situation occurs during sliding between large commensurate flat surfaces of A on B. In that case, c is given by the elastic constants of A at the interface [11.12], hence can not be changed independently. Since c is rather large in many materials, zero friction should be observed for moderate applied loads in the absence of wear and plastic deformations. For a multi-atom "tip" which is *commensurate* with the substrate, the tip–substrate potential is proportional to the number of tip atoms at the interface, n, as is the critical value c_{crit} for nonzero friction. In this case, the effective FFM spring depends both on the external spring and the elastic response of the tip material. The inverse value of c_{crit} is given by the sum of the inverse values of the corresponding spring constants. For a large tip which is *incommensurate* with the substrate, no friction should occur [11.3].

The average friction force $\langle F_f \rangle$ as a function of the load F_{ext} and the force constant c is shown as a contour plot in Fig. 11.7. Clearly, the applicable load range is limited by the underlying assumption of contact without wear. This figure illustrates that not only the friction force F_f, but also the friction coefficient $\mu = \langle F_f \rangle / F_{ext}$, depend strongly both on the interaction potential between the two materials in contact and on the intrinsic force constant c of the friction force microscope. This clearly makes the friction force dependent on the

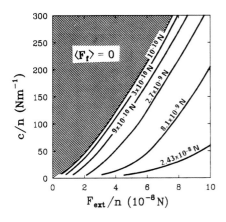

Fig. 11.7. Contour plot of the average friction force $\langle F_f \rangle$ as a function of the load F_{ext} and the force constant c. The calculations are for a monatomic (or a larger commensurate n-atom) Pd tip on graphite. The force constant and all forces are normalized by the number of tip atoms n in contact with the substrate [11.13]

construction parameters of the FFM. There is also one advantage in this fact – c can be chosen in such a way that non-zero friction occurs even at small loads F_{ext}.

11.2.2 Rolling Friction

We have seen that sliding friction is related to a hysteresis in the $x_t(x_M)$ relationship depicted in Fig. 11.6c. As I will discuss in the following, *rolling friction* is related to a hysteresis in the $z_t(z_M)$ relationship occurring during a vertical approach of an SFM tip to a surface. The corresponding instability has briefly been discussed in Chapter 4 (Volume I, pages 108 ff) and is illustrated

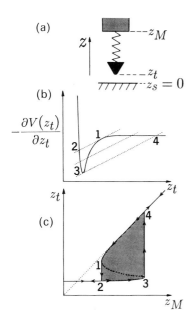

Fig. 11.8. (a) Schematic picture of a scanning force microscope. **(b)** Force on the tip resulting from the interaction with the substrate (*solid line*) and from the elongated spring (*dotted line*). **(c)** Equilibrium tip position z_t as a function of the microscope suspension z_M. The hatched area is proportional to the energy dissipated during one approach-retraction cycle

in Fig. 11.8. The cantilever can be conveniently represented by a vertical spring with a force constant c (Fig. 11.8a). The substrate-induced force on the tip F_t is compensated by the elongated spring,

$$F_t(z_M) = -c(z_t - z_M) .$$ (11.8)

Here, z_M is the equilibrium tip position in the absence of the substrate. The total potential energy V_{tot} of the system consists of the tip–substrate interaction $V(z_t - z_s) = V(z_t)$ (note that the topmost substrate layer is at $z_s = 0$) and the energy stored in the vertical spring,

$$V_{tot}(z_t, z_M) = V(z_t) + \tfrac{1}{2} c(z_t - z_M)^2 .$$ (11.9)

For a given position z_M of the tip suspension, the equilibrium position of the tip z_t is obtained by minimizing V_{tot} with respect to z_t. We obtain

$$\frac{\partial V_{tot}}{\partial z_t} = \frac{\partial V(z_t)}{\partial z_t} + c(z_t - z_M) = 0$$ (11.10)

or, with (11.8),

$$F_t = -cz_t + cz_M = \frac{\partial V(z_t)}{\partial z_t} .$$ (11.11)

A graphical solution of (11.11) is exhibited in Fig. 11.8b and the resulting relation $z_t(z_M)$ is displayed in Fig. 11.8c.

Let us consider a tip approaching the substrate from $z \to \infty$. The force on the tip F_t will be zero first, resulting in $z_t = z_M$, as shown in Fig. 11.8b, c. As the tip slowly approaches the point labeled "1", a deflection towards the surface occurs due to the attractive tip–substrate interaction. At the point labeled "1", the spring can no longer compensate the strong tip–substrate attraction, and the tip jumps to a point labeled "2". The resulting kinetic energy of the tip is dissipated into heat or plastic deformations at the tip–substrate interface. Should the microscope suspension z_M approach the substrate further, the tip will first experience a weakly attractive, then a repulsive interaction with the substrate and will undergo no instabilities. Upon retracting the microscope suspension z_M, the tip will first probe the strongly attractive part of the potential between points "2" and "3". At point "3", the force in the stretched spring can no longer be compensated by the tip–substrate attraction, and the tip jumps to the point labeled by "4". Upon further retracting the microscope suspension z_M, the tip will experience a decreasing attractive interaction from the substrate, resulting in $z_t \to z_M$ with no further instabilities. This is quite analogous to the results presented in Fig. 4.6 for the case of a stationary tip suspension z_M and a moving sample.

Since according to (11.8), $F_t = -c(z_t - z_M)$ is the force on the tip, the hatched area in Fig. 11.8c is proportional to the energy dissipated during an

(a)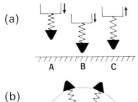

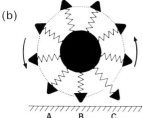
(b)

(c)

Fig. 11.9. Approach-retraction cycle of (**a**) a single SFM tip and (**b**) an array of tips attached to a cylinder rolling on a substrate. The relation of these models to rolling friction between a solid cylinder and a substrate, shown in (**c**), is discussed in the text

approach-retraction cycle of an SFM tip. This A–B–C–A cycle, which is illustrated in Fig. 11.9a for a single SFM tip, occurs during a revolution by $\Delta\phi$ of a rolling cylinder which has a set of SFM tips attached to the surface, as shown in Fig. 11.9b. The latter is, as a matter of fact, a reasonable model for the energy dissipation at the surface of a solid cylinder, once we associate SFM tips with surface atoms. The corresponding cylinder, which can be imagined as covered by sticky tape and rolling on a substrate, is shown in Fig. 11.9c. The analogy between the cylinder and the SFM is based on the fact that the surface layer of the cylinder interacts with the substrate by Lennard–Jones type potentials, and is kept in place by harmonic forces, same as the SFM tip in Fig. 11.8a.

We conclude that the microscopic origin of *rolling friction* without wear lies in the hysteresis in the $z_t(z_M)$ relation in Fig. 11.8c. Consequently, the microscopic mechanism which transfers macroscopic rotational energy from the cylinder into microscopic degrees of freedom (heat) can be studied quantitatively by measuring the energy dissipated during a single approach-retraction cycle of an SFM tip.

11.3 Predictive Calculations of the Friction Force

11.3.1 Tip–Substrate Interactions in Realistic Systems: Pd on Graphite

As I discussed above, a quantitative study of sliding or rolling friction is possible, once the interaction potential V at the interface between material A and material

B is known accurately enough. In this section, I want to illustrate how for a model system, namely a Pd FFM tip interacting with a graphite substrate, this interaction can be calculated from first principles.

Graphite is an ideal substrate for the study of friction since the binding energy of adsorbed atoms and the variations thereof along the surface are negligibly small near the equilibrium adsorbate–substrate separation. This is specifically true for the Pd-graphite interaction, which is depicted in Fig. 11.10a for the on-top (T) and the sixfold hollow (H) site as a function of the Pd-graphite separation [11.9]. The adsorption energies E_{ad} of Pd atoms (representing the Pd monolayer) in the two adsorption sites on graphite have been defined as $E_{ad} = E_{total}(Pd/graphite) - E_{total}(Pd) - E_{total}(graphite)$. The first-principles total energy calculations for this system have been performed using the Density Functional Formalism within the Local Density Approximation (LDA) [11.17] and the *ab initio* pseudopotential local orbital method [11.18]. The details of the calculation have been discussed in [11.9]. The surface of hexagonal graphite has been represented by a 4-layer slab and the adsorbate by a monolayer of Pd atoms in registry with the substrate (1 Pd atom per surface Wigner–Seitz cell of graphite). The valence charge density of the system is shown in Fig. 11.10b.

For a realistically large and sharp SFM tip, the tip–substrate interaction is likely to be modified by the long-range van der Waals force which is not reproduced correctly by LDA. The van der Waals force between an extended conical tip and a flat surface is estimated using the expression

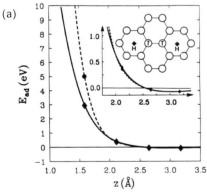

(a)

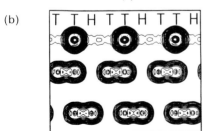

(b)

Fig. 11.10. (a) Pd adsorption energy E_{ad} as a function of the adsorption height z above the surface of hexagonal graphite. The solid lines connecting the data points given by ◆ are for the sixfold hollow (H) sites, and the dashed lines connecting the data points given by ● are for the on-top (T) sites. An enlarged section of the graph near equilibrium adsorption is shown in the inset. A second inset shows the adsorption geometry and a possible trajectory of the Pd layer along x in top view. (b) Valence charge density of the Pd/graphite system. The results of the LDA calculation are for the on-top adsorption site near the equilibrium adsorption distance z_{eq}, and are shown in the xz-plane perpendicular to the surface. The ratio of two consecutive charge density contours $\varrho(n + 1)/\varrho(n)$ is 1.2 [11.9, 16]

$F_{vdW}(z) = A_H \times \tan^2 \alpha/(6z)$, where α is half the opening angle of the tip cone [11.19]. In this expression, A_H is the Hamaker constant and z is the distance between the conical tip and the surface. As typical values for a metal tip, one can assume $\alpha = 30°$ and $A_H = 3 \times 10^{-19}$ J. For tip–substrate distances $z > 3$ Å, the van der Waals forces are very small, typically $F_{vdW} < 10^{-10}$ N. At smaller distances, these forces can be neglected when compared to the closed-shell and internuclear repulsion which are both described correctly within LDA. Since each of these regions is dominated by only one type of interaction, the total tip–substrate force F_{ext} can be approximated as a superposition of the force described by LDA and the van der Waals force. It turns out that the van der Waals forces are not very important for the interpretation of experimental results, since they do not show atomic resolution [11.20] and are easily compensated in the experiment by adjusting the force on the cantilever which supports the tip.

From Fig. 11.10a we see that the equilibrium adsorption bond strength of ≤ 0.1 eV is very weak and much smaller than the cohesive energy of the tip material (Pd metal) or the graphite substrate, which is an important prerequisite for friction with no wear. Near the equilibrium adsorption height $z \approx 3$ Å, the corrugation of the graphite charge density is negligibly small due to Smoluchowski smoothing [11.21] and the position-dependence of the adsorption bond strength is $\ll 0.1$ eV [11.22], which should result in a very small friction coefficient. This calculation indicates that at bond lengths $z \leq 2$ Å, the hollow site is favored with respect to the on-top site. At $z \approx 2$ Å, the adsorption energies are nearly the same and, at larger distances, it is the on-top site which is slightly favored by < 0.05 eV. This is consistent with the dominant interaction changing from closed-shell repulsion (which strongly favors the hollow site at very small adsorption bond lengths) to a weak chemisorption bond (which is stabilized by the hybridization with p_z orbitals in the on-top site). As I will discuss later, the change of the preferential adsorption site from the on-top to the hollow site as a function of the applied load leads to an anomaly in the friction coefficient μ.

A prerequisite for the calculation of the friction force is the precise knowledge of the Pd adsorption energy $E_{ad}(x, z)$ along the whole graphite surface. Since the corresponding *ab initio* calculations are computationally very expensive, it is useful to find simpler ways to determine this quantity. Several simple potentials have been used for this purpose so far [11.23–25]. Since small inaccuracies in the interaction potentials have a large effect on the friction force, a safer way is to parametrize existing LDA results in a way which makes an evaluation of E_{ad} very easy everywhere. This can be achieved by approximating E_{ad} by a local function which depends only on the total charge density of the graphite host at the Pd adsorption site [11.26],

$$E_{ad}(\mathbf{r}) = E_{ad}(\varrho(\mathbf{r})) \ . \tag{11.12}$$

This form of the interaction potential is inspired by the density functional formalism [11.17] and the embedded atom method [11.27] and hence is

282   D. Tománek

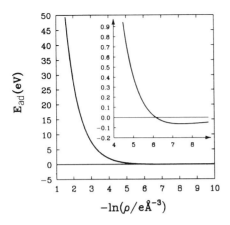

Fig. 11.11. Relation between the Pd adsorption energy $E_{ad}(r)$ and the total charge density of graphite $\varrho(r)$ at the adsorption site r, given by (11.13). An enlarged section of the graph near equilibrium adsorption is given in the inset [11.26]

expected to be quite general, not restricted to the Pd-graphite system. A convenient parametrization of $E_{ad}(\varrho(r))$ is

$$E_{ad}(\varrho(r)) = \varepsilon_1(\varrho/\varrho_0)^{\alpha_1} - \varepsilon_2(\varrho/\varrho_0)^{\alpha_2} \ . \tag{11.13}$$

In the case of Pd on graphite, $\varepsilon_1 = 343.076$ eV, $\varepsilon_2 = 2.1554$ eV, $\alpha_1 = 1.245$, $\alpha_2 = 0.41806$, and $\varrho_0 = 1.0\,e/\text{Å}^3$. The dependence of E_{ad} on ϱ, obtained using the parametrized form in (11.13), is depicted in Fig. 11.11.

In many cases, the total charge density can be well approximated by a superposition of atomic charge densities,

$$\varrho(r) = \sum_n \varrho_{at}(r - R_n) \ . \tag{11.14}$$

This parametrization is especially convenient in case of deformed surfaces where an LDA calculation is difficult due to reduced symmetry. On flat graphite

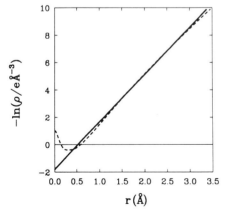

Fig. 11.12. Radial plot of the charge density of a carbon atom $\varrho_{at}(r)$, based on LDA (*dashed line*). The solid line shows the parametrized form of the charge density, given by (11.15) [11.26]

surfaces, the maximum difference between the LDA charge density and the superposition of atomic charge densities is only a few percent.

Finally, the LDA charge density of the substrate (carbon) atoms can be conveniently expressed as

$$\varrho_{at}(r) = \varrho_C e^{-\beta r} , \tag{11.15}$$

where $\varrho_C = 6.0735 \, e/\text{Å}^3$ and $\beta = 3.459 \, \text{Å}^{-1}$. As shown in Fig. 11.12, this fit represents the LDA results very well.

11.3.2 Atomic-Scale Friction in Realistic Systems: Pd on Graphite

In the following, I will discuss the atomic-scale friction between a monatomic Pd FFM tip and graphite, based on the "maximum friction microscope" model illustrated in Fig. 11.4a. I will show how the interaction potential between a Pd monolayer and graphite (Pd in registry with the substrate), which has been given in Sect. 11.3.1, can be used to estimate the friction force (and friction coefficient μ) as a function of the applied load.

For a microscopic understanding of the friction process, let us first consider the motion of the Pd layer along the graphite surface, under the influence of an external load per atom [11.28] f_{ext} which is normal to the surface. Following [11.9], we consider a straight trajectory along the surface x direction connecting nearest neighbor sixfold hollow sites on graphite, which are separated by Δx and connected by a bridge site.

As mentioned in the description of the "maximum friction microscope" in Sect. 11.2.1 and shown in Fig. 11.4a, the horizontal positions of the tip and the microscope suspension are the same, $x_t = x_M = x$. The potential energy V of the system along this trajectory has two main components. The first consists of variations of the tip–surface interaction (or adsorption bond energy) $V_{int}(x, z_t) = E_{ad}(x, z_t)$. The second is given by the work against the external load f_{ext} applied on the apex atom of the tip, due to the variations of the tip–substrate distance (or adsorption bond length). Hence,

$$V(x, f_{ext}) = E_{ad}(x, z_{t, min}(x)) + f_{ext} z_{t, min}(x) - V_0(f_{ext}) . \tag{11.16}$$

Here, the potential energy has been set to zero at the hollow site by defining

$$V_0(f_{ext}) = E_{ad}(x_H, z_{t, min}(x)) + f_{ext} z_{t, min}(x_H) . \tag{11.17}$$

The equilibrium tip height $z_{t, min}(x)$ along the trajectory can be determined from

$$f_{ext} = -\frac{\partial}{\partial z} E_{ad}(x, z) . \tag{11.18}$$

In Fig. 11.13a, $V(x)$ is shown for different external loads. We find that the variations of V are dominated by the mechanical component and only partly compensated by the site-dependence of the adsorption energy. As a result of the

(a)

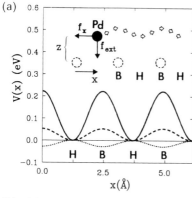

(b)

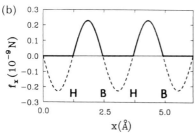

Fig. 11.13. (a) Potential energy $V(x)$ of the Pd-graphite system as a function of the position of the Pd layer along the surface x-direction, for external forces $f_{ext} = 3 \times 10^{-9}$ N (*dotted line*), 6×10^{-9} N (*dashed line*) and 9×10^{-9} N (*solid line*). The inset shows the adsorption geometry and trajectory of the Pd layer in side view. (b) Atomic-scale structure of the force along the surface f_x (*dashed line*) and the friction force $f_f = \max(f_x, 0)$ (*solid line*) for $f_{ext} = 9 \times 10^{-9}$ N [11.9]

variations of V along x, there is a position-dependent force f_x along the x direction. In analogy to (11.3), this force is given by

$$f_x(x, f_{ext}) = \frac{\partial}{\partial x} V(x, f_{ext})$$
(11.19)

and shown in Fig. 11.13b. The maximum value of f_x describes the static friction governing the onset of stick–slip motion. Non-zero average sliding friction comes from the nonconservative "sticking mechanism" of this particular FFM, which is illustrated in Fig. 11.4a and which results in zero horizontal force on the tip for $\Delta x/2 < x < \Delta x$ (see (11.3)). In order to estimate the friction force along the trajectory, we note that the energy loss due to friction W_f along Δx can not exceed the activation energy corresponding to the largest change of V, hence

$$W_f \leq \Delta V_{max} \ .$$
(11.20)

Let us now assume that at Δx, the entire energy stored in the spring gets transferred into surface phonons and electron–hole pairs [11.15], as indicated in Fig. 11.4a. Then, both sides of (11.20) will be equal. The horizontal force on the tip will show atomic-scale structure and will not average to zero, as indicated in Figs. 11.5b and 11.13b. This has been observed recently using the FFM [11.2].

The energy W_f dissipated in friction along the trajectory Δx can be used to define the average friction force $\langle f_f \rangle$ as

$$W_f = \langle f_f \rangle \Delta x \ , \tag{11.21}$$

or, using (11.20) for W_f,

$$\langle f_f \rangle = \frac{1}{\Delta x} \Delta V_{max} \ . \tag{11.22}$$

The friction coefficient μ, defined in (11.1), can now be estimated as

$$\mu = \frac{\langle f_f \rangle}{f_{ext}} = \frac{\Delta V_{max}}{f_{ext} \Delta x} \ . \tag{11.23}$$

In Fig. 11.14, μ is shown as a function of f_{ext}. We find a general increase of μ with increasing external force, in contradiction to the general notion that μ is nearly independent of the load. The minimum in $\mu(f_{ext})$ near $f_{ext} = 5 \times 10^{-9}$ N is caused by the switching of the minima in $V(x)$ from H to B, depicted in Fig. 11.13a [11.29].

The above estimates of μ have been obtained for an infinitely rigid substrate, an assumption which holds only within a limited load range. Theoretical results of [11.16, 30], which will be summarized in Sect. 11.4, indicate that if the external force (per atom) exceeds 10^{-8} N, the graphite surface is very strongly deformed [11.31] and likely to be ruptured [11.6]. Since no plastic deformations have been observed in the SFM/FFM studies [11.2], the applied forces were probably in the region $f_{ext} < 10^{-8}$ N. For these values of f_{ext}, the above results for the friction coefficient of $\mu \approx 10^{-2}$ agree with the experimental value [11.2,6].

In order to obtain a meaningful comparison with observable friction forces, we have to make further assumptions about the macroscopic tip–substrate

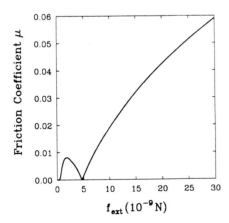

Fig. 11.14. Microscopic friction coefficient μ as a function of the external force per atom f_{ext} [11.9]

interface and the elastic response of the substrate to external forces. In the simplest case, we consider an atomically flat interface, where n atoms are in contact with the substrate, and neglect elastic deformations. Then, the external force per atom f_{ext} is related to the total external force F_{ext} by

$$f_{ext} = \frac{1}{n} F_{ext} \;. \tag{11.24}$$

In Fig. 11.15 we use the calculated $\mu(f_{ext})$ to plot the total friction force F_f for such a perfectly flat interface consisting of 1500 Pd atoms. Since μ increases with increasing value of f_{ext}, the F_f versus F_{ext} relationship is non-linear, which has also been observed in the SFM experiment [11.2].

At this point, it is important to address the validity range of the above calculations. Obviously, the implicit assumption of an infinitely rigid substrate will remain realistic only for limited loads F_{ext} applied on the tip, especially in the case of graphite. Also, there is a critical load for each system which marks the onset of plastic deformations within either the substrate or the tip. The latter point will be discussed in more detail in the following section. Here, I would like to discuss a simple modification of the above results for friction in the case of an extended tip and an elastic substrate.

In the case of large external forces and an elastic substrate such as graphite, linear elasticity theory predicts [11.32] the substrate deformations to be proportional to $F_{ext}^{1/3}$. In the case of a spherical tip [11.2], the tip–substrate interface area and the corresponding number of atoms in contact at the interface are proportional to $F_{ext}^{2/3}$. Then, the force per atom f_{ext} is proportional to $F_{ext}^{1/3}$. Hence for increasing external forces, variations of the effective force per atom and of μ are strongly reduced due to the increasing interface area. This is illustrated by the dashed line in Fig. 11.15, which is based on the assumption that $n = 1500$ tip atoms are in contact with the substrate at $F_{ext} = 10^{-6}$ N. These results are in good agreement with the FFM results for a large nonspecific tungsten tip with a radius $R = 1500$ Å–3000 Å on graphite [11.2], but show

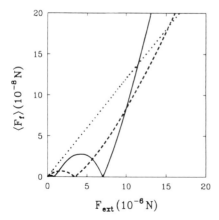

Fig. 11.15. Macroscopic friction force F_f as a function of the external force F_{ext} for a large object. The solid line describes a "flat" object, the surface of which consists of 1500 atoms in contact with a rigid substrate. The dashed line describes a large spherical tip and also considers the effect of elastic substrate deformations on the effective contact area. The dotted line corresponds to a constant friction coefficient $\mu = 0.012$ [11.9]

a slightly larger increase of the friction force than that observed for the range of external forces investigated.

It is instructive to discuss some consequences of the above theory for atomic-scale friction. At rough interfaces, the friction coefficient μ generally increases due to the onset of plastic deformations or wear associated with bond breaking at the interface. It is interesting to note that even in the case of no wear, the above theory would predict an increased value of μ at a rough interface, since the number of atoms n in contact with the substrate is smaller in that case, which would lead to an increase of f_{ext} and hence of μ. Also, with increasing relative velocity between the bodies in contact, the coupling between macroscopic and internal microscopic degrees of freedom (phonons, electron–hole pairs) gets less efficient. Then, $W_f < \Delta V_{max}$ in (11.20), which should result in a decrease of $\langle f_f \rangle$ and μ.

11.4 Limits of Non-destructive Tip–Substrate Interactions in Scanning Force Microscopy

The operating load range of the SFM (and hence the FFM) is limited in two ways. If the applied load on the tip is too small, atomic-scale modulations of the tip–substrate distance z_t and of the horizontal force on the tip f_x will lie below the detection limit. If the load is too large, the substrate and/or the tip will be destroyed. The optimum operating range can be predicted, once the tip–substrate interaction potentials and the elastic properties of the tip and the substrate are known.

Some of these questions have been addressed previously in calculations of the interaction between an infinite "periodic" carbon or aluminium tip and a rigid surface [11.33]. Other calculations have considered the interaction between a single SFM tip and an elastic surface represented by a semi-infinite continuum [11.30] or by a model system of finite thickness [11.25, 34]. In the following, I will show, how this optimum load operating range can be determined in a parameter-free calculation. The numerical results will be for a monatomic and a multi-atom Pd SFM tip interacting with graphite, a system which has been discussed in [11.16].

I will start with a discussion of the minimum load required to observe atomic-scale features in SFM images and/or non-zero friction (Fig. 11.7) on a *rigid* substrate. Figure 11.16a shows the expected SFM corrugation Δz during a horizontal xy scan of the graphite surface by a monatomic Pd tip, for $f_{ext} = 10^{-8}$ N. The top view of the geometry is shown schematically in the inset of Fig. 11.16b for a monatomic Pd tip (left) and a three-atom Pd tip (right) at the hollow ("H") site. The hatched cricles correspond to Pd atoms at the tip apex.

Figure 11.16b shows the SFM corrugation $\Delta z(x)$ for different loads f_{ext}. The tip trajectory along the surface x-direction, shown by arrows in the inset of Fig. 11.16b, contains the "T" and "H" sites and yields the largest corrugation. As

288   D. Tománek

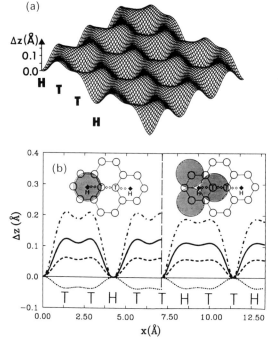

Fig. 11.16. (a) Surface corrugation Δz experienced by a monatomic Pd SFM tip scanning the xy surface plane of rigid graphite under the applied SFM load (per atom) $f_{ext} = 10^{-8}$ N. **(b)** Δz (with respect to the "H" site) along the surface x direction for a "sharp" 1-atom tip *(left)* and a "dull" 3-atom tip *(right)*. The applied loads are $f_{ext} = 10^{-9}$ N *(dotted line)*, 5×10^{-9} N *(dashed line)*, 10^{-8} N *(solid line)*, and 2×10^{-8} N *(dash-dotted line)*. The inset shows the geometry of the tip-graphite system in top view. The SFM tip is shown above the hollow site and the shaded area represents Pd atoms [11.16]

discussed in Sect. 11.3.1, the favored surface site changes with changing load. For the sake of simple comparison, Δz (hollow site) is set to zero in Fig. 11.16b, which gives a sign change of Δz near $f_{ext} = 2.5 \times 10^{-9}$ N. This calculation indicates that in order to achieve observable atomic resolution in the SFM in the constant force mode, which requires $\Delta z \gtrsim 0.05$ Å, the load on the tip must be $f_{ext} \gtrsim 5 \times 10^{-9}$ N. Since the corrugation Δz along a trajectory connecting adjacent "T" sites is very small (Fig. 11.16b), the observation of individual carbon atoms is unlikely, as has been confirmed by experiment [11.35]. Depending on the force constant c in the "realistic friction microscope", a similar minimum load is expected for the onset of non-zero sliding friction (Fig. 11.7) during the scan.

For an n-atom tip which is commensurate with the substrate, the average load per SFM tip atom is $f_{ext} = F_{ext}/n$ and the equilibrium tip height $z_{t, min}$ can be estimated using (11.18). It should be noted that, under certain conditions, such a "dull" multi-atom tip can still produce atomic corrugation for f_{ext} similar to a monatomic tip. This is the case for an ideally aligned tip with a close-packed (111) surface, since the unit cells of Pd and graphite are nearly identical in this case, as shown in the inset of the right-hand side Fig. 11.16b. Thus, for an incompressible substrate, the "sharpness" of the SFM tip need not play a decisive role in the resolution. Of course, a strong suppression of height corrugations and the (horizontal) friction force is expected in the case of large tips which are incommensurate with the substrate. This has been discussed in [11.3] and can be understood intuitively in the simple-minded haircomb model of sliding friction in Fig. 11.1a.

The range of applicable loads F_{ext} is limited by the condition that substrate distortions near the SFM tip should remain in the elastic region. Since full-scale LDA calculations of local SFM-induced distortions of a semi-infinite graphite surface are practically not feasible, the following approach has proven to be quite useful. For small loads, the relaxation of carbon atoms at the graphite surface due to the SFM tip can be determined using continuum elasticity theory [11.36], with elastic constants obtained from *ab initio* calculations [11.30]. This continuum approach is applicable in the linear response regime and has been successfully used previously to calculate local rigidity, local distortions and the healing length of graphite near an SFM tip and near intercalant impurities [11.30,36]. In a second step, the total charge density of the distorted graphite substrate is reconstructed from a superposition of atomic charge densities (see (11.14)) which can be calculated by LDA. Based on this total charge density, the tip–substrate interaction and the equilibrium tip position for a given applied load can be obtained using (11.12).

In the continuum elasticity calculation, the semi-infinite system of graphite layers is characterized by the interlayer spacing d, the in-plane C–C bond length d_{C-C}, the flexural rigidity D, the transverse rigidity K (proportional to C_{44}) and c-axis compressibility G (proportional to C_{33}) [11.30,36]. The LDA calculations for undistorted graphite yield $d = 3.35$ Å and $d_{C-C} = 1.42$ Å, in excellent agreement with experimental and previous theoretical results [11.37,38]. The continuum calculation is further based on the elastic constants $D = 7589$ K, $K = 932$ K Å^{-2} and $G = 789$ K Å^{-4} which have been obtained from calculated graphite vibration modes [11.30] and the experiment [11.37]. The continuum elasticity theory, as applied to SFM experiments on graphite, is discussed in more detail in Chap. 10.

The total charge density of the graphite surface, distorted by a Pd SFM tip, is shown in Fig. 11.17. A comparison of charge density contours with results of the self-consistent calculation in Fig. 11.10b proves a posteriori the applicability of the linear superposition of atomic charge densities. Fig. 11.17a shows that the substrate distortions in response to a monatomic SFM tip at a load $F_{ext} = f_{ext} = 5 \times 10^{-9}$ N are already substantial, while the corrugations of $\Delta z \approx 0.06$ Å, shown in Fig. 11.16b, are marginally detectable. For larger applied loads $F_{ext} = f_{ext} > 5.0 \times 10^{-9}$ N, which would lead to sizeable corrugations in Δz, the local distortions of graphite are very large. Even though linear response theory is not applicable in this force range, the estimated substrate distortions, shown in Fig. 11.17b, indicate the possibility of tip-induced *plastic* deformations in this load range. As shown in Fig. 11.17c, similar large (and possibly plastic) deformations are expected for a tip with a multi-atom apex, even though the load per atom f_{ext} may be relatively small.

In general, we expect plastic deformations to occur whenever the interlayer distance in graphite approaches the value of the intralayer C–C distance. According to linear response theory, this occurs for $f_{ext} \gtrsim 5 \times 10^{-9}$ N. Under

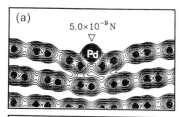

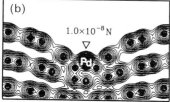

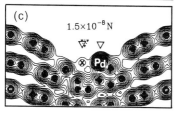

Fig. 11.17. Total charge density ϱ of a Pd SFM tip interacting with the elastic surface of graphite near the hollow site. Contours of constant ϱ are shown in the xz-plane perpendicular to the surface. The results are for loads **(a)** $F_{ext} = f_{ext} = 5 \times 10^{-9}$ N and **(b)** $F_{ext} = f_{ext}$ 1×10^{-8} N applied on a monatomic tip, and **(c)** for $F_{ext} = 3f_{ext} = 1.5 \times 10^{-8}$ N applied on a 3-atom tip. The ratio of two consecutive charge density contours $\varrho(n + 1)/\varrho(n)$ is 1.4. The location of the applied load acting on the Pd atoms at the apex of the tip is indicated by ∇ [11.16]

these conditions, an irreversible rehybridization of carbon orbitals from graphitic sp^2 to diamond-like sp^3 bonding is likely to occur below the tip apex. An estimate of the critical SFM force for this plastic deformation, which does not rely on the continuum elasticity theory, has been obtained in a first-principles calculation [11.39] of the graphite–diamond transition as a function of external pressure along the graphite c axis. These results, corresponding to an "infinitely extended tip", indicate a critical force per surface atom of $f_{ext} = 10^{-9}$ N for this transition. Due to the large flexural rigidity of graphite, this force increases by half an order of magnitude for a one-atom tip, in agreement with the value quoted above.

A realistic SFM tip is more complex than the model tip discussed above and could consist of a micro-tip of one or few atoms on top of a larger tip. A substantial portion of this larger tip could, through the "cushion" of a contamination layer [11.31] or a graphite flake [11.40, 41], distribute the applied load more evenly across a large substrate area, reduce the large curvature near the tip (Fig. 11.17) and increase the minimum separation between graphite layers. This effect would increase the upper limit of applicable loads f_{ext} compatible with elastic substrate deformations and make nondestructive imaging of surface topographic features and of sliding friction on the atomic scale possible [11.35].

Acknowledgements. I thank Professors H. Thomas, J. Sethna and H.-J. Güntherodt, as well as Dr. G. Overney, for stimulating discussions. Partial support by the Office of Naval Research under contract No. N00014-90-J-1396 is acknowledged.

References

11.1 A summary of the present knowledge in tribology can be found in Adhesion and Friction, ed. by M. Grunze, H.J. Kreuzer: Springer Ser. in Surf. Sci., Vol. 17 (Springer, Berlin Heidelberg 1989)

11.2 C. Mathew Mate, Gary M. McClelland, Ragnar Erlandsson, Shirley Chiang: Phys. Rev. Lett. **59**, 1942 (1987)

11.3 G.M. McClelland: In [11.1]

11.4 J. Spreadborough: Wear **5**, 18 (1962)

11.5 R.D. Arnell, J.W. Midgley, D.G. Teer: Proc. Inst. Mech. Eng. **179**, 115 (1966)

11.6 J. Skinner, N. Gane, D. Tabor: Nat. Phys. Sci. **232**, 195 (1971)

11.7 G. Binnig, C.F. Quate, Ch. Gerber: Phys. Rev. Lett. **56**, 930 (1986): ibid. Appl. Phys. Lett. **40**, 178 (1982); G. Binnig, Ch. Gerber, E. Stoll, T.R. Albrecht, C.F. Quate: Europhys. Lett. **3**, 1281 (1987)

11.8 Uzi Landman, W.D. Luedtke, A. Nitzan: Surf. Sci. **210**, L177 (1989); Uzi Landman, W.D. Luedtke, M.W. Ribarsky: J. Vac. Sci. Technol. A **7**, 2829 (1989)

11.9 W. Zhong, D. Tománek: Phys. Rev. Lett. **64**, 3054 (1990)

11.10 L.D. Landau, E.M. Lifshitz: Course of Theoretical Physics, Vol. 1: Mechanics (Pergamon, Oxford 1960) p. 122

11.11 The optimum selection of C is characterized by strong A–C and B–C bonds, while C itself should have a low shear modulus.

11.12 G.A. Tomlinson: Phil. Mag. S. 7, Vol. 7, 905 (1929)

11.13 D. Tománek, W. Zhong, H. Thomas: Europhys. Lett. **15**, 887 (1991)

11.14 E. Meyer, R. Overney, L. Howald, D. Brodbeck, R. Lüthi, H.-J. Güntherodt: In *Fundamentals of Friction*, ed. by I. Singer, H. Pollock: NATO-Advanced Study Institute Series (Kluwer, Dordrecht 1992)

11.15 J.E. Sacco, J.B. Sokoloff, A. Widom: Phys. Rev. B **20**, 5071 (1979)

11.16 W. Zhong, G. Overney, D. Tománek: Europhys. Lett. **15**, 49 (1991)

11.17 W. Kohn, L.J. Sham: Phys. Rev. **140**, A1133 (1965)

11.18 C.T. Chan, D. Vanderbilt, S.G. Louie: Phys. Rev. B **33**, 2455 (1986); C.T. Chan, D. Vanderbilt, S.G. Louie, J.R. Chelikowsky: Phys. Rev. B **33**, 7941 (1986)

11.19 M. Anders, C. Heiden: (submitted for publication)

11.20 C. Horie, H. Miyazaki: Phys. Rev. B **42**, 11757 (1990)

11.21 R. Smoluchowski: Phys. Rev. **60**, 661 (1941)

11.22 This effect is also responsible for the large value of the surface diffusion constant on graphite and apparent small sensitivity of surface friction to adsorbed films, as discussed in [11.6]

11.23 J.B. Pethica, W.C. Oliver: Physica Scripta T **19**, 61 (1987); I.P. Batra, S. Ciraci: Phys. Rev. Lett. **60**, 1314 (1988)

11.24 S. Gould, K. Burke, P.K. Hansma: Phys. Rev. B **40**, 5363 (1989)

11.25 F.F. Abraham and I.P. Batra: Surf. Sci. **209**, L125 (1989)

11.26 D. Tománek, W. Zhong: Phys. Rev. B **43**, 12623 (1991)

11.27 M.S. Daw, M.I. Baskes: Phys. Rev. B **29**, 6443 (1984)

11.28 In the following, I denote microscopic forces per atom by f and macroscopic forces applied on large objects by F

11.29 Near $f_{ext} = 5 \times 10^{-9}$ N, the potential V is essentially constant across the graphite surface, which leads to $\mu \approx 0$ (Fig. 11.14). Small deviations from this behavior can arise due to tip-induced substrate deformations and small inaccuracies in the energy formula given in (11.12)

11.30 D. Tománek, G. Overney, H. Miyazaki, S.D. Mahanti, H.J. Güntherodt: Phys. Rev. Lett. **63**, 876 (1989); ibid. **63**, 1896(E) (1989)

11.31 H.J. Mamin, E. Ganz, D.W. Abraham, R.E. Thomson, J. Clarke: Phys. Rev. B **34**, 9015 (1986)

11.32 L.D. Landau, E.M. Lifshitz: Course of Theoretical Physics, Vol. 7: Theory of Elasticity, (Pergamon, Oxford 1986) p. 53

11.33 S. Ciraci, A. Baratoff, I.P. Batra: Phys. Rev. B **41**, 2763 (1990); I.P. Batra, S. Ciraci: J. Vac. Sci. Technol. A **6**, 313 (1988)

11.34 U. Landman, W.D. Luedtke, N.A. Burnham, R.J. Colton: Science **248**, 454 (1990)

11.35 E. Meyer, H. Heinzelmann, P. Grütter, Th. Jung, Th. Weiskopf, H.-R. Hidber, R. Lapka, H. Rudin, H.-J. Güntherodt: J. Microsc. **152**, 269 (1988)

11.36 S. Lee, H. Miyazaki, S.D. Mahanti, S.A. Solin: Phys. Rev. Lett. **62**, 3066 (1989)

11.37 H. Zabel: In *Graphite Intercalation Compounds*, ed. by H. Zabel, S.A. Solin: Topics in Current Phys. Vol. 14 (Springer, Berlin Heidelberg 1989).

11.38 C.T. Chan, K.M. Ho, W.A. Kamitakahara: Phys. Rev. B **36**, 3499 (1987); C.T. Chan, W.A. Kamitakahara, K.M. Ho, P.C. Eklund: Phys. Rev. Lett. **58**, 1528 (1987)

11.39 S. Fahy, S.G. Louie, M.L. Cohen: Phys. Rev. B **34**, 1191 (1986)

11.40 J.B. Pethica: Phys. Rev. Lett. **57**, 3235 (1986)

11.41 G. Overney, D. Tománek, W. Zhong, Z. Sun, H. Miyazaki, S.D. Mahanti, H.J. Güntherodt: J. Phys.: Cond. Mat. **4**, 4233 (1992)

12. Theory of Non-contact Force Microscopy

U. Hartmann

With 34 Figures

Different forces which may be present in non-contact scanning force microscopy are theoretically analyzed with respect to their typical magnitude and range. It is shown that van der Waals forces provide an ever-present contribution to long-range probe–sample interactions. If a liquid is present in the intervening gap between probe and sample, it is found that ionic double-layer forces may play an important role. If the probe is in very close proximity to the substrate, the discrete structure of intervening liquids may lead to characteristic solvation forces. For liquids being present as thin adsorbed films on top of the substrate, capillary forces turn out to be the source of very strong long-range probe–sample interactions.

12.1 Methodical Outline

A general theory concerning the long-range probe–sample interactions effective in non-contact Scanning Force Microscopy (SFM), i.e., at probe–sample separations well beyond the regime of overlap of the electron wave functions, is a rather ambitious project. Even in the absence of externally applied electro- and magnetostatic interactions the approach has to account for various inter-molecular and surface forces which are, however, ultimately all of electromagnetic origin. Figure 12.1 gives a survey of the different components which generally contribute to the total probe–sample interaction.

In the absence of any contamination on probe and sample surface, i.e., under clean UHV conditions, an ever-present long-range interaction is provided by van der Waals forces. In this area theory starts with some well-known results from quantum electrodynamics. In order to account for the typical geometry involved in an SFM, i.e., a sharp probe opposite to a flat or curved sample surface, the Derjaguin geometrical approximation is used which essentially reduces the inherent many-body problem to a two-body approach.

Under ambient conditions surface contaminants, e.g., water films, are generally present on probe and sample. Liquid films on solids often give rise to a surface charging and thus to an electrostatic interaction between probe and sample. The effect of these ionic forces is treated by classical Poisson–Boltzmann statistics, where the particular probe–sample geometry is again accounted for by employing the Derjaguin approximation. If the probe–sample separation is

Springer Series in Surface Sciences, Vol. 29
Scanning Tunneling Microscopy III Eds.: R. Wiesendanger · H.-J. Güntherodt
© Springer-Verlag Berlin Heidelberg 1993

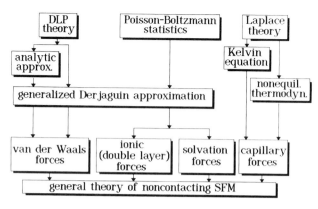

Fig. 12.1. Schematic of the approach toward a general theory of non-contact scanning force microscopy

reduced to a few molecular diameters, liquids can no longer be treated by a pure continuum approach. The discrete molecular structure gives rise to solvation forces which are due to the long-range ordering of liquid molecules in the gap between probe and sample.

Finally, capillary condensation is a common phenomenon in SFM under ambient conditions. In this area the well-known Laplace equation provides an appropriate starting basis. Capillary action is then treated in terms of two extreme approaches: while the first is for liquid films strictly obeying a thermodynamic equilibrium behavior represented by the Kelvin equation, the second approach is for liquids which are actually not in thermodynamic equilibrium.

It must be emphasized that the general situation in non-contacting SFM is governed by a complex interplay of all the aforementioned contributions. The situation is further complicated by the fact that not all of these contributions are simply additive. The following detailed discussion strictly relies on a macroscopic point of view. All material properties involved are treated in terms of isotropic bulk considerations and even properties attributed to individual molecules are consequently deduced from the overall macroscopic behavior of the solids or liquids composed by these molecules. The considerations concerning the presence of liquids in SFM are of course not restricted to "parasitic" effects due to contaminating films but in particular also apply to the situation where the SFM is completely operated in a liquid immersion medium or where just the properties of a liquid film, e.g., of a polymeric layer on top of a substrate, are of interest.

12.2 Van der Waals Forces

12.2.1 General Description of the Phenomenon

Macroscopic van der Waals (VdW) forces arise from the interplay of electromagnetic field fluctuations with boundary conditions on ponderable bodies.

These field fluctuations result from zero-point quantum vibrations as well as from thermal agitation of permanent electronic multipoles and extend well beyond the surface of any absorbing medium – partly as traveling waves, partly as exponentially damped "evanescent" waves. According to this particular picture *Lifshitz* calculated the mutual attraction of two semi-infinite dielectric slabs separated by an intervening vacuum gap [12.1]. Since the Lifshitz "random field approach" involves a solution of the full Maxwell equations rather than of the simpler Laplace equation, retardation effects are accounted for in a natural way. The well-known fundamental results of the *Eisenschitz* and *London* [12.2] and *Casimir* and *Polder* [12.3] theories are obtained as specific cases of this general approach.

Since the VdW interaction between any two bodies occurs through the fluctuating electromagnetic field it stands to reason that the following alternative viewpoint could be developed: as schematically shown in Fig. 12.2 for the typical probe–sample arrangement involved in SFM, the fluctuating electromagnetic field can be considered in terms of a distribution of virtual photons associated with probe and sample. Now, if both come into close proximity, an exchange of these virtual photons occurs, giving rise to a macroscopic force between probe and sample. This alternative viewpoint is actually the basis for a treatment of the problem by methods of quantum field theory. Using the formidable apparatus of Matsubara–Fradkin–Green function technique of quantum statistical mechanics, *Dzyaloshinskii, Lifshitz* and *Pitaevskii* (DLP) rederived the Lifshitz two-slab result and extended the approach to the presence of any intervening medium filling the gap between the dielectric slabs [12.4].

Subsequently, several other approaches to the general problem of electromagnetic interaction between macroscopic bodies, all more or less equivalent, have been developed by various researchers [12.5]. In the present context the most important aspect common to all this work is the following: on a microscopic level the origin of the dispersion forces between two molecules is linked to a process which can be described by the induction of polarization on one due to the instantaneous polarization field of the other. However, this process is seriously affected by a third molecule placed near the two. The macroscopic consequence is that VdW forces are in general highly non-additive. For example, if two perfectly conducting bodies (a perfect conductor may be considered as the

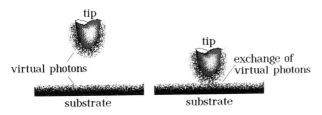

Fig. 12.2. Distribution of virtual photons associated with probe and sample. At close proximity an exchange of virtual photons takes place giving rise to van der Waals interactions

limit of a London superconductor, as the penetration depth approaches zero) mutually interact via VdW forces, only bounding surface layers will contribute to the interaction, while the interiors of the bodies are completely screened. Thus, the interaction can certainly not be characterized by straightforward pairwise summation of isotropic intermolecular contributions, at least not in this somewhat fictitious case.

However, it is precisely the explicit assumption of the additivity of two-body intermolecular pair potentials which is the basis of the classical *Hamaker* approach [12.6]. Granted additivity, the interaction between any two macroscopic bodies which have well-defined geometric shapes and uniform molecular densities can be calculated by a simple double-volume integration. In spite of its apparent limitations the Hamaker approach has the virtue not only of ease in comprehension, but works over a wider range than would at first be thought possible. The most conspicuous result is that the additivity approach yields the correct overall power law dependence of the VdW interaction between two arbitrarily shaped macroscopic bodies on the separation between them. Although not rigorously proved, this appears to hold for the London limit, i.e., for a completely non-retarded interaction, as well as for the Casimir limit, i.e., complete radiation field retardation [12.7]. From the field point of view, the geometrical boundary conditions associated with the SFM's probe–sample arrangement lead to tremendous mathematical difficulties in a rigorous calculation of VdW interactions, especially if retardation is included. Actually, several rather involved mathematical detours by various authors have shown that the key problem of a precise calculation of the magnitude of VdW forces as a function of separation of interacting bodies which exhibit curved surfaces can be solved fairly unambiguously only in some elementary cases involving spherical configurations [12.5].

Due to all the aforementioned difficulties it appears quite clear why a rigorous treatment of VdW interactions in SFM has not yet been presented. On the one hand, field theories are extremely complicated and tend to obscure the physical processes giving rise to the probe–sample forces. On the other hand, although two-body forces generally provide the dominant contribution, the explicit assumption of pairwise molecular additivity of VdW interactions of the many-particle system simply does not hold. The corrections due to many-body effects are generally essential in order to estimate whether the VdW interaction of a given tip–sample arrangement is within or well beyond the experimentally accessible regime.

In what follows below, a treatment of VdW interactions in non-contact SFM is proposed which is based on elements of both the quantum field DLP theory and the Hamaker additivity approach. While some basic results from field theory provide an appropriate starting point, a characterization of material dielectric contributions, and a final analysis of the limitations of the developed framework, the additivity approach allows to account in a practical way, in terms of reasonable approximations, for the particular geometrical boundary conditions involved. In this sense the resulting model can best be referred to as a "renormalized Hamaker approach".

12.2.2 The Two-Slab Problem: Separation of Geometrical and Material Properties

The DLP theory [12.4] gives the exact result for the electromagnetic interaction of two dielectric slabs separated by a third dielectric material of arbitrary thickness:

$$f(z) = -8\pi^2 \frac{kT}{c^3} \sum_{m=0}^{\infty} \left(\frac{1}{2}\right)_0 \varepsilon_3^{3/2}(iv_m) v_m^3 \int_1^{\infty} dp\, p^2 \left(\frac{\alpha(iv_m, p)}{\exp[\eta(v_m, iv_m, p)] - \alpha(iv_m, p)}\right.$$

$$\left. + \frac{\beta(iv_m, p)}{\exp[\eta(v_m, iv_m, p)] - \beta(iv_m, p)}\right), \tag{12.1}$$

where

$$v_m = 2\pi m\, kT/h \ , \tag{12.2a}$$

$$\alpha(iv_m, p) = \frac{\gamma_{13}(iv_m, p) - p\,\gamma_{23}(iv_m, p) - p}{\gamma_{13}(iv_m, p) + p\,\gamma_{23}(iv_m, p) + p} \ , \tag{12.2b}$$

$$\beta(iv_m, p) = \frac{\gamma_{13}(iv_m, p) - p\varepsilon_1(iv_m)/\varepsilon_3(iv_m)}{\gamma_{13}(iv_m, p) + p\varepsilon_1(iv_m)/\varepsilon_3(iv_m)} \frac{\gamma_{23}(iv_m, p) - p\varepsilon_2(iv_m)/\varepsilon_3(iv_m)}{\gamma_{23}(iv_m, p) + p\varepsilon_2(iv_m)/\varepsilon_3{}^3(iv_m)} \ , \tag{12.2c}$$

$$\gamma_{j3}(iv_m, p) = \sqrt{p^2 - 1 + \varepsilon_j(iv_m)/\varepsilon_3(iv_m)} \ , \tag{12.2d}$$

with $j = 1, 2$, and

$$\eta(v_m, iv_m, p) = 4\pi\, v_m \sqrt{\varepsilon_3(iv_m)}\, p\, z/c \ . \tag{12.2e}$$

In this somewhat complex expression $f(z)$ stands for the "VdW pressure", i.e., the force per unit surface area exerted on the two slabs as a function of their separation z. kT is the thermal agitation energy, c the speed of light, and h Planck's constant. p is simply an integration constant and α, β, γ, and η are functions of p and the characteristic frequencies v_m. The three media involved are completely characterized by their dielectric permittivities ε_j, with $j = 1, 2, 3$, where "3" corresponds to the intervening medium. The summation in (12.1) entails calculating the functions ε_j at discrete imaginary frequencies iv_m, where $(1/2)_0$ means that only the first term of the sum has to be multiplied by $1/2$. The dielectric permittivities at imaginary frequency are related to the imaginary parts of the dielectric permittivities taken at real frequency by the well-known Kramers–Kronig relation

$$\varepsilon_j(iv_m) = 1 + \frac{2}{\pi} \int_0^{\infty} d\xi \frac{\xi\varepsilon_j''(\xi)}{\xi^2 + v_m^2} \ . \tag{12.3}$$

The imaginary parts of the complex permittivities $\varepsilon_j(\xi) = \varepsilon_j'(\xi) + i\varepsilon_j''(\xi)$ entering (12.3) are always positive and determine the dissipation of energy as a function of field frequency. The values of ε_j at purely imaginary arguments which enter (12.1, 2) are thus real quantities which decrease monotonically from their electrostatic limits ε_{j0} to 1 for $v_m \to \infty$. Separation of the entropic and quantum mechanical contributions involved in (12.1) can now simply be performed by considering the zero frequency ($m = 0$) and the non-zero frequency contributions separately. In order to ensure convergence of the integral it is wise to follow the transformation procedure originally given by *Lifshitz* [12.1]. With $y = mp$ one obtains

$$f(z) = -64\pi^5 \frac{(kT)^4}{(hc)^3} \sum_{m=0}^{\infty} \left(\frac{1}{2}\right)_0 \varepsilon_3^{3/2} (iv_m) \int_m^{\infty} dy\, y^2 \left(\frac{\alpha_m(iv_m, y)}{\exp[\kappa(iv_m, y)] - \alpha_m(iv_m, y)}\right.$$

$$\left. + \frac{\beta_m(iv_m, y)}{\exp[\kappa(iv_m, y)] - \beta_m(iv_m, y)}\right) , \tag{12.4}$$

where

$$\alpha_m(iv_m, y) = \frac{m\,\gamma_{13m}(iv_m, y) - y}{m\,\gamma_{13m}(iv_m, y) + y} \frac{m\,\gamma_{23m}(iv_m, y) - y}{m\,\gamma_{23m}(iv_m, y) + y} , \tag{12.5a}$$

$$\beta_m(iv_m, y) = \frac{m\,\gamma_{13m}(iv_m, y) - y\,\varepsilon_1(iv_m)/\varepsilon_3(iv_m)}{m\,\gamma_{13m}(iv_m, y) + y\,\varepsilon_1(iv_m)/\varepsilon_3(iv_m)}$$

$$\times \frac{m\,\gamma_{23m}(iv_m, y) - y\,\varepsilon_2(iv_m)/\varepsilon_3(iv_m)}{m\,\gamma_{23m}(iv_m, y) + y\,\varepsilon_2(iv_m)/\varepsilon_3(iv_m)} , \tag{12.5b}$$

$$m\,\gamma_{j3m}(iv_m, y) = \sqrt{y^2 + m^2 \left[\varepsilon_j(iv_m)/\varepsilon_3(iv_m) - 1\right]} , \tag{12.5c}$$

$$\kappa(iv_m, y) = 8\pi^2\, kT\sqrt{\varepsilon_3(iv_m)}\, yz/hc . \tag{12.5d}$$

For $m = 0$ one thus has $\alpha_0 = 0$ and $\beta_0 = \Delta_{130}\Delta_{230}$, where the latter quantity is determined by the electrostatic limit of

$$\Delta_{j3}(iv) = \frac{\varepsilon_j(iv) - \varepsilon_3(iv)}{\varepsilon_j(iv) + \varepsilon_3(iv)} , \tag{12.6}$$

given for $v = 0$. Using the definite integral

$$\text{constant}\int_0^{\infty} dx \frac{x^q}{\exp(x) - \text{constant}} = q! \sum_{k=1}^{\infty} \frac{(\text{constant})^k}{k^{q+1}} , \tag{12.7}$$

one finally obtains

$$f_e(z) = -\frac{H_e}{6\pi z^3} , \tag{12.8a}$$

where

$$H_e = \frac{3}{4} kT \sum_{m=1}^{\infty} \frac{\Delta_{130}^m \Delta_{230}^m}{m^3} \tag{12.8b}$$

incorporates all material properties in terms of the three static dielectric constants ε_{j0}. $f_e(z)$ characterizes the purely entropic contribution to the total VdW pressure given by (12.1) and involves a simple inverse power law dependence on the separation z of the two slabs. The zero frequency force is due to the thermal agitation of permanent electric dipoles present in the three media and includes Debye and Keesom contributions. For reasons of consistency with the following treatment of the quantum mechanical dispersion contribution, the material properties are all incorporated into the so-called "entropic Hamaker constant" given by (12.8b). It should be noted that the latter quantity cannot exceed a value of $[3\zeta(3)/4]kT$ (ζ denotes Riemann's zeta function) which is about 3.6×10^{-21} J or 22.5 meV for $T = 300$ K. The maximum value is obtained for $\varepsilon_{10}, \varepsilon_{20} \to \infty$ and $\varepsilon_{30} = 1$, i.e., for two perfect conductors interacting across vacuum. According to (12.8b) the entropic Hamaker constant becomes negative if the static dielectric constant of the intervening medium is just in between those of the two dielectric slabs. This then leads via (12.8a) to a repulsion of the slabs.

To discuss the dispersion force contribution resulting from zero-point quantum fluctuations the non-zero frequency terms ($m > 0$) in (12.1) have to be evaluated. According to (12.2a) the discrete frequencies are given by $4.3n \times 10^{13}$ Hz at room temperature. Since this is clearly beyond typical rotational relaxation frequencies of a molecule, the effective dielectric contributions according to (12.3) are solely determined by electronic polarizabilities. Absorption frequencies related to the latter are usually located somewhere in the UV region. However, with respect to this regime the ν_m's are very close together. Thus, since one has from (12.2a) $dm = (h/2\pi kT)d\nu$, one applies the transformation

$$kT \sum_{m=1}^{\infty} \to \frac{h}{2\pi} \int_{\nu_1}^{\infty} d\nu \tag{12.9}$$

to (12.1) and obtains

$$f(z) = -4\pi \frac{h}{c^3} \int_{\nu_1}^{\infty} d\nu \, \nu^3 \, \varepsilon_3^{3/2}(i\nu) \int_1^{\infty} dp \, p^2 \left(\frac{\alpha(i\nu, p)}{\exp[\eta(\nu, i\nu, p)] - \alpha(i\nu, p)} \right.$$
$$\left. + \frac{\beta(i\nu, p)}{\exp[\eta(\nu, i\nu, p)] - \beta(i\nu, p)} \right), \tag{12.10}$$

where $\alpha(i\nu, p)$, $\beta(i\nu, p)$, and $\eta(\nu, i\nu, p)$ are given by (12.2b–e), however, now for a continuous electromagnetic spectrum. Since ν_1 according to (12.2a) is much smaller than the prominent electronic absorption frequencies the spectral integration in (12.10) can be performed from zero to infinity.

Following the DLP approach [12.4] the asymptotic VdW pressure $f(z \to 0)$ is given from (12.10) by

$$f_n(z) = -\frac{H_n}{6\pi z^3}, \quad \text{with} \tag{12.11}$$

$$H_n = \frac{3}{8\pi} h \int_0^\infty dx\, x^2 \int_0^\infty dv \frac{\Delta_{13}(iv)\, \Delta_{23}(iv)}{\exp(x) - \Delta_{13}(iv)\, \Delta_{23}(iv)} \ . \tag{12.12a}$$

Using the identity given in (12.7) one can rewrite this as

$$H_n = \frac{3}{4\pi} h \sum_{m=1}^\infty \frac{1}{m^3} \int_0^\infty dv\, \Delta_{13}^m(iv)\, \Delta_{23}^m(iv) \ . \tag{12.12b}$$

In almost all cases of practical interest, where some experimental results have to be compared with theory, restriction to the first term of the above sum should be sufficient, where corrections due to higher-order terms are always less than $1 - 1/\zeta(3) \approx 16.7\%$ of the $m = 1$ term. Equation (12.11) characterizes the dispersion contribution to the total VdW pressure acting on the two slabs in the London limit, i.e., in the absence of radiation field retardation at small separation z. The inverse power law dependence is exactly the same as in (12.8a). From the *Hamaker* point of view this is not so surprising since intermolecular Debye, Keesom, and London forces all exhibit the same dependence on the separation of two molecules, $\sim 1/r^7$ [12.8]. However, contrary to the entropic Hamaker constant given by (12.8b) the "non-retarded Hamaker constant" according to (12.12b) now involves the detailed dielectric behavior of the three media through the complete electromagnetic spectrum. Since H_n is thus related to dynamic electronic polarizabilities, while H_e is related to zero frequency orientational processes, there is generally no close relation between both quantities.

In the opposite limit of large separation between the two dielectric slabs the asymptotic VdW pressure $f(z \to \infty)$ obtained from (12.10) is given according to the DLP result [12.4] by

$$f_r(z) = - H_r/z^4 \ , \quad \text{where} \tag{12.13}$$

$$H_r = \frac{hc}{64\pi^3 \sqrt{\varepsilon_3(0)}} \int_0^\infty du\, u^3 \int_1^\infty \frac{dp}{p^2} \left(\frac{\alpha(0,p)}{\exp(u) - \alpha(0,p)} + \frac{\beta(0,p)}{\exp(u) - \beta(0,p)} \right) \ . \tag{12.14a}$$

$\alpha(0, p)$ and $\beta(0, p)$ are again given by (12.2b–d), however, now in the static limit of the electronic polarizability. Using (12.7) one can rewrite the above as

$$H_r = \frac{3hc}{32\pi^3 \sqrt{\varepsilon_3(0)}} \sum_{m=1}^\infty \frac{1}{m^4} \int_1^\infty \frac{dp}{p^2} \left[\alpha^m(0,p) + \beta^m(0,p) \right] \ . \tag{12.14b}$$

Equation (12.13) characterizes the VdW pressure due to zero-point quantum fluctuations in the Casimir limit, i.e., for total radiation field retardation. A glance at (12.11) shows that, as in the case of two interacting molecules [12.3], retardation leads to an increase of the power law index by unity. However, the material properties now enter through (12.14b) in terms of dielectric permittivities $\varepsilon_j(0)$, $j = 1, 2, 3$, depending on the electronic polarizabilities in the electrostatic limit. Thus, $\varepsilon_j(0)$ must not be confused with orientational contributions ε_{j0} determining the entropic Hamaker constant in (12.8b). $H_r[\varepsilon_1(0), \varepsilon_2(0), \varepsilon_3(0)]$ is called the "retarded Hamaker constant".

In spite of the fact of having already performed a tour de force of rather lengthy calculations one is still at a point where one only has the VdW pressure acting upon two semi-infinite dielectric slabs separated by a third dielectric medium of arbitrary thickness. However, this is actually still the only geometrical arrangement for which a rigorous solution of equations of the form of (12.1) has been presented, which is equally valid at all separations and for any material combination. Without fail this means that the adaption of the above results to the SFM configuration must involve several serious manipulations of the basic results obtained from field theory.

A certain problem in handling the formulae results from the convolution of material and geometrical properties present in the integrand of the complete dispersion force solution in (12.10). A separation of both, as in the case of the entropic component given by (12.8a), is only obtained for the London and Casimir limits characterized by (12.11, 13), respectively. However, a straightforward interpolation between both asymptotic regimes is given by

$$f(z) = -\frac{H_n}{6\pi} \frac{\tanh(\lambda_{132}/z)}{z^3} , \tag{12.15}$$

where

$$\lambda_{132} = 6\pi H_r / H_n \tag{12.16}$$

is a characteristic wavelength which indicates the onset of retardation. λ_{132} is determined by the electronic contributions to the dielectric permittivities via the quotient of non-retarded Hamaker constant, according to (12.12b), and retarded constant, according to (12.14b). The above approximation is based on the assumption that H_n and H_r have the same sign. It turns out that this assumption does not hold for any material combination of the two slabs and the intervening medium (see concluding remarks in Sect. 12.2.5). It is fairly obvious that (12.15, 16) combined immediately yield the London and Casimir limit. This simple analytical approximation of the complex exact result (12.10), provides an accuracy which is more than sufficient for SFM applications. If entropic contributions are included the total VdW, pressure then is given by

$$f(z) = -\frac{1}{6\pi}\left(H_e + H_n \tanh\frac{\lambda_{132}}{z}\right)\frac{1}{z^3} . \tag{12.17}$$

This latter result shows that, while retardation causes a transition from an initial $1/z^3$ to a $1/z^4$ distance dependence of the dispersion contribution, the interaction is dominated by entropic contributions at very large separations giving again a $1/z^3$ inverse power law [12.9]. However, as will be shown below, this phenomenon is well beyond the regime which is accessible to SFM.

12.2.3 Transition to Renormalized Molecular Interactions

The macroscopic DLP theory [12.4] can be used to derive the effective inter-action of any two individual molecules within two dielectric slabs exhibiting a macroscopic VdW interaction. Accounting for an intervening dielectric medium of permittivity $\varepsilon_3(iv)$ the intermolecular force is given in the non-retarded limit by

$$F_n(z) = -A/z^7 \ , \tag{12.18a}$$

where z is the intermolecular distance and

$$A = \frac{9h}{8\pi^3\varepsilon_0^2} \int_0^\infty dv \frac{\alpha_1^*(iv)\,\alpha_2^*(iv)}{\varepsilon_3^2(iv)} \ . \tag{12.18b}$$

$\alpha_j^*(iv)$ are the dynamic electronic "excess polarizabilities" of the two interacting molecules in the immersion medium. For $\varepsilon_3 = 1$, i.e., interaction in vaccum, the $\alpha_j^*(iv)$'s become the ordinary polarizabilities $\alpha_j(iv)$ of isolated molecules and (12.18) are identical with the well-known London formula [12.2]. On the other hand, the retarded limit gives [12.4, 10]

$$F_r(z) = -B/z^8 \ , \quad \text{with} \tag{12.19a}$$

$$B = \frac{161hc}{128\pi^4\varepsilon_0^2} \frac{\alpha_1^*(0)\,\alpha_2^*(0)}{\varepsilon_3^{5/2}(0)} \ , \tag{12.19b}$$

where the electronic contributions now have to be considered in their electro-static limits. For $\varepsilon_3 = 1$ and $\alpha_j^*(0) = \alpha_j(0)$ the above result coincides with the classical *Casimir–Polder* result [12.3].

Since the above results have been derived from the macroscopic DLP theory [12.4] the excess electronic polarizabilities reflect molecular properties that are generally not directly related to the behavior of the isolated molecule but rather to its behavior in an environment composed by all molecules of the macroscopic arrangement under consideration, e.g., of the two-slab arrangement. The above molecular constants A and B thus involve an implicit renormalization with respect to the dielectric and geometrical properties of the complete macroscopic environment. This means in particular that $\alpha_j^*(iv)$ is not solely determined by the overall dielectric permittivities of all three media involved, but varies if for a given material combination only the geometry of the system is modified. Consequently, if $\alpha_j^*(iv)$ is considered in this way it involves corrections for many-body effects. Using the intermolecular interactions given in (12.18, 19) within the *Hamaker* approach [12.6], which involves volume integration of these pairwise interactions to obtain the macroscopic VdW force, yields the correct result if A and B are renormalized in an appropriate way.

If, e.g., the excess dielectric polarizability $\alpha_j^*(iv)$ of a sphere of radius R and permittivity $\varepsilon_j(iv)$,

$$\alpha_j^*(iv) = 4\pi\varepsilon_0\,\varepsilon_3(iv)\,_2\Delta_{j3}(iv)\,R^3 \ , \tag{12.20a}$$

with

$$2\Delta_{j3}(iv) = \frac{\varepsilon_j(iv) - \varepsilon_3(iv)}{\varepsilon_j(iv) + 2\varepsilon_3(iv)} , \qquad (12.20b)$$

is introduced into (12.18, 19) for two spherical particles separated by a distance d one obtains the accurate result for the macroscopic dispersion interaction of the particles in the London and Casimir limits, respectively:

$$F_n(d) = - \frac{H_n}{6\pi} \frac{R_1^3 R_2^3}{d^7} , \qquad (12.21a)$$

where the non-retarded Hamaker constant is given by

$$H_n = 108h \int_0^{\infty} dv \, _2\Delta_{13}(iv) \, _2\Delta_{23}(iv) \qquad (12.21b)$$

and

$$F_r(d) = - H_r \frac{R_1^3 R_2^3}{d^8} , \qquad (12.22a)$$

with the retarded Hamaker constant

$$H_r = \frac{161hc}{8\pi^2 \sqrt{\varepsilon_3(0)}} \, _2\Delta_{13}(0) \, _2\Delta_{23}(0) . \qquad (12.22b)$$

For an arbitrary geometrical configuration consisting of two macroscopic bodies with volumes V_1 and V_2 the Hamaker approach is given by the sixfold integral

$$F_{n,r} = \int_{V_1} dv_1 \int_{V_2} dv_2 \, f_{n,r} , \qquad (12.23)$$

where $F_{n,r}$ denotes the non-retarded or retarded macroscopic dispersion force and $f_{n,r}$ is the renormalized two-body intermolecular contribution according to (12.18, 19). Equation (12.23) applied to the two-slab arrangement yields:

$$F_n(d) = - \frac{\pi \varrho_1 \varrho_2 A}{36 d^3} \quad \text{and} \qquad (12.24a)$$

$$F_r(d) = - \frac{\pi \varrho_1 \varrho_2 B}{70 d^4} , \qquad (12.24b)$$

where ϱ_1 and ϱ_2 are the molecular densities. Comparison of (12.24a and 11) as well as of (12.24b and 13) yields the effective molecular constants A and B in terms of their "two-slab renormalization":

$$\varrho_1 \varrho_2 A = 6H_n/\pi^2 \ , \quad \text{and} \tag{12.25a}$$

$$\varrho_1 \varrho_2 B = 70H_r/\pi \ . \tag{12.25b}$$

Using (12.12b, 18b) one obtains from (12.25a) with reasonable accuracy

$$\varrho \, \alpha_j^*(iv) = 2\varepsilon_0 \, \varepsilon_3(iv) \, \Delta_{j3}(iv) \tag{12.26}$$

for the effective excess dynamic polarizability of an individual molecule "j", where $\Delta_{j3}(iv)$ is defined in (12.6). Employing this result in a threefold Hamaker integration and using (12.20), the non-retarded interaction between a small particle or a molecule "2" and a semi-infinite dielectric slab "1" is given by

$$F_n(d) = -\frac{H_n}{6\pi} \frac{R_2^3}{d^4} \ , \quad \text{with} \tag{12.27a}$$

$$H_n = \frac{9}{2} h \int_0^\infty dv \, \Delta_{13}(iv) \, _2\Delta_{23}(iv) \ . \tag{12.27b}$$

The corresponding result for the retarded interaction can easily be derived from the original DLP work [12.4]:

$$F_r(d) = -H_r \frac{R_2^3}{d^5} \ , \quad \text{with} \tag{12.28a}$$

$$H_r = \frac{3hc}{4\pi^2 \sqrt{\varepsilon_3(0)}} \, _2\Delta_{23}(0) \left[\frac{1}{3} + \frac{\varepsilon_1(0)}{\varepsilon_3(0)} + \frac{4\varepsilon_3^{3/2}(0) - [\varepsilon_1(0) + \varepsilon_3(0)]\sqrt{\varepsilon_1(0)}}{2[\varepsilon_1(0) - \varepsilon_3(0)]\sqrt{\varepsilon_3(0)}} \right.$$
$$- \frac{\varepsilon_3^3(0) + \varepsilon_1(0)\,\varepsilon_3^2(0) + 2\varepsilon_1(0)\,[\varepsilon_1(0) - \varepsilon_3(0)]^2}{2\{\varepsilon_3(0)[\varepsilon_1(0) - \varepsilon_3(0)]\}^{3/2}} \, \text{arsinh} \sqrt{\frac{\varepsilon_1(0)}{\varepsilon_3(0)} - 1}$$
$$\left. + \frac{\varepsilon_1^2(0)\,\varepsilon_3^{3/2}(0)}{\sqrt{\varepsilon_1(0) + \varepsilon_2(0)}} \left(\text{arsinh} \sqrt{\frac{\varepsilon_1(0)}{\varepsilon_3(0)}} - \text{arsinh} \sqrt{\frac{\varepsilon_3(0)}{\varepsilon_1(0)}} \right) \right] . \tag{12.28b}$$

The result holds for arbitrary dielectric constants $\varepsilon_j(0)$. If especially $\varepsilon_1(0)$ is sufficiently small (≤ 5) the above result simplifies to

$$H_r = \frac{23hc}{40\pi^2 \sqrt{\varepsilon_3(0)}} \, \Delta_{13}(0) \, _2\Delta_{23}(0) \ . \tag{12.28c}$$

If one has a metallic half-space, $\varepsilon_1(0) \to \infty$, the retarded Hamaker constant is simply given by

$$H_r = \frac{3hc}{4\pi^2 \sqrt{\varepsilon_3(0)}} \, _2\Delta_{23}(0) \ . \tag{12.28d}$$

While (12.25) are ultimately the basis for the renormalized Hamaker approach used in the following, (12.27, 28) play a role in the modeling of processes of molecular-scale surface manipulation involving physisorption of large non-polar molecules (Sect. 12.2.9). Equations (12.22, 28) are finally used to check the limits of the presented theory as provided by size effects (Sect. 12.2.8).

In order to analyze the behavior of a large molecule near a substrate surface it is convenient to extend the somewhat empirical interpolation given by (12.15) and (12.16) to the particle-substrate dispersion interaction. Equations (12.27, 28) can then be combined as

$$F(d) = -\frac{H_n}{6\pi} R_2^3 \frac{\tanh(\lambda_{132}/d)}{d^4} \, , \qquad (12.29a)$$

with the retardation wave length given by

$$\lambda_{132} = 6\pi \frac{H_r}{H_n} \, . \qquad (12.29b)$$

This approach is valid for $d \gg R$.

12.2.4 The Effect of Probe Geometry

In order to model the probe–sample interaction in SFM the general expression for the VdW pressure previously obtained now has to be adapted to the particular geometrical boundary conditions involved. Since the actual meso-scopic geometry of the employed sharp probes, i.e., the shape at nanometer scale near the apex region, is generally not known in detail, it is convenient to analyze the effect of probe geometry by considering some basic tip shapes exhibiting a rotational symmetry. Additionally accounting for a certain curvature of the sample surface one obtains the geometrical arrangement shown in Fig. 12.3.

The force between the two curved bodies can be obtained in a straightforward way by integrating the interaction between the circular regions of infinitesimal area $2\pi x \, dx$ on one surface and the opposite surface, which is assumed to be locally flat and a distance $\zeta = d + z_1 + z_2$ away. The error involved in this approximation is thus due to the assumption of local flatness of one surface – usually of the sample surface since the probing tip should be much sharper.

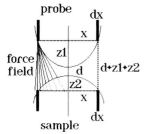

Fig. 12.3. Basic geometry in the Derjaguin approximation

However, since the VdW interaction according to (12.17) exhibits an overall $1/z^3$ distance dependence at small separations, those contributions of the force field in Fig. 12.3 involving increasing distances to the probe's volume element under consideration exhibit a rapid damping with respect to near-field contributions. This effect is further enhanced by radiation field retardation gradually leading to a $1/z^4$ inverse power law for large distances as given by the $z \to \infty$ limit of (12.17).

According to Fig. 12.3 the VdW force between probe and sample is given by

$$F(d) = 2\pi \int_{\zeta=d}^{\infty} dx\, x\, f(\zeta) \ , \tag{12.30}$$

where $f(\zeta)$ is simply the previously obtained VdW pressure between two slabs separated by an arbitrary medium of local thickness ζ, with $\zeta(x) = z_1(x) + z_2(x) + d$, and d is the distance between the apices of probe and sample. The relation between the cross-sectional radius x and the vertical coordinate z is given by

$$x = z \tan \phi \ , \tag{12.31a}$$

for a cone with half-cone angle ϕ,

$$x = R_< \sqrt{z/R_>} \ , \tag{12.31b}$$

for a paraboloid with semi-axes $R_<$ and $R_>$,

$$x = \sqrt{2R_> z - z^2}\, R_</R_> \underset{R_> \gg z}{\cong} R_< \sqrt{2z/R_>} \ , \tag{12.31c}$$

for an ellipsoid with the above semi-axes. To unify calculations it is convenient to define an effective measure of curvature by

$$R = \begin{cases} \tan \phi & \text{(cone)} \\ R_<^2/2R_> & \text{(paraboloid)} \\ R_<^2/R_> & \text{(ellipsoid)} \end{cases} . \tag{12.32}$$

Combining (12.31, 32) one immediately obtains

$$x\, dx = \begin{cases} \mathbb{R}\,[\zeta - d] & \text{(cone)} \\ 1 & \text{(paraboloid, ellipsoid)} \end{cases} \mathbb{R}\, d\zeta \ , \tag{12.33a}$$

with

$$\mathbb{R} = R_1 R_2/(R_1 + R_2) \ , \tag{12.33b}$$

where R_1 and R_2 characterize the curvature of probe and sample, respectively.

Inserting the above substitutions for $x\,dx$ into (12.30) yields

$$F(d) = 2\pi\, \mathbb{R}^2 \int\limits_d^\infty d\zeta\, \omega(\zeta) \tag{12.34a}$$

for two opposite conical surfaces and

$$F(d) = 2\pi\, \mathbb{R}\, \omega(d) \tag{12.34b}$$

for two opposite paraboloidal or ellipsoidal surfaces, respectively.

$$\omega(d) = \int\limits_d^\infty d\zeta\, f(\zeta) \tag{12.35}$$

is the total VdW energy per unit area of two flat surfaces separated by the same distance d as the apices of probe and local sample protrusion (Fig. 12.3). For the special case of two interacting spheres with radii $R_1, R_2 \gg d$ the above treatment is known as the *Derjaguin* approximation [12.11]. It should be emphasized that it is not necessary to explicitly specify the type of interaction $f(\zeta)$ which enters (12.30). The Derjaguin formulae (12.34) are thus valid for any type of interaction law, whether attractive, repulsive, or oscillating.

In order to check the effect of probe geometry in detail the dispersion pressure given by (12.15, 16) is inserted into (12.34, 35). One thus obtains for the conical arrangement

$$F(d) = -\frac{H_n}{3}\, \mathbb{R}^2\, \frac{d}{\lambda_{132}^2} \int\limits_0^{\lambda_{132}/d} dx \ln(\cosh x) \ . \tag{12.36}$$

In the non-retarded and retarded limits, where $f(\zeta)$ in (12.34, 35) follows a simple inverse power law $\sim 1/\zeta^k$, with $k = 3, 4$, the identity

$$\int\limits_d^\infty d\zeta\, (\zeta - d) f(\zeta) = \left(\frac{k-1}{k-2} - 1\right) d\, \omega(d) \tag{12.37}$$

used in (12.30) yields the particulary simple results

$$F_n(d) = -\, H_n\, \mathbb{R}^2 / 6d \quad \text{and} \tag{12.38a}$$

$$F_r(d) = -\, \pi\, H_r\, \mathbb{R}^2 / 3d^2 \tag{12.38b}$$

as non-retarded and retarded limits of (12.36). According to (12.34b) the dispersion force for two opposing paraboloidal or ellipsoidal surfaces is given by

$$F(d) = -\frac{H_n}{3}\, \frac{\mathbb{R}}{\lambda_{132}} \left[\frac{\ln[\cosh(\lambda_{132}/d)]}{d} - \frac{1}{\lambda_{132}} \int\limits_0^{\lambda_{132}/d} dx \ln(\cosh x)\right] \ . \tag{12.39}$$

Expansion of the ln(cosh) terms for large and small arguments leads to the non-retarded and retarded limits given by

$$F_n(d) = -H_n \, \mathbb{R}/6d^2 \quad \text{and} \tag{12.40a}$$

$$F_r(d) = -2\pi \, H_r \, \mathbb{R}/3d^3 \; . \tag{12.40b}$$

If the sample locally exhibits an atomically flat surface (12.33b) simplifies to

$$\lim_{R_2 \to \infty} \mathbb{R} = R \; , \tag{12.41}$$

where, according to (12.32), R is the effective radius of apex curvature for a paraboloidal or ellipsoidal probe and $R = \tan \phi$ for a conical probe.

Apart from describing the probe–sample dispersion interaction (12.39) also characterizes the adsorption of a large non-polar molecule or small particle on an atomically flat substrate surface. Directly at the surface one has $R \gg d$ and the non-retarded dispersion interaction is given by (12.40a). On the other hand, if the particle is initially far away from the substrate, $R \ll d$, the interaction is given by (12.29). This involves a non-retarded transition of the dispersion force from a $1/d^2$ dependence at small distances to $1/d^4$ dependence at large distances and finally a transition to $1/d^5$ at very large distances which is due to retardation. If the retardation wavelength λ_{132} of the particle-substrate arrangement is assumed to be independent of the particle's distance from the surface, i.e., if λ_{132} is the same in (12.15, 29a), then the particle-substrate dispersion interaction is modeled by

$$F(d) = 2\pi \, R \, \omega(d) \tanh\left(\frac{\lambda_{132}^R}{d}\right)^2 \; , \tag{12.42a}$$

where R characterizes the dimension of the particle according to (12.32) and $\omega(d)$ is the specific energy obtained for the two-slab system as given in (12.35). The non-retarded $1/d^2$ to $1/d^4$ transition is determined by the transition length

$$\lambda_{132}^R = R \sqrt{\frac{1}{\pi} \frac{H_n \text{ from (12.27b)}}{H_n \text{ from (12.12b)}}} \; , \tag{12.42b}$$

which is more or less close to R. It can easily be verified that (12.42) satisfies the limiting results given in (12.40a) for $d \ll R$ and in (12.29) for $d \gg R$. Equations (12.42) allow the modeling of particle or molecule physisorption processes if the involved dispersion interactions are governed by bulk dielectric properties.

Figure 12.4a shows the decrease of the dispersion force for increasing working distance for a conical probe according to (12.36) and for a paraboloidal or ellipsoidal probe according to (12.39). The curve for the interaction of two slabs is given by (12.15) and the physisorption curve for a small particle or large molecule with a flat surface is obtained from (12.42), where $\lambda_{132}^R/\lambda_{132} = 0.1$ was used as a somewhat typical non-retarded transition length. For refer-

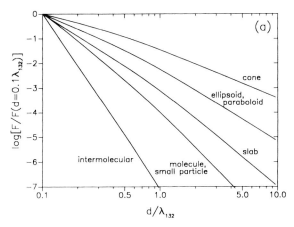

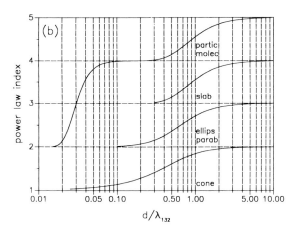

Fig. 12.4. Dispersion interaction for some fundamental arrangements. (a) Shows the normalized force as a function of separation, where λ_{132} denotes the material-dependent retardation wavelength. (b) Shows the corresponding differential power law index to be used in an inverse power law ansatz for the distance dependence of the dispersion force

ence, the interaction between small particles with radii R_1 and R_2 – according to (12.21a, 22a) modeled by $F(d) = -(H_n/6\pi)\, R_1^3 R_2^3 \tanh(\lambda_{132}/d)/d^7$, with λ_{132} given by (12.16) – is also indicated. If one performs the transformation $(H_n/6\pi)R_1^3 R_1^3 \rightarrow A$ and $\lambda_{132} \rightarrow B/A$, with the molecular constants A and B given by (12.18b, 19b), the latter curve corresponds to the intermolecular dispersion force and includes retardation effects.

Even if all above formulae for the dispersion force involve the same material-dependent retardation wavelength λ_{132} the gradual onset of retardation effects is clearly determined by the mesoscopic geometry of the interacting bodies [12.12]. This phenomenon is clarified by considering the differential power law index,

$$k(d) = -\frac{d\partial}{\partial d}\ln F(d)\ , \tag{12.43}$$

which has to be applied if the VdW force $F(d)$ is approximated for a given distance d by an inverse power law of type $F(d) \sim 1/d^{k(d)}$. Application of (12.43) yields for the two-slab arrangement the simple result

$$k(d) = -d\frac{\partial f(d)/\partial d}{f(d)} ,\tag{12.44a}$$

where $f(d)$ is the VdW pressure according to (12.15). For a paraboloidal or ellipsoidal SFM probe one obtains from (12.34b)

$$k(d) = d f(d)/\omega(d) ,\tag{12.44b}$$

where $f(d)$ and $\omega(d)$ are the VdW pressure according to (12.15) and the VdW energy per unit surface area according to (12.35) obtained from the two-slab arrangement. The result for the conical probe is obtained from (12.34a):

$$k(d) = d \omega(d)/\int_d^\infty d\zeta\, \omega(\zeta) .\tag{12.44c}$$

The above results describe in detail the geometry-dependent transition to retardation for an extremely blunt probe (cylindrical, i.e., two-slab arrangement) for a realistic probe type (paraboloidal or ellipsoidal) and for the limit of an atomically sharp probe (conical). The distance dependence of the differential power law index according to (12.44) is shown in Fig. 12.4b. Additionally, the physisorption behavior of a small particle or large molecule onto a flat surface, obtained by applying (12.43) to (12.42a), is indicated.

In the present context the most important result from Fig. 12.4 is that VdW forces drop with an $1/d^2$ inverse power law for the most realistic probe geometries, i.e., for paraboloidal or ellipsoidal apices, in the non-retarded limit. This clearly indicates a long-range probe–sample interaction in comparison with the exponential decrease of a tunneling current or the short-range manifestation of interatomic repulsive forces resulting from electronic orbital overlap. Consequently, the spatial resolution which can be obtained in VdW microscopy should be determined by the mesoscopic probe dimensions at nanometer scale near the probe's apex and of course by the probe–sample separation [12.12]. An estimation of the lateral resolution is obtained by determining the probe diameter $\Delta(d)$ corresponding to the center of interaction. The latter is determined by the maximum contribution of the integrand in (12.30), i.e., by

$$\frac{\partial}{\partial x}(x f[\zeta(x)]) = 0 .\tag{12.45}$$

Accounting for the $1/\zeta^k$ dependence of $f_n(\zeta)$ this can be replaced by the more convenient form

$$1 - \frac{k}{\zeta}x\frac{\partial\zeta}{\partial x} = 0 .\tag{12.46}$$

According to (12.11) the non-retarded VdW pressure involves $k = 3$. Using the geometrical relations given in Fig. 12.3 one obtains the remarkably simple results

$$\Delta_n(d) = 2\sqrt{(2/5)\,\mathbb{R}\,d} \tag{12.47a}$$

for the resolution of a paraboloidal or ellipsoidal probe and

$$\Delta_n(d) = \mathbb{R}\,d \tag{12.47b}$$

for a conical probe. $\mathbb{R}$ is related by (12.32, 33b) to the effective dimensions of probe apex and local sample protrusion. For an atomically flat sample $\mathbb{R} \to R$ simply characterizes the sharpness of the probe. While the minimum resolvable lateral dimension for a conical probe is proportional to the working distance d, paraboloidal or ellipsoidal probes exhibit a square root dependence on the working distance. While (12.47) gives the lateral resolution in VdW microscopy in terms of simple analytical results, more elaborate solutions can only be obtained numerically [12.13, 14]. Apart from quantifying the microscope's resolution (12.47) additionally tells us that the resolution is independent of material properties in the considered non-retarded limit. This is, however, not valid if retardation becomes effective. In this case the power law index k in (12.46) becomes distance- and material-dependent.

Finally it should be emphasized that the solutions obtained largely analytically for forces, power law indices, and lateral resolutions by using the approximate result for the dispersion pressure in (12.15) can be numerically obtained in an exact way by using the DLP result from (12.10), whenever the dispersion pressure $f(d)$ for the two-slab arrangement is needed. Consequently, the corresponding specific dispersion energy in (12.35) has to be calculated by directly integrating (12.10). The result is then

$$\omega(d) = \frac{h}{c^2} \int_0^\infty dv\, v^2\, \varepsilon_3^2(iv) \int_1^\infty dp\, p\, (\ln\{1 - \alpha(iv, p)\exp[-\eta(v, iv, p)]\}$$
$$+ \ln\{1 - \beta(iv, p)\exp[-\eta(v, iv, p)]\})\,, \tag{12.48}$$

where α, β and η are defined as in (12.10) for $z \equiv d$. The entropic component always additionally present is rigorously given by (12.8). Integration of the latter equation immediately yields the entropic VdW energy $\omega_e(d)$ per unit surface area. Equations (12.8, 10, 48) then provide the general framework for a rigorous numerical calculation of probe–sample forces via (12.34), of power law indices via (12.44), and of lateral resolutions at any working distance via (12.45). However, the major advantage of the analytical treatment presented above is that it emphasizes the physical processes giving rise to VdW interaction in SFM, while the rigorous numerical treatment ultimately based on (12.1) tends to obscure the basic physical aspects due to considerable mathematical complexities.

12.2.5 Dielectric Contributions: The Hamaker Constants

Apart from the probe and sample geometry considered above, the magnitude of VdW forces in SFM is determined by the detailed dielectric properties of probe, sample, and an immersion medium which may be present in the intervening gap. The real dielectric permittivities taken at imaginary frequencies, $\varepsilon_j(iv)$, enter the two-slab dispersion pressure in (12.15) via the non-retarded Hamaker constant H_n and via the retardation wavelength λ_{132}. The latter quantity is determined according to (12.16) by the ratio of H_n to the retarded Hamaker constant H_r. For a given probe–sample geometry H_n and H_r thus completely determine the magnitude of the resulting force as well as the onset of retardation effects. The following discussion is devoted to a calculation of the Hamaker constants in terms of only two characteristic material properties: the optical refractive index and the effective electronic absorption wavelength.

The energy absorption spectrum of any medium for frequencies from zero through to the ultraviolet (UV) regime is characterized by

$$\varepsilon_j(iv) = 1 + \sum_l \frac{c_{jl}}{1 + v/v_{jl}} + \sum_m \frac{f_{jm}}{1 + (v/v_{jm})^2 + \gamma_{jm} v/v_{jm}^2} , \qquad (12.49)$$

where the first non-unitary term describes the effect of possible Debye rotational relaxation processes, and the second models absorption using a Lorentz harmonic oscillator model of the dielectric [12.5]. The characteristic constants $c_{jl}, v_{jl}, f_{jm}, v_{jm}$ are given in pertinent tables of dielectric data. The damping coefficients γ_{jm} associated with the Lorentz oscillations are rather difficult to determine and are not known in most cases. However, since for dielectrics the widths of the absorption spectra are always small compared with the absorption frequencies, i.e., $\gamma_m \ll v_m$, the term $\gamma_m v/v_m^2$ can be dropped to a satisfactory approximation in (12.49). If the static dielectric constant is denoted by ε_{j0}, and if one only has one prominent rotational absorption peak for $v = v_{j,\text{rot}}$ and one prominent electronic absorption peak for $v = v_{j,e}$ the above may be written as

$$\varepsilon_j(iv) = 1 + \frac{\varepsilon_{j0} - n_j^2}{1 + v/v_{j,\text{rot}}} + \frac{n_j^2 - 1}{1 + (v/v_{j,e})^2} , \qquad (12.50)$$

where n is the optical refractive index. While v_{rot} is typically given by microwave or lower frequencies, v_e is located in the UV regime, and for most materials of practical interest one has $v_e \approx 3 \times 10^{15}$ Hz. If there are m individual electronic absorption frequencies, (12.50) has to be replaced by

$$\varepsilon_j(iv) = 1 + \frac{\varepsilon_{j0} - n_1^2}{1 + v/v_{\text{rot}}} + \sum_{l=2}^{m+1} \frac{n_{j,l-1}^2 - n_{j,l}^2}{1 + (v/v_{j,l-1})^2} , \qquad (12.51)$$

where $n_{j,m+1} \equiv 1$.

In the far-UV and soft X-ray regime, all matter responds like a free electron gas [12.15] and the response function changes to

$$\varepsilon_j(iv) = 1 + (v_e/v)^2 , \qquad (12.52)$$

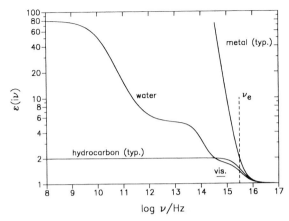

Fig. 12.5. Dielectric permittivity on the imaginary frequency axis as a function of real frequency for water, typical hydrocarbons, and typical metals. $\nu_e = 3 \times 10^{15}$ Hz is taken as the prominent electronic absorption frequency. The visible regime is indicated for reference

where ν_e is now the free electron gas plasma frequency. This latter expression also characterizes approximately the dielectric permittivity of a metal from zero through the visible to the soft X-ray regime. In the intermediate regime between far UV and soft X-ray, there is little knowledge of $\varepsilon(i\nu)$. However, some reasonable interpolation schemes may be constructed [12.5].

According to the above result, matter may roughly be subdivided into three classes of dielectric behavior, as shown in Fig. 12.5. For water, the simplest Debye rotational relaxation and some closely spaced infrared (IR) bands lead to variations of $\varepsilon(i\nu)$ below the UV regime. Thus, $\varepsilon(i\nu)$ has to be evaluated according to (12.51) conveniently using effective values for refractive indices and absorption frequencies in the IR and UV regime [12.5], respectively. On the other hand, typical hydrocarbons (liquid or crystallized) exhibit a constant $\varepsilon(i\nu)$ from zero frequency through to the optical regime. The complex absorption spectrum in the near-UV regime is conveniently summarized by taking mean values corresponding to the first ionization potential [12.5]. In this case $\varepsilon_j(i\nu)$ is simply approximated by (12.50), where only Lorentz harmonic contributions have to be considered. The third class of dielectric behavior belongs to metals. In this case $\varepsilon_j(i\nu)$ is simply given by (12.52), where typical plasma frequencies are $3–5 \times 10^{15}$ Hz [12.8] and $\varepsilon_{j0} \to \infty$.

According to (12.6, 12b) the non-retarded Hamaker constant H_n is determined by the dielectric response functions of probe, sample and intervening medium given according to (12.50–52). Figure 12.6 shows the spectral contributions to the VdW interaction for some material combinations of practical interest. The dispersion force in the non-retarded regime is directly proportional to the area under a curve. The Hamaker constant is obviously most sensitive to spectral features between about 1 and 10–20 eV. This involves, e.g., the widths of typical band gaps in semiconductors. The maximum H_n is found for two typical metals interacting across vacuum. If one metal is replaced by mica, a representative dielectric, H_n becomes considerably smaller and the maximum of the

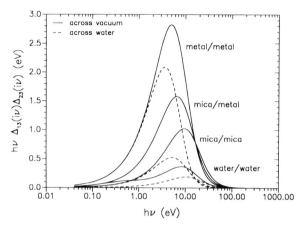

Fig. 12.6. Spectral contributions to the non-retarded Hamaker constant H_n for some material combinations of practical interest. H_n is directly proportional to the area under a curve

corresponding curve in Fig. 12.6 is slightly shifted to higher spectral energies. If both interacting materials are mica, H_n further decreases and the maximum is further shifted to the right. Two water films interacting across saturated air exhibit a still smaller H_n, where the maximum is now slightly shifted to lower energies with respect to the mica/mica interaction. The long low-energy tail of the water/water curve is due to the IR spectral contributions as discussed above. Filling the intervening gap between the two interacting media with water considerably reduces the H_n values, respectively. This is not due to a Debye orientational process of the highly polar water molecules, but results from the dynamic electronic contribution $\varepsilon_3(iv) > 1$ to the response functions $\Delta_{13}(iv)$ and $\Delta_{23}(iv)$ defined in (12.6). The maxima of the curves in Fig. 12.6 are slightly shifted to lower energies with respect to the vacuum values.

If each of the three media involved is characterized with respect to its dielectric permittivity by (12.50) and if all media exhibit approximately the same electronic absorption frequency v_e, the non-retarded Hamaker constant H_n according to (12.12b) can be evaluated analytically [12.8], where sufficient accuracy is obtained if only the first term of the sum is considered. Since possible low-frequency rotational processes corresponding to the first term in (12.50) are represented by the entropic constant H_e given according to (12.8b), H_n is simply given in terms of the three optical refractive indices n_j ($j = 1, 2, 3$) for probe, sample and immersion medium, and the absorption frequency v_e:

$$H_n = \frac{3}{8\sqrt{2}} \frac{(n_1^2 - n_3^2)(n_2^2 - n_3^2)}{(n_1^2 + n_3^2)\sqrt{n_2^2 + n_3^2} + (n_2^2 + n_3^2)\sqrt{n_1^2 + n_3^2}} h v_e \ . \tag{12.53}$$

The detailed behavior of H_n as a function of the refractive indices is shown in Fig. 12.7. Let n_2 be the smaller index with respect to the probe–sample ensemble.

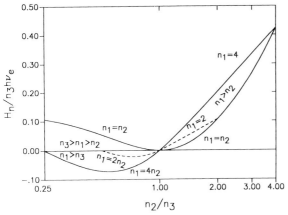

Fig. 12.7. Non-retarded Hamaker constant for dielectric systems as a function of optical refractive indices of probe, sample, and intervening medium (n_3). v_e denotes the prominent electronic absorption frequency

A reasonable range covering almost all dielectrics is given by $1/4 \leq n_2/n_3 \leq 4$, where n_3 is the index of the intervening medium. If $n_2/n_3 > 1$, H_n is always positive and is usually given by a point in between the curves for $n_1 = n_2$ and $n_1 = 4$. Most dielectric materials, however, lie in between the curves $n_1 = n_2 \leq 2$ and $n_1 = 2$. If the optical refractive index of the immersion medium matches that of either probe or sample, H_n becomes zero and the non-retarded force vanishes. For $n_2/n_3 < 1$ one has to distinguish between two regimes: If $n_1 < n_3$, H_n is again positive and is located somewhere in between the curve $n_1 = n_2$ and the abscissa which corresponds to $n_3 = n_1$. However, if $n_1 > n_3$, H_n becomes negative and, according to (12.11), the non-retarded dispersion pressure becomes repulsive. Almost all dielectrics in this regime lie in between the abscissa and the curve $n_1 = 4n_2$, while most of them are limited by the curve $n_1 = 2n_2$.

The main conclusions from the above analysis of H_n are: (i) If probe and sample materials are interchanged, the non-retarded dispersion force remains the same. (ii) For interactions in vacuum or for identical probe and sample materials the non-retarded force is always attractive. (iii) If probe and sample are made from dielectrics with different refractive indices n_1 and n_2, and if there is an intervening medium with $n_3 > 1$, the non-retarded force may be attractive ($n_3 < n_1, n_2$ or $n_3 > n_1, n_2$), vanishing (n_3 matches either n_1 or n_2 or both of them), or repulsive ($n_1 < n_3 < n_2$ or $n_1 > n_3 > n_2$). (iv) According to Fig. 12.7 most of the dielectric material combinations produce a Hamaker constant $H_n \lesssim h v_e/10$. With a typical value of $v_e = 3 \times 10^{15}$ Hz one has $H_n \lesssim 2 \times 10^{-19}$ J (1.2 eV). H_n has to be added to the entropic Hamaker constant H_e defined in (12.8b) to obtain the total non-retarded VdW force given for the $z \ll \lambda_{132}$ limit of (12.17). Depending on the static values ε_{j0} ($j = 1, 2, 3$), the entropic force component may also be attractive, vanishing, or repulsive, where signs and magnitudes of H_n and H_e are generally not correlated. However, the maximum value found for H_e (Sect. 12.2.2) is only 1.5% of the limiting value given above for H_n.

This clearly implies that entropic VdW forces play only a minor role in SFM applications [12.16].

If either the probe or the sample is made from a typical metal, characterized by (12.52), the non-retarded Hamaker constant is given by

$$H_n = \frac{3}{8\sqrt{2}} \frac{n_2^2 - n_3^2}{n_2^2 + n_3^2 + \sqrt{n_2^2 + n_3^2}} h v_e \ , \tag{12.54}$$

where v_e is the prominent absorption frequency of the system, n_2 the optical refractive index of the remaining dielectric, and n_3 that of the intervening medium. Equation (12.54) permits an estimate of the maximum repulsive dispersion force that can be obtained: $n_2 \ll n_3$ yields $H_n = -(3/8\sqrt{2})n_3^2 h v_e/(n_3^2 + n_3)$ which gives for large n_3 and $v_e = 3 \times 10^{15}$ Hz the upper limit $|H_n| \lesssim 5 \times 10^{-19}$ J (3.1 eV). For two dielectrics with different absorption frequencies v_{e1} and v_{e2}, interacting across vacuum, one obtains

$$H_n = \frac{3}{8\sqrt{2}} h \frac{v_{e1} v_{e2} (n_1^2 - 1)(n_2^2 - 1)}{v_{e1}(n_1^2 + 1)\sqrt{n_2^2 + 1} + v_{e2}(n_2^2 + 1)\sqrt{n_1^2 + 1}} \ , \tag{12.55}$$

which was already presented in [12.8]. If either probe or sample is a metal one finds

$$H_n = \frac{3}{8\sqrt{2}} h \frac{v_{e1} v_{e2} (n_2^2 - 1)}{v_{e1} \sqrt{n_2^2 + 1} + v_{e2}(n_2^2 + 1)} \ , \tag{12.56}$$

and, if both probe and sample are metallic,

$$H_n = \frac{3}{8\sqrt{2}} h \frac{v_{e1} v_{e2}}{v_{e1} + v_{e2}} \ , \tag{12.57}$$

which reduces to the particularly simple result $H_n = (3/16\sqrt{2}) h v_e$ [12.8] if both metals have the same electronic absorption frequency. Assuming a free electron gas plasma frequency of 5×10^{15} Hz one obtains $H_n \lesssim 9 \times 10^{19}$ J (5.4 eV) as a realistic upper limit for metallic probe–sample arrangements. The dependence of the non-retarded Hamaker constants on the electronic absorption frequencies of probe and sample is shown in Fig. 12.8 in detail. For given values of v_{e1} and v_{e2} the resulting H_n is always highest if both probe and sample are metals. The metal/dielectric arrangement yields lower values depending on the optical refractive index n_2 of the dielectric (either probe or sample). The lowest values of H_n are obtained if probe and sample are dielectric, where the magnitude of the non-retarded dispersion force now depends on n_1 and n_2. Anyway, an increase of the absorption frequencies v_{e1} and v_{e2} always leads to an increase of H_n.

The above results obtained for H_n are only part of the whole story. The total VdW pressure according to (12.17) is completely characterized if, apart from H_e and H_n, also the retarded Hamaker constant H_r according to (12.14b) is

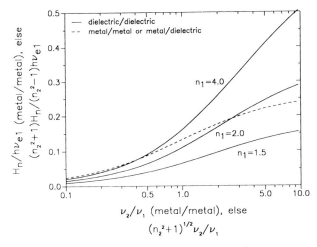

Fig. 12.8. Non-retarded Hamaker constant H_n as a function of the prominent absorption frequencies v_1 and v_2 of the probe–sample combination. n_1 and n_2 denote the optical refractive indices if dielectric materials are involved

calculated. H_r depends on the static electronic limits $\varepsilon_j(0)$ of the dielectric response functions of probe, sample and intervening medium. Since the relative magnitudes of $\varepsilon_j(0)$ ($j = 1, 2, 3$) are neither in general related to the overall behavior of the functions $\varepsilon_j(iv)$ [(12.12b) via (12.6)] over the complete electromagnetic spectrum nor to the quasistatic orientational contributions ε_{j0} [(12.8b) via (12.6)], H_r is a priori not closely related to H_n and H_e with respect to sign and magnitude. Apart from the magnitude of the dispersion pressure in the retarded limit given by (12.13), H_r determines together with H_n via the retardation wavelength (12.16) the onset of retardation effects.

The electrostatic limits of the electronic permittivity components are given from (12.51) by

$$\varepsilon_j(0)_{\text{total}} - (\varepsilon_{j0} - n_{j1}^2) = n_{j1}^2 \rightarrow \varepsilon_j(0)_{\text{electronic}} . \tag{12.58}$$

As for most hydrocarbons (Fig. 12.5) the n_{j1}'s often equal the usual optical refractive indices n_j. However, as in the case of water, which is of particular practical importance for many SFM experiments, n_{j1} is sometimes determined by lower-frequency (IR) absorption bands. Introduction of generalized refractive indices n_j ranging from unity to infinity in (12.14b) permits a unified analysis of H_r for all material combinations, i.e., metals and dielectrics. The resulting values of H_r, as depending on the individual refractive indices $n_{j1} = n_j$, are shown in Fig. 12.9. Let n_2 be the smaller index for the probe–sample system under consideration, n_3 is the index of the intervening immersion medium. If $n_2 > n_3$ (Fig. 12.9a) H_r is always positive and its magnitude is given by a point in between the curves for $n_1 = n_2$ and $n_1 \rightarrow \infty$. For $n_1, n_2 \rightarrow \infty$ (two interacting

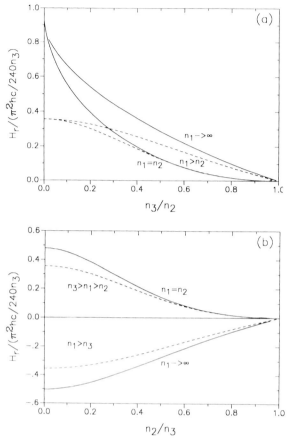

Fig. 12.9. Retarded Hamaker constant H_r as a function of effective refractive indices (infrared and visible). (**a**) shows the positive values of H_r, where both probe and sample have larger indices than the intervening medium (n_3). (**b**) shows the situation if the indices of probe and sample are smaller than that of the immersion medium. The dotted lines indicate results from the low-permittivity analytical approximation. Both numerical and analytical results correspond to the first term of the infinite series involved

metal slabs) one obtains from (12.14b)

$$H_r = \frac{\pi}{480}\frac{hc}{n_3} , \tag{12.59}$$

which gives, according to (12.13), a retarded dispersion pressure which is completely independent of the nature of the employed metals; a property that does not hold for small distances, where the dispersion force according to (12.11, 12b) depends on higher-frequency contributions to the dielectric response functions which are generally different for different metals. For $n_3 = 1$, (12.59) coincides with the well-known *Casimir* and *Polder* result [12.3]. If only typical dielectric materials are involved, (12.14b) may be evaluated analytically [12.10] by expanding $\alpha(0, p)$ and $\beta(0, p)$ for small $\varepsilon_j(0)/\varepsilon_3(0)$, $j = 1, 2$. The approximate result is

$$H_r = \frac{23}{320\pi^3}\frac{hc}{n_3}\sum_{l=1}^{\infty}\frac{1}{l^4}\,\Delta_{13}^l(0)\,\Delta_{23}^l(0) , \tag{12.60}$$

with Δ_{j3} according to (12.6). The validity of this approximation, depending on the magnitudes of n_1/n_3 and n_2/n_3, can be obtained from Fig. 12.9. For two interacting metals, i.e., for $\Delta_{13}(0)$, $\Delta_{23}(0) \rightarrow 1$, the low-permittivity approximation still predicts the correct order of magnitude for H_r. More precisely, (12.60) yields $69\zeta(4)/(2\pi^4) \approx 38\%$ of the correct value given by (12.59). If one has $n_j/n_3 = 1$ at least for one of the quotients ($j = 1, 2$), H_r becomes zero and the retarded dispersion force vanishes. If $n_2 < n_3$ (Fig. 12.9b), one has to distinguish between two regimes: if also $n_1 < n_3$, H_r is again positive and is located in between the abscissa, corresponding to $n_1 = n_3$, and the curve $n_1 = n_2$. The maximum value of this latter curve is obtained for $n_1/n_3, n_2/n_3 \rightarrow 0$. The low-permittivity approximation, which underestimates the exact value, yields in this case again 38% of the value given by (12.59). On the other hand, if $n_1 > n_3$, H_r becomes negative and is given by some point in between the abscissa ($n_1 = n_3$) and the curve for $n_1 \rightarrow \infty$. The approximation for the minimum of this latter curve, obtained for $n_2/n_3 \rightarrow 0$, i.e., $\Delta_{13}, \Delta_{23} \rightarrow 1$ in (12.60), gives a magnitude of 38% of the value in (12.59) which is, according to Fig. 12.9, an underestimate of the exact value. The maximum repulsive dispersion force that can be obtained for any material combination is obtained from the condition $\partial H_r/\partial n_3 = 0$, where $n_1 \rightarrow \infty$ is an obvious boundary condition to achieve high H_r values. The use of (12.60) yields $n_2 = n_3(\sqrt{5} - 2)^{1/2}$ and a maximum repulsive retarded dispersion force with a magnitude of about 22% of the value in (12.59). This is again slightly underestimated with respect to the exact value numerically obtained from (12.14b).

It should be emphasized that the entropic Hamaker constant H_e scales with kT; the non-retarded constant with $h\nu_e$; and the retarded constant with hc. The absolute maximum obtained for H_r is for two metals interacting across vacuum and amounts according to (12.59) to $H_r = 1.2 \times 10^{-27}$ Jm (7.4 eV nm). Comparison of (12.53, 60) confirms that the above statements (i)–(iii) characterizing the behavior of the non-retarded force can be directly extended to the retarded force, however, where one now has to consider the low-frequency indices n_{j1} from (12.51) instead of the ordinary optical indices n_j. If there is no absorption in the IR regime the situation is simple and $n_{j1} = n_j$ as in the case of hydrocarbons. However, strong IR absorption, as in the case of water, considerably complicates the situation: the relative weight of different frequency regimes, (IR, visible, and UV) becomes a sensitive function of separation between probe and sample. At small distances (non-retarded regime) the interaction is dominated by UV fluctuations. With increasing distance these contributions are progressively damped leading to a dominance of visible and then IR contributions. For very large separations the interaction would finally be dominated by Debye rotational relaxation processes. This complicated behavior may in principle be characterized by treating the different spectral components according to (12.50) additively in terms of separate Hamaker constants and retardation wavelengths. In the present context, the major point is, that due to a missing correlation between the magnitudes of n_{j1} and n_j, H_n and H_r may have different signs, i.e., the VdW force may be attractive at small probe–sample separation and exhibits a retardation-induced transition to repulsion at larger separations or vice versa. In this case the simple analytical approximation of the DLP theory given in (12.17) breaks down. However, even in this case it is possible to keep the concept of separating geometrical and dielectric contributions. The DLP result from (12.1) may now be modeled by

$$f(z) = -\frac{1}{6\pi} \left\{ H_e + H_n \tanh\left[\frac{|\lambda_{132}|}{z}\left(\frac{|\lambda_{132}|}{z} - 1\right)\right]\right\} \frac{1}{z^3}, \tag{12.61}$$

where the definitions of H_e, H_n, H_r, and λ_{132} remain totally unchanged. This formally implies the occurence of negative retardation wavelengths which are obtained according to (12.16) if H_n and H_r have different signs. Equation (12.61) exhibits the same behavior as (12.17) for the $z \lessgtr \lambda_{132}$ limits, but additionally demonstrates a retardation-induced transition between the attractive and repulsive regimes at $z = \lambda_{132}$ as can be seen in Fig. 12.10. A completely blunt (cylindrical) SFM probe would detect a force exactly corresponding to the curves obtained for the VdW pressure. However, according to (12.34b), more realistic probe models (paraboloidal, ellipsoidal) predict a measurement of forces being proportional to the specific VdW energy given via an integration of (12.61). This implies that the transition distance, measured with a paraboloidal or ellipsoidal probe, is somewhat smaller than that measured with a cylindrical probe (λ_{132}). The smallest transition separation is, according to (12.34a), obtained for a conical probe. The intriguing conclusion is that for a probe–sample interaction which does not involve a monotonic distance dependence, the force measured at a given probe–sample separation may be attractive for one probe and repulsive for another with different apex geometry.

The retardation wavelength λ_{132} defined in (12.16) depends on the dielectric response functions of probe, sample and intervening medium. Retardation effects of the radiation field between probe and sample become noticeable if the probe–sample separation is comparable with λ_{132}. The retardation wavelength

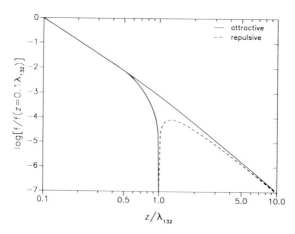

Fig. 12.10. Dispersion pressure for the two-slab configuration as a function of separation. If the system exhibits strong infrared absorption a retardation-induced transition from attraction to repulsion (or vice versa) may occur. An overall attractive (or repulsive) interaction occurs if non-retarded and retarded Hamaker constants have the same sign

is thus closely related to the prominent absorption wavelength $\lambda_e = c/v_e$ of the material combination which is usually about 100 nm, i.e., within the UV regime. The actual onset of retardation effects, manifest in a gradual increase of the differential power law index k according to (12.43), is then for a given material combination determined by the probe geometry (Sect. 12.2.4). In the following some simple analytical results for λ_{132} are presented which allow a straight-forward verification of the relevance of retardation effects for most material combinations of practical importance to SFM.

Combining (12.53, 60) one obtains the retardation wavelength for a solely dielectric material combination. First-order approximation yields

$$\lambda_{132} = \frac{23\sqrt{2}}{20\pi^2} \frac{1}{n_{31}} \left(\prod_{l=1}^{2} \frac{n_l^2 + n_3^2}{n_{l1}^2 + n_{31}^2} \frac{n_{l1}^2 - n_{31}^2}{n_l^2 - n_3^2} \right) \left(\frac{1}{\sqrt{n_1^2 + n_3^2}} + \frac{1}{\sqrt{n_2^2 + n_3^2}} \right) \frac{c}{v_e} .$$

(12.62)

If the system does not exhibit effective IR absorption, i.e., $n_{j1} = n_j$ ($j = 1, 2, 3$), the product in parenthesis reduces to unity and λ_{132} is solely determined by the ordinary optical refractive indices and the prominent electronic UV absorption frequency. If the probe or the sample is metallic (12.54, 60) yield the approximate result

$$\lambda_{132} = \frac{23\sqrt{2}}{20\pi^2} \frac{1}{n_{31}} \left(\frac{n_2^2 + n_3^2}{n_{21}^2 + n_{31}^2} \frac{n_{21}^2 - n_{31}^2}{n_2^2 - n_3^2} \right) \left(1 + \frac{1}{\sqrt{n_2^2 + n_3^2}} \right) \frac{c}{v_e} ,$$

(12.63)

where the product in parenthesis again becomes unity in the absence of IR absorption. If dielectric probe and sample have different absorption frequencies and if they interact across vacuum, (12.55, 60) approximately give

$$\lambda_{132} = \frac{23\sqrt{2}}{20\pi^2} \left(\prod_{l=1}^{2} \frac{n_l^2 + 1}{n_{l1}^2 + 1} \frac{n_{l1}^2 - 1}{n_l^2 - 1} \right) \left(\frac{1}{v_{e1}\sqrt{n_1^2 + 1}} + \frac{1}{v_{e2}\sqrt{n_2^2 + 1}} \right) c ,$$

(12.64)

which again simplifies for $n_{l1} = n_l$. If either the probe or the sample is metallic one obtains from (12.56, 60)

$$\lambda_{132} = \frac{23\sqrt{2}}{20\pi^2} \left(\frac{n_2^2 + 1}{n_{21}^2 + 1} \frac{n_{21}^2 - 1}{n_2^2 - 1} \right) \left(\frac{1}{v_{e1}} + \frac{1}{v_{e2}\sqrt{n_2^2 + 1}} \right) c ,$$

(12.65)

with the aforementioned simplification for $n_{21} = n_2$. If a purely metallic probe–sample system interacts across vacuum, (12.57, 59) yield the exact result

$$\lambda_{132} = \frac{\pi^2\sqrt{2}}{30} \left(\frac{1}{v_{e1}} + \frac{1}{v_{e2}} \right) c ,$$

(12.66)

which only involves the free electron gas plasma frequencies as characteristics of the metals. If one in particular has $c/v_{e1} = c/v_{e2} = \lambda_e$, λ_{132} amounts to 93% of λ_e. A glance at (12.62–65) shows that this can be considered as an upper limit for any material combination with $n_{j1} = n_j$ ($j = 1, 2, 3$), i.e., for arrangements, where IR absorption only plays a minor role. On the other hand, large values of λ_{132} are obtained according to (12.16) if H_n is nearly vanishing and H_r is determined by IR absorption.

Figure 12.11 shows typical values of λ_{132} obtained in an accurate way by numerically solving Eqs. (12.12b, 14b). The maximum value for λ_{132} in a solely dielectric probe–sample arrangement is about 31% of λ_e. For a metal/dielectric combination this value amounts to 37%. Both values are considerably lower than the aforementioned value which may be obtained for a metal/metal combination of probe and sample. Typical values of λ_{132} are 20–35% of λ_e if one does not have a purely metallic arrangement.

The rigorous solution for λ_{132} as a function of the prominent UV absorption frequencies involved is shown in Fig. 12.12. The minimum value of λ_{132} for a metal/metal arrangement is about 46% of $\lambda_{e1} = c/v_{e1}$ if $v_{e2} \rightarrow v_{e1}$. For a metal/dielectric or dielectric/dielectric combination λ_{132} can be much smaller depending on the optical refractive indices involved.

Systems with extremely low retardation wavelength may be constructed according to (12.62). Suitable material combinations consist of a dielectric probe, sample, and immersion medium with an appropriate choice of refractive indices; minimizing λ_{132}. Most effective would be a match of the IR indices, n_{31} of the immersion medium with n_{11} and/or n_{21} of the probe–sample combination. Unfortunately, reliable IR data are not available for most materials. For material combinations which do not exhibit pronounced IR absorption the ordinary optical indices n_j ($j = 1, 2, 3$) should all be as large as possible, where a highly refractive immersion medium (n_3) is especially effective. In this way, retardation wavelengths smaller than 10 nm are generated, which opens the way

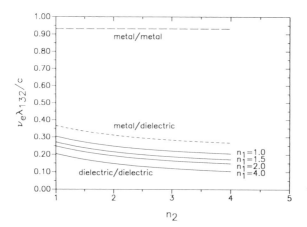

Fig. 12.11. Retardation wavelength λ_{132} as a function of the optical refractive indices of probe and/or sample interacting across vacuum. v_e is the prominent electronic absorption frequency of the system for which the absence of infrared absorption bands is assumed. The upper limit provided by the metal–metal arrangement is indicated for reference

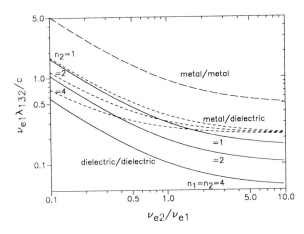

Fig. 12.12. Retardation wavelength as a function of the prominent ultraviolet absorption frequencies ν_{e1} and ν_{e2} of probe and sample. n_1 and n_2 denote the ordinary optical refractive indices if dielectric materials are involved. The curves are valid for systems without effective absorption bands in the infrared regime

for an experimental confirmation of radiation field retardation effects by SFM (Sect. 12.2.6).

12.2.6 On the Observability of Van der Waals Forces

The framework for calculating VdW forces for any material combination and any probe geometry as a function of probe–sample separation is now complete. The material properties of a certain system are characterized by the three Hamaker constants H_e, H_n, and H_r according to (12.8b, 12b, 14b). This includes the determination of the retardation wavelength λ_{132} via (12.16). The total VdW pressure $f(z)$ for the two-slab arrangement is then given by (12.17) or (12.61) in terms of a reasonable approximation. For relevant probe geometries the VdW interaction is characterized by (12.34b) which involves the probe's effective radius of curvature. An estimate of the resulting lateral resolution is obtained from (12.47a).

Figure 12.13 shows the typical order of magnitude of the two-slab VdW pressure as well as the material-dependent onset of retardation effects for some representative material combinations. The dielectric data used for these model calculations are given in Table 12.1. In the regime from 1 to 100 nm separation the pressure drops by six to seven orders of magnitude in ambient air (or vacuum). As mentioned before, two typical metals yield the strongest possible interaction. Mica, representing a typical dielectric material, yields a pressure of about 25% of the metal value in the non-retarded regime and of about 7% in the retarded limit, respectively. Crystallized hydrocarbons and water are most frequently the sources of surface contaminations. These media only exhibit a small VdW pressure with respect to the metal limit. Consequently, if initially clean metal surfaces become contaminated by films of hydrocarbons or water the VdW interaction may decrease by 80–90% or more for a given width of the intervening air gap. If the complete intervening gap between two mica surfaces is

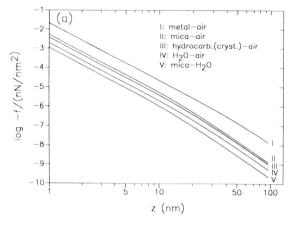

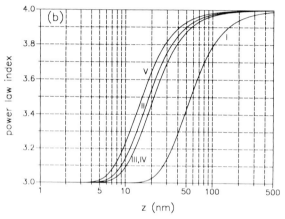

Fig. 12.13. (a) shows the two-slab van der Waals pressure as a function of separation for some representative material combinations. **(b)** shows the corresponding retardation-induced increase of the differential power law indices

Table 12.1 Dielectric data used for the calculations. For reasons of comparison the present data is deduced from the basic data given in [12.8, 17]. For water infrared absorption contributions have been neglected

	H_n $(10^{-20}\,J)$	H_e $(10^{-20}\,J)$	H_r $(10^{-29}\,Jm)$	λ_{132} (nm)
metal/air/metal	40	0.30	130	61
mica/air/mica	10	0.17	9.3	20
H_2O/air/H_2O	3.7	0.29	4.5	23
hydroc./air/hydroc.	7.1	0.04	8.7	23
mica/H_2O/mica	2.0	0.21	2.0	17

filled with water the VdW pressure drops with respect to the air (or vacuum) value by about 80%.

The onset of retardation effects also critically depends on the system composition. The metal system yields the highest λ_{132}. The hydrocarbon and water values are about the same. The retardation wavelength for two mica slabs in air (vacuum) is reduced by about 17% if the intervening gap is filled with water.

As an example of direct practical relevance Fig. 12.14 shows the VdW interaction between a realistic metal probe (paraboloidal or ellipsoidal) with a mesoscopic radius of apex curvature of 100 nm and two different atomically flat substrates; a typical metal and mica, representing a typical dielectric. Assuming an experimental sensitivity of 10 pN which is not unrealistic for present-day UHV–SFM systems, forces should be detectable up to about 20 nm for the metal sample and up to about 10 nm for mica. Radiation field retardation becomes effective just near these probe–sample separations. The entropic limit, according to (12.8a) with $H_e = 3.6 \times 10^{-21}$ J indicates that thermally agitated VdW forces could only be measured at working distances $\lesssim 1$ nm. In the dynamic or "ac" mode of SFM, the vertical force derivative $F'(d) = \partial F / \partial d$ is detected. An accessible experimental sensitivity may be given by 10 μN/m. This extends the measurable regime up to about 70 nm for the metal sample and up

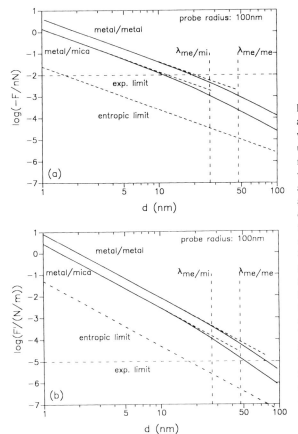

Fig. 12.14. Van der Waals interaction of a 100 nm metal probe with a metal and a mica substrate under clean vacuum conditions, respectively. The retardation wavelengths λ for the metal/metal and metal/mica configurations are indicated. The entropic limit determines the absolute room-temperature maximum for thermally agitated interaction contributions. Deviations from a linear decrease of the curves with increasing probe–sample separation reflect the gradual onset of retardation effects. The indicated experimental limits are accessible by state-of-the-art instruments. (a) Shows the forces measured upon static operation of the force microscope and (b) the vertical force derivative, detected in the dynamic mode

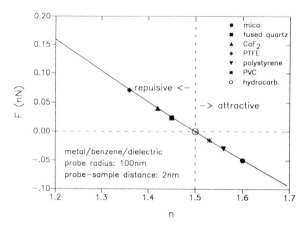

Fig. 12.15. Van der Waals force between a metal probe operated in a benzene immersion and various dielectric substrates at a fixed probe–sample distance. n denotes the sample's ordinary optical refractive index. For purposes of comparison refractive indices and absorption indices have been chosen according to [12.8]

to about 50 nm for mica. According to Fig. 12.12(b) this clearly involves the onset of retardation effects.

Performance of SFM in an immersion medium generally offers the possibility to choose material combinations yielding attractive, repulsive or just vanishing VdW interactions between probe and substrate. Assuming a metal probe operated in a benzene immersion at a fixed probe–sample separation, Fig. 12.15 gives the resulting VdW forces for various dielectric substrates as a function of the ordinary optical refractive index n of the sample according to (12.53). While PolyTetraFluoroEthylene (PTFE), CaF_2, and fused quartz with $n < 1.5$ produce repulsive non-retarded VdW forces, PolyVinylChloride (PVC), polystyrene, and mica with $n > 1.5$ yield attractive forces. Crystallized hydrocarbons just match the index of benzene, $n = 1.5$, and the VdW force reduces to the small entropic contribution.

12.2.7 The Effect of Adsorbed Surface Layers

The analytical solutions for the VdW pressure of the two-slab configuration given in (12.17, 61) allow straightforward extension to multilayer configurations. Figure 12.16 shows the basic geometry for two slabs "1" and "2", both with an adsorbed surface film "4". An arbitrary intervening substance is denoted by "3". By extending previous results found for the non-retarded interaction [12.5, 10] to an analysis for arbitrary separations d and film thicknesses t_{41} and t_{42}, the VdW pressure is given by

$$f_{ad}(z) = f_{434}(z) - f_{341}(z + t_{41}) - f_{342}(z + t_{42}) + f_{142}(z + t_{41} + t_{42}) .$$

$$(12.67)$$

The subscripts of type klm denote the material combination which actually has to be considered to calculate an individual term of the above expression. k and

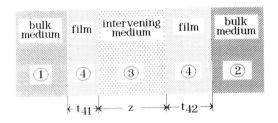

Fig. 12.16. Basic geometry of the four-slab arrangement used to analyze the interaction of two bulk media which have surfaces covered with adsorbed layers

m denote the opposite slabs, respectively, and l the intervening medium. The solution of this four-slab problem is thus reduced to the calculation of four "partial" VdW pressures involving four sets of Hamaker constants. Since these partial pressures have different entropic, non-retarded, and retarded magnitudes and varying retardation wavelengths, the distance dependence of f_{ad} is generally much more complex than that for the two-slab arrangement, in particular, if there are f_{klm}-terms showing a retardation-induced transition between attractive and repulsive regimes according to (12.61). However, it follows immediately from (12.67) that $f_{ad}(z) \to f_{434}(z)$ for $t_{41}/z, t_{42}/z \to \infty$; for large thicknesses of the adsorbed surface layers the VdW pressure is solely determined by the interaction of the layers across the intervening medium. On the other hand, if $t_{41}/z, t_{42}/z \to 0$, one immediately finds $f_{ad}(z) \to f_{132}(z) \equiv f(z)$ which is simply the

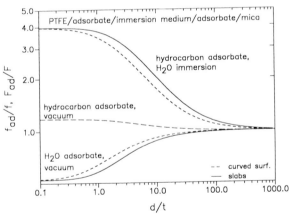

Fig. 12.17. Model calculation showing the effect of adsorbed hydrocarbon (liquid or crystallized) or water layers on the interaction between PolyTetraFluoroEthylene (PTFE) and a mica surface. The quotients f_{ad}/f and F_{ad}/F denote the force ratios obtained for adsorbate-covered surfaces with respect to clean surfaces, for planar and paraboloidally or ellipsoidally curved surfaces, respectively. The adsorbate thickness t is assumed to be the same on PTFE and mica. d denotes the width of the intervening gap, either for vacuum or water immersion. The curved- and planar-surface curves for hydrocarbon adsorbate in vacuum cannot be distinguished within the accuracy of the plot. The dashed curves would be detected with a typical probe in dc-mode force microscopy, while the solid lines reflect ac data

solution of the two-slab problem according to Eq. (12.17) or (12.61). In the latter case the interaction is dominated by the interaction of the two bulk media across the intervening medium.

Figure 12.17 exemplarily shows the considerable differences of the VdW interactions which occur if initially clean PolyTetraFluoroEthylene (PTFE) and mica surfaces adsorb typical hydrocarbons (liquid or crystallized) or water. The adsorption of hydrocarbons slightly increases the vacuum forces. However, if the intervening gap is filled with water the magnitude of the VdW forces increases by about a factor of four with respect to the interaction of clean surfaces across water. Water adsorption in air considerably reduces the forces with respect to clean surfaces. In all cases involving adsorbed surface layers the bulk interaction value is approached not before the intervening gap exceeds the layer thickness by two to three orders of magnitude. This clearly emphasizes the fact that VdW interactions are highly surface sensitive: even only a monolayer adsorbed on a substrate considerably modifies the probe–sample interaction with respect to the clean substrate up to separations of several nanometers. The situation is additionally complicated by the fact that the difference in VdW force measured between clean and coated substrate surfaces also depends on the probe geometry (Fig. 12.17). This intriguing phenomenon is due to the integral equations (12.34) determining the probe–sample force from the two-slab pressure.

12.2.8 Size, Shape, and Surface Effects: Limitations of the Theory

The rigorously macroscopic analysis of VdW interactions in SFM implicitly exhibits some apparent shortcomings which are ultimately due the particular mesoscopic, i.e., nanometer-scale physical properties of sharp probes and corrugated sample surfaces exhibiting deviations from ordinary bulk physics. To obtain an upper quantitative estimate for those errors resulting from size and shape effects it is convenient to apply the present formalism to some particular worst case configurations for which exact results from quantum field theory are available for comparison. Two such arrangements which have been subject to rigorous treatments are two interacting spheres and a sphere interacting with a semi-infinite slab. These configurations do reflect insofar worst case situations as a sphere of finite size emphasizes geometrical errors involved in the Derjaguin approximation as well as shape-induced deviations from a simple bulk dielectric behavior. Size and shape effects occur when the probe–sample separation becomes comparable with the effective mesoscopic probe radius as defined in (12.32). Since realistic probe radii are generally of the order of the retardation wavelengths defined by (12.16), the following analysis is restricted to the retarded limit of probe–sample dispersion interaction to obtain an upper boundary for the involved errors. However, extension of the treatment to arbitrary probe–sample separations is straightforward.

According to the basic Hamaker approach given in (12.23), the dispersion interaction between a sphere of radius R_1 and a single molecule at a distance

z from the sphere's center is simply given by

$$f_m(z) = -\pi \varrho_1 B \frac{1}{z} \int_{z-R_1}^{z+R_1} dr \frac{R_1^2 - (z-r)^2}{r^7} , \qquad (12.68)$$

where ϱ_1 is the molecular density within the sphere and B the molecular interaction constant given by (12.19b). The interaction between two spheres separated by a distance d is then given by

$$F(d) = \frac{\pi \varrho_2}{d+2R_2} \int_{d+R_2}^{d+3R_2} dz\, f_m(z)\, z\, [R_2^2 - (d+2R_2-z)^2] , \qquad (12.69)$$

where R_2 and ϱ_2 are radius and molecular density of the second sphere and $f_m(z)$ is taken from (12.68). The interaction between a sphere and a semi-infinite slab is obtained without problems by analytically evaluating the above integrals and letting one of the radii go to infinity. However, from reasons clarified below, the limiting behavior for $d \gg R$ is more interesting in the present context. For two identical spheres ($R_1 = R_2 = R$, $\varrho_1 = \varrho_2 = \varrho$) one obtains

$$F(d) = -\frac{16}{9}\pi^2 \varrho^2 B \frac{R^6}{d^8} \qquad (12.70)$$

and for the sphere–slab configuration ($R_1 = R, R_2 \to \infty, \varrho_1 = \varrho_2 = \varrho$)

$$F(d) = -\frac{8}{105}\pi^2 \varrho^2 B \frac{R^3}{d^5} . \qquad (12.71)$$

Both above results have already been derived in (12.22a, 28a). If one now assumes that the screening of the radiation field by the near-surface molecules is the same as for the two-slab configuration, the microscopic quantity $\varrho^2 B$ is related to the macroscopic Hamaker constant by (12.25b). Especially for ideal metals, which may be considered as the limit of a London superconductor, as the penetration depth approaches zero, one obtains via (12.59) for an interaction in vacuum

$$F(d) = -\frac{7\pi^2}{27} hc \frac{R^6}{d^8} \qquad (12.72)$$

from (12.70), and

$$F(d) = -\frac{\pi^2}{90} hc \frac{R^3}{d^5} \qquad (12.73)$$

from (12.71). However, these results are not completely correct since the surface screening of the radiation field is affected by the actual curvature of the interacting surfaces. The correct result for the two-sphere configuration is obtained by using the Hamaker constant given in (12.22b). For perfectly conducting spheres, as considered in the present case, one has, apart from the electric polarizability, to account for the magnetic polarizability which provides an additional contribution of 50% of the electric component to the total polarizability [12.18]. Appropriate combination of electric and magnetic dipole photon contributions yields $_2\Delta_{13}(0)_2\Delta_{23}(0) = \Delta^2(0) = \Delta_E^2(0) + \Delta_M^2(0) + (14/23)\Delta_{EM}(0)$ [12.19], where $\Delta_E(0) = 1$ and $\Delta_M(0) = 1/2$ are the pure electric and magnetic contributions, respectively, and $\Delta_{EM}(0) = 1/2$ is due to an interference of electric and magnetic dipole photons. Inserting $\Delta^2(0) = 143/92$ into (12.22b) then ultimately leads to

$$F(d) = -\frac{1001}{32\pi^2} hc \frac{R^6}{d^8} , \qquad (12.74)$$

which has already been previously derived by a more involved treatment [12.19]. Comparison with (12.72) shows that the two-slab renormalization underestimates the sphere–sphere VdW force by about 19% which is due to the reduced screening of the curved surfaces. For the sphere–slab arrangement the rigorous result is obtained by using the Hamaker constant given in (12.28d) for a perfectly conducting metal sphere. Using $_2\Delta_{23}(0) = 3/2$ this yields

$$F(d) = -\frac{9}{8\pi^2} hc \frac{R^3}{d^5} , \qquad (12.75)$$

which is in agreement with a previous result [12.20] obtained by different methods of theory. A comparison with (12.73) yields a slight underestimate of about 4% due to the two-slab renormalization.

At very small separations, $d \ll R$, the Derjaguin approach according to (12.40b) yields the correct results

$$F(d) = -\frac{\pi^2}{1440} hc \frac{R}{d^3} , \qquad (12.76)$$

for the sphere–sphere interaction, and

$$F(d) = -\frac{\pi^2}{720} hc \frac{R}{d^3} , \qquad (12.77)$$

for the sphere–slab interaction, where in both cases the Hamaker constant according to (12.59) has to be used; radiation field screening is about the same for planar and very smoothly curved surfaces. Comparison of (12.74) and (12.75) with (12.76) and (12.77) shows that the dispersion force changes from $1/d^3$ to a $1/d^8$ dependence for the two spheres, and from a $1/d^3$ to a $1/d^5$ dependence for the sphere–slab arrangement. Both cases are correctly modeled by the Hamaker approach according to (12.69), as shown in Fig. 12.18. For $d \gtrsim R/10$ the Derjaguin approximations exhibit increasing deviations from the Hamaker curves. Deviations in radiation field screening with respect to the two-slab

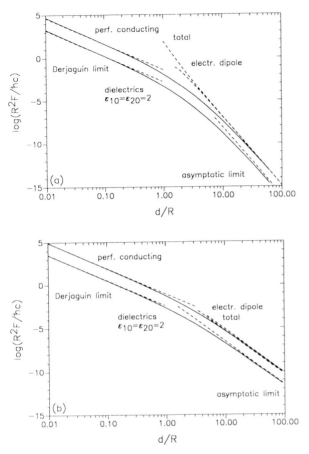

Fig. 12.18. Retarded vacuum dispersion force between two spheres (**a**) and between a sphere and a semi-infinite slab (**b**). R denotes the sphere's radius, F the magnitude of the attractive force, and d the surface-to-surface distance. The upper curves correspond to perfectly conducting constituents, while the lower ones characterize identical dielectrics. The dashed lines indicate results from short- and long-distance approximations

configuration gradually occur and reach the aforementioned asymptotic values when the Hamaker curves approach the asymptotic limit.

Figure 12.18 additionally includes results of the above comparative study for interacting dielectrics. In this case, surface screening is much less pronounced as for perfectly conducting bodies. Thus, the Hamaker approach with two-slab renormalization yields almost accurate results at any interaction distance and for arbitrarily curved surfaces. The major conclusion that can be drawn from this worst case scenario is that the maximum error due to surface screening of a probe with unknown electric and magnetic form factors amounts, at large distances, to 19% for an arbitrarily corrugated sample surface and to 4% for an atomically flat substrate. At ordinary working distances, $d \ll R$, and for dielectrics geometry-modified screening is completely negligible [12.21].

Another shortcoming of the present theory is that it implicitly neglects multipole contributions beyond exchange of dipole photons. In general, for probe–sample separations greater than about one nanometer the exchange of dipole photons overshadows that due to dipole–quadrupole and higher multipole exchange processes. However, for smaller separations, as present in contact-mode SFM, and for some particular material combinations also at larger separations, multipole interactions assume increasing importance. For the retarded interaction of perfectly conducting spheres, the total force including the interference between electric and magnetic quadrupole photons [12.19] is shown in Fig. 12.18 in addition to the pure electric dipole contribution. However, in most cases these corrections are of little relevance in the present context.

A much more serious obstacle for a rigorous characterization of VdW interactions in SFM results from the explicit assumption of isotropic bulk dielectric permittivities of probe, sample, and immersion medium. Especially if probe and sample are in close proximity, this assumption is subject to dispute, in view of the microstructure of solids [12.22] and liquids (Sect. 12.4) and the existence of particular surface states. This may require specific corrections to account for the particular molecular-scale surface dielectric permittivities. However, unfortunately only little reliable information on specific surface dielectric properties has become available so far. Further progress is also needed in appropriately treating the behavior of real metal tips and substrates. Especially for sharp probes and substrate protrusions the delocalized electrons, moving under the influence of the radiation field fluctuations, require a specific non-local microscopic treatment [12.23].

12.2.9 Application of Van der Waals Forces to Molecular-Scale Analysis and Surface Manipulation

Manipulation of substrate surfaces by using scanned probe devices assumes an increasing importance. In particular the deposition of individual molecules or small particles provides an approach to study microscopic electronic or mechanic properties and to achieve positional control of interaction processes. First experimental results [12.24] imply that VdW forces may play an important role within this field. Figure 12.19 shows some proposals for the analysis and manipulation of small particles or molecules by a systematic employment of VdW interactions.

A small particle or molecule physisorbed on a flat substrate may be moved in close contact to the substrate by a "sliding process", as shown in the upper left image of Fig. 12.19. The VdW bonds between particle and substrate and between particle and tip have to ensure on the one hand the fixing of the particle between tip and substrate during sliding and, on the other, the anchoring to the substrate during final withdrawal of the tip. A liquid environment permits the

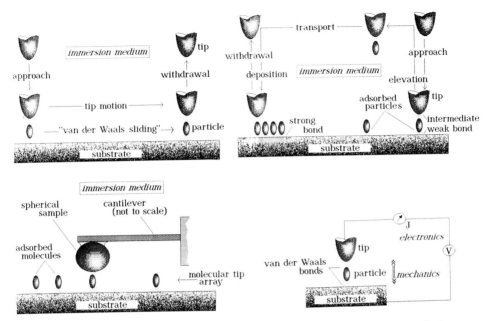

Fig. 12.19. Manipulation (*upper row*) and analysis (*lower row*) of small particles and molecules by systematically employing van der Waals interactions

variation of the non-retarded Hamaker constant for the tip–substrate interaction over a wide range preferably according to (12.53) or (12.54). PolyTetraFluoroEthylene (PTFE) may be considered as a promising universal substrate material since its optical refractive index ($n = 1.359$ [12.8]) is lower than that of several liquids yielding repulsive interactions with respect to most tip materials. Especially for water immersion, $n = 1.333$, the interaction between PTFE and any tip material should almost vanish. PTFE can be easily modified to render its surface hydrophilic.

Another surface manipulation process (Fig. 12.19, upper right) involves the elevation of the particle if the tip–particle VdW bond is stronger than that between particle and substrate. The particle can thus be transported over larger distances and obstacles. Deposition is performed at a place where the substrate–particle interaction is stronger than the tip–particle interaction. For this purpose the tip should be dielectric and the place of position of the PTFE substrate should have a higher optical refractive index than the tip. As was shown in Sect. 12.2.7, already a monolayer coverage on top of the PTFE substrate can considerably increase the particle–substrate Hamaker constant and can rise the interaction above that between tip and particle. A liquid immersion again allows to control the perturbative tip–substrate interaction.

The lack of really reproducible, well-characterized and mesoscopically (nanometer scale) sharp probes in standard SFM systems is the most apparent obstacle for high resolution VdW measurements. The "molecular tip array" (MTA) philosophy developed by *Drexler* [12.25] could alleviate this problem. The proposed geometry is shown in the lower left of Fig. 12.19. The sample, shaped as a small bead, is attached to the microscope's cantilever by adhesion. Small particles or macromolecules are adsorbed on the flat substrate surface. Since the bead's radius R^* is assumed to be very large compared to the effective molecular radius, which is according to (12.32) given by $R = 2R_<^2/R_>$ for an ellipsoidal molecule with semi-axes $R_<$ and $R_>$, an individual molecule underneath the bead images the gently curved surface upon raster scanning the substrate with respect to the bead. The lateral resolution follows directly from (12.47a) and amounts to

$$\Lambda_n(d) = 2R_< \sqrt{2d/5R_>} \ . \tag{12.78}$$

While the interaction between bead and molecule at arbitrary separations may be obtained from (12.42), the total non-retarded VdW force for separations being large compared with the effective molecular radius, i.e., $d \gg R_<^2/R_>$, is according to (12.27)

$$F_n(d) = -\frac{H_n}{6\pi} \frac{R_<^6}{R_>^3 d^4} \ . \tag{12.79}$$

For $d \ll R_<^2/R_>$ (12.40a) yields

$$F_n(d) = -\frac{1}{6} H_n \frac{R_<^2}{R_> d^2} \ . \tag{12.80}$$

The substantial bead radius raises the issue of unwanted surface forces. The bead–substrate interaction is according to (12.40a)

$$F_n^*(d) = -\frac{1}{6} H_n^* \frac{R^*}{(2R_> + d)^2} \ , \tag{12.81}$$

with a Hamaker constant perferably according to (12.53) or (12.54). The ratio of this "parasitic" force to the imaging force is, for $d \gg R_<^2/R_>$,

$$\frac{F_n^*}{F_n}(d) < \pi \frac{H_n^*}{H_n} \frac{R^* R_>^3 d^2}{R_<^6} \ . \tag{12.82}$$

For a close bead-molecule separation, $d \ll R_<^2/R_>$, one obtains

$$\frac{F_n^*}{F_n}(d) < \frac{1}{4} \frac{H_n^*}{H_n} \frac{R^* d^2}{R_<^2 R_>} \ . \tag{12.83}$$

The above quotients may be considered as "noise-to-signal ratios" and should be much smaller than unity.

The considerable potential of MTA imaging is emphasized if one somewhat quantifies the above design analysis. For simplicity, arbitrary spherical macro-molecules with $R_< = R_> = 1$ nm are assumed. A molecule–bead separation of $d = 1$ nm separates the VdW interaction from short-range forces due to orbital overlap. Under these conditions (12.78) yields a lateral resolution of $\Delta_n = 1.3$ nm for the VdW imaging of the bead's surface. Using a somewhat typical Hamaker constant of 1.5×10^{-19} J (Sect. 12.2.5), the force according to (12.80) amounts to $F_n = 25$ pN, which is within reach of present technology. Suppression of para-sitic forces F_n^* requires, according to (12.83), a Hamaker constant H_n^* which is less than 4% of H_n. This may easily be achieved by using PTFE substrates in combination with an aqueous immersion.

Potential tip structures may predominantly include single-chain proteins, proteins with bound partially-exposed ligands, or nanometer-scale crystalline particles [12.25]. The considerable capabilities of modern organic synthesis and biotechnology offer broad freedom in molecular tip design. MTA technology would permit the quasi-simultaneous use of a broad variety of tips scattered across the substrate. This may include tips of different composition, electric charge, magnetization, and orientation. A tip density of more than $1000/\mu m^2$ has been considered as reasonable [12.25]. First results in obtaining suitable metallic bead–cantilever systems are reported in [12.26].

Apart from VdW imaging the surface of a spherical sample at ultra-high spatial resolution, the MTA technology may be well-suited to obtain a deeper insight into molecular electronics and mechanics (Fig. 12.19, lower right). Using an SFM with conducting tip–cantilever system, simultaneous tunneling and force measurements may be performed on a single molecule. This may help to clarify the process of tunneling through localized electronic states in organic molecules by detecting the tip–molecule–substrate tunneling current I as a func-tion of tunneling voltage V and force exerted on the molecule.

12.2.10 Some Concluding Remarks

The present analysis emphasizes that VdW forces play an important role in SFM. At probe–sample separations less than a few nanometers the force drops according to $F_n(d) = -(H_n/6)R/d^2$, where R is the probe's effective radius of curvature, H_n the non-retarded Hamaker constant, and d the probe–sample separation. A somewhat representative value for the force at $d = 1$ nm is $|F_n|/R = 10$ mN/m. While the interaction is always attractive in dry air or vacuum, it may be attractive, repulsive, or even vanishing if the gap between probe and sample is filled with a liquid medium. Thermally activated processes generally play a minor role, and it is in almost all cases sufficient to analyze the dispersion part of the forces.

Non-contacting VdW microscopy is capable of providing information on surface dielectric permittivities at sub-100 nm resolution. The technique is sensitive to even monolayer coverages of a substrate. Important future fields of application are the investigation of liquid/air (vapour) interfaces [12.27] and the imaging of soft (biological) samples including individual macromolecules. As in contact-mode SFM (AFM), where VdW forces have a substantial influence on the net force balance, and thus on the probe–sample contact radius, the long-range interactions may play a role in other non-contacting modes of operation, i.e., in electric and magnetic force microscopy, if these are performed at low working distances ($\approx$ 1 nm). However, in this latter context VdW forces may be reduced in a well-defined way by covering the sample surface with a suitable dielectric and/or using an adapted liquid immersion medium.

Finally, some open questions with respect to the general subject of VdW interactions in SFM should be listed:

(i) In what way may the effective dielectric permittivities deviate from the assumed anisotropic bulk dielectric properties, especially for sharp metal tips?

(ii) Is the present rigorously macroscopic treatment satisfactory down to probe–sample separations which involve electron-orbital overlap, or is a special non-local microscopic treatment needed?

(iii) May VdW forces be externally stimulated in a measurable way by electromagnetic irradiation preferably at wavelengths between IR and UV? Such an excitation, beyond zero-point fluctuations, would permit the performance of "scanning force spectroscopy" as a technique to sense the spectral variation of surface dielectric permittivities.

(iv) Do excited surface states, i.e., surface plasmons [12.28] have a measurable effect on the probe–sample VdW interaction?

These questions are considered as some major future challenges for elaborate SFM experiments on VdW forces. Additional questions are concerned with the delicate interplay of VdW forces with other interactions to be discussed in the following.

12.3 Ionic Forces

12.3.1 Probe–Sample Charging in Ambient Liquids

Situations in which VdW forces solely determine the probe–sample interaction in SFM are in general restricted to an operation under clean vacuum conditions. Under ambient conditions, which are often present in SFM experiments long-range electrostatic forces are frequently additionally involved, and the interplay of these latter and VdW forces has important consequences. If wetting films are present on probe and sample or if the intervening gap is filled with a liquid, surface charging may come about essentially in two ways [12.8]: (i) by ionization or dissociation of ionizable surface groups and (ii) by adsorption of ions onto initially uncharged surfaces. Whatever the actual mechanism, the equilibrium final surface charge is balanced by a diffuse atmosphere of counterions close to the surfaces, resulting in the so-called "double layer", see Fig. 12.20. The electrostatic interaction between probe and sample is closely related to the counterion concentration profile.

Since the intervening homogeneously dielectric gap between two semi-infinite equally charged slabs is field-free, the counterions do not experience an attractive electrostatic force toward the surfaces. The ionic concentration profile is solely determined by interionic electrostatic repulsion and the entropy of mixing, while the amount of surface charge only controls the total number of counterions. For simplicity an identical charge density σ is assumed on the opposite surfaces and one also assumes electroneutrality of the complete two-slab arrangement. The resulting non-linear second-order Poisson–Boltzmann differential equation [12.8] leads to a general form of the so-called contact value theorem,

$$f(d) = kT\left[\varrho(d) - \varrho(\infty)\right] \ . \tag{12.84}$$

The ionic excess osmotic pressure is, for a given thermal activation energy, simply proportional to the excess counterion density, present directly in front of the surfaces of the charged slabs. $\varrho(\infty)$ is the ionic surface density for an isolated charged surface. Since the intervening liquid does not contain a bulk electrolytic reservoir one has

$$\varrho(\infty) = \frac{\sigma^2}{2\varepsilon\varepsilon_0\,kT} \ , \tag{12.85}$$

where ε is the static dielectric constant of the immersion fluid. The two-slab counterion surface concentration amounts to

$$\varrho(d) = \frac{\sigma^2}{2\varepsilon\varepsilon_0\,kT} + \frac{2\varepsilon\varepsilon_0\,kT}{[e\,n\,\lambda(d)]^2} \ , \tag{12.86}$$

where n denotes the ionic valency. The characteristic length λ depends, for

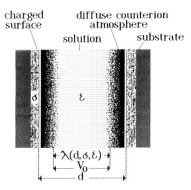

charged surface · diffuse counterion atmosphere · solution · substrate

Fig. 12.20. Diffuse counterion atmosphere near the surfaces of two slabs which exhibit a certain surface charge density σ. The intervening gap of thickness d contains a solution with a static dielectric constant ε. V_0 and λ denote voltage and separation between fictitious centric planes of the near-surface counterion profiles

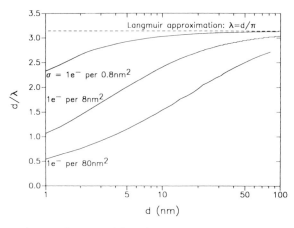

Fig. 12.21. Characteristic separation length λ for monovalent counterions in water as a function of separation of two surfaces exhibiting an equal charge density σ. The latter quantity is given in electrons per surface area, where $1e^-/0.8\,\text{nm}^2 = 0.2\,\text{C/m}^2$ represents a typical value for a fully ionized surface

a given valency and for given values of σ and ε, on the separation d of the slabs and has to fulfill the following conditions:

$$\frac{2kT}{ne\lambda} \tan\frac{d}{2\lambda} + \frac{\sigma}{\varepsilon\varepsilon_0} = 0 \ . \tag{12.87}$$

Insertion of (12.85, 86) into (12.84) shows that $f(d)$ in (12.84) may be associated with the pressure in the interspace of a simple parallel-plate capacitor with $C/s = \varepsilon\varepsilon_0/z$. If one applies a certain voltage V_0 this pressure amounts to $f(z) = -\varepsilon\varepsilon_0 V_0^2/2z^2$. The analogy holds if the virtual separation of the capacitor plates is given by $z = |n|\lambda(d)$, while one has to apply an imaginary voltage

$$V_0 = \mathrm{i}\, 2kT/e \ , \tag{12.88}$$

which results in a repulsive interaction between the plates. It should be noted that this elementary voltage is completely independent of the type of counterions. The magnitude amounts to 52.5 mV at room temperature. The two-slab ionic pressure is now simply given by

$$f(d) = -\frac{1}{2}\varepsilon\varepsilon_0\left[\frac{V_0}{n\,\lambda(d)}\right]^2 \ . \tag{12.89}$$

$|n|\lambda$ thus obviously represents the separation of the fictitious capacitor plates which may be associated with the maxima of the near-surface counterion concentration profiles (Fig. 12.20). Figure 12.21 shows the dependence of λ for monovalent ions in water on the separation d between two slabs for three different surface charge densities, computed according to (12.87). λ decreases with decreasing d and increasing σ. For high surface-charge densities and at large separations λ becomes proportional to d. In this limit the pressure becomes

$$f(d) = -\frac{\pi^2}{2}\varepsilon\varepsilon_0\left[\frac{V_0}{nd}\right]^2 \ , \tag{12.90}$$

which is known as the Langmuir relation. This relation may be used to calculate the equilibrium thickness and disjoining pressure of wetting films on probe and sample in SFM systems. In the opposite limit $d \to 0$ (12.87) yields $\lambda \to \sqrt{-(\varepsilon\varepsilon_0 \, kT/ne\, \sigma)} \, d$, and the pressure according to (12.89) is given by

$$f(d) = i \, V_0 \, \sigma/nd \; , \tag{12.91}$$

which describes a real and repulsive pressure since V_0 is imaginary and σ and n have opposite signs. This simple equation is considered to be of particular importance for SFM experiments involving deionized immersion liquids and a moderate charging of probe and sample surface.

However, the total interaction between probe and sample in a liquid environment must of course also include the VdW force. Unlike the ionic force, VdW interactions are largely insensitive to variations of the counterion concentration, while they are highly sensitive to those surface reactions ultimately leading to the ionic forces, i.e., dissociation or adsorption processes (Sect. 12.2.7). Thus, for any given probe–sample–immersion configuration, the total interaction is obtained by simple linear superposition of VdW and ionic contributions. The comparison of (12.11, 91) shows that the VdW force generally exceeds the ionic force at small separations of the interacting surfaces, while, according to (12.90), the ionic force is dominant at large separations. If the VdW force is attractive, this results in a transition from repulsive to attractive interactions if the probe approaches the sample, as shown in Fig. 12.22. Even for highly charged surfaces, the VdW force causes deviations from the simple ionic double layer behavior

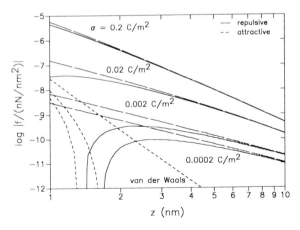

Fig. 12.22. Interplay of ionic and van der Waals pressure as a function of separation between two planar surfaces interacting in pure water. Surface charging is assumed to result from a monovalent ionization process. The long-dashed lines correspond to the pure repulsive ionic force. A typical non-retarded Hamaker constant of 10^{-20} J yields an attractive van der Waals interaction following the short-dashed straight line. The resulting total pressure is given by the solid lines which show a zero-axis crossing for the two lower charge densities

up to surface separations of more than a nanometer. For low surface-charge densities both contributions may interplay throughout the whole regime that is interesting for SFM experiments. If the VdW force is attractive, the total pressure generally changes from repulsion to attraction below 10 nm separation of the surfaces.

If two slabs are finally forced into molecular contact, the pressure pushing the trapped counterions toward the surfaces dramatically increases according to (12.91). The high ionic pressure may initiate "charge regulation processes", e.g., readsorption of counterions onto original surface sites. As a result, the surface charge density exhibits a reduction with decreasing distance between the slabs. The ionic force thus falls below the value predicted by (12.91). However, charge regulation is expected to be of only little importance in non-contacting SFM since probe–sample separations are generally well above the molecular diameter. Moreover, for a sharp tip close to a flat substrate, charge regulation would be restricted to the tip's very apex, while the major part of the interaction comes about from longer-range contributions. Thus, (12.89) should be a good basis to calculate the actual ionic probe–sample interaction via the framework developed in Sect. 12.2.4.

12.3.2 The Effect of an Electrolyte Solution

The treatment in Sect. 12.3.1 was based on the assumption that the immersion medium is a pure liquid, i.e., that it only contains a certain counterion concentration just compensating the total surface charge of probe and sample. This assumption is generally not strictly valid for SFM systems involving wetting films on probe and sample or liquid immersion: pure water at pH 7 contains 10^{-7} M (1 M = 1 mol/dm^3 corresponds to a number density of $6 \times 10^{26}/m^3$) of H_3O^+ and OH^- ions. Many biological samples exhibit ion concentrations about 0.2 M resulting from dissociated inorganic salts. A bulk reservoir of electrolyte ions has a profound effect on the ionic probe–sample interaction.

For an isolated surface, covered with a charge density σ and immersed in a monovalent electrolyte solution of bulk concentration ϱ_b, the surface electrostatic potential is given by

$$\psi_0(\sigma, \varrho_b) = -i\, V_0 \operatorname{arsinh} \frac{\sigma}{\sqrt{8\, e\, \lambda_D\, \varrho_b}}, \tag{12.92}$$

which is a convenient form of the Grahame relation [12.29]. The imaginary potential difference V_0 is defined in (12.88) and

$$\lambda_D(\varrho_b) = \frac{1}{e}\sqrt{\frac{\varepsilon\varepsilon_0\, kT}{\varrho_b}} \tag{12.93}$$

denotes the Debye length. The dependence of ψ_0 and λ_D on the bulk electrolytic concentration is shown in Fig. 12.23.

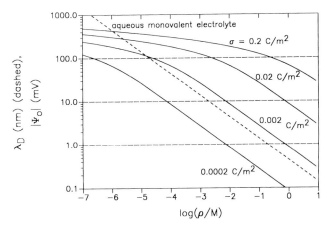

Fig. 12.23. Debye length and surface potential of an isolated charged surface as a function of bulk electrolytic concentration

If $\lambda_D \sigma / \varepsilon\varepsilon_0 \ll - \mathrm{i}\, V_0/2$, then (12.92) yields

$$\psi_0(\sigma, \varrho_b) = \sigma \lambda_D(\varrho_b)/\varepsilon\varepsilon_0 \; , \tag{12.94}$$

which represents a parallel-plate capacitor, the plates of which exhibit a surface charge density σ and are separated by λ_D. A glance at Fig. 12.23 shows that the above proportionality between surface potential and charge density holds up to surface potentials of about $-\mathrm{i}V_0/2$. The Debye length characterizes the separation of the effective centric plane of the counterion profile from the charged surface, as shown in Fig. 12.24. According to the Gouy–Chapman approach

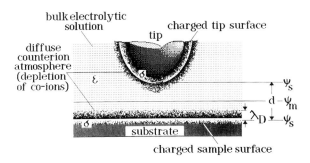

Fig. 12.24. Diffuse counterion atmosphere near probe and substrate which both exhibit a surface charge density σ. The intervening gap of width d contains an electrolytic solution with a static dielectric constant ε. The Debye length λ_D characterizes the separation of the centric planes of the counterion clouds from the surfaces of probe and sample, respectively. ψ_s and ψ_m denote the surface and midplane potentials

[12.29] the electrostatic potential at any separation z from the isolated surface is given by

$$\psi(z) = -\,i\,2V_0\,\text{artanh}[\exp(-z/\lambda_D)\tanh(i\,\psi_0/2V_0)]\;, \tag{12.95a}$$

which reduces to

$$\psi(z) = -\,i\,2V_0\tanh(i\,\psi_0/2V_0)\exp(-z/\lambda_D) \tag{12.95b}$$

for $z \gg \lambda_D$ and/or $\psi_0 \ll -\,i2V_0$. Especially for low surface potentials, $\psi_0 \ll -\,iV_0$, this latter relation may be represented by the Debye–Hückel approximation

$$\psi(z) = \psi_0\exp(-z/\lambda_D)\;. \tag{12.95c}$$

The ionic pressure between two equally charged surfaces may now be calculated according to (12.84). However, it is more convenient to use the contact value theorem in the alternative form [12.8]

$$f(z) = kT[\varrho_m(z) - \varrho_b]\;, \tag{12.96}$$

which is the excess osmotic pressure of the ions in the midplane over the bulk pressure. Since the bulk ionic concentration ϱ_b is known, the problem is reduced to the calculation of the midplane ionic concentration ϱ_m which is related to the midplane potential $\psi_m(z)$ by the Boltzmann ansatz. This leads to

$$f(z) = 4kT\,\varrho_b\sinh^2[i\,\psi_m(z)/V_0]\;. \tag{12.97a}$$

In the "weak overlap approximation" $\psi_m(z)$ is found by a linear superposition of the potentials of the isolated surfaces produced at $z/2$, i.e.,

$$\psi_m(z) = 2\psi(z/2)\;, \tag{12.97b}$$

where $\psi(z/2)$ is given by (12.95). The resulting ionic pressure according to the full weak overlap approximation, i.e., (12.97) combined with (12.95a), is shown in Fig. 12.25 for various surface charge densities and two electrolyte concentrations. At any surface charge density more dilute electrolytes with long Debye screening lengths according to (12.93) lead to a stronger repulsion between the slabs than concentrated electrolytes. The difference in magnitude and decay of the electrolytic ionic pressure with respect to the pure liquid results shown in Fig. 12.22 is striking.

While dynamic-mode SFM essentially detects the pressure according to (12.97), the actual ionic force exerted on a probe of given radius R, according to (12.32), has to be obtained via the Derjaguin formulae according to (12.34, 35). Calculations considerably simplify if a small midplane potential can be assumed, i.e., $\psi_m \ll -\,iV_0$. This latter condition is satisfied if $d \gg 2\lambda_D$ and/or $\psi_0 \ll -\,iV_0$, where d is the probe–sample separation and ψ_0 the potential of an isolated surface which is assumed to approximately represent the real surface potential

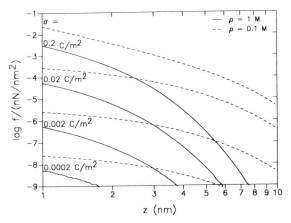

Fig. 12.25. Repulsive ionic pressure between two slabs with equal surface charge density σ in an aqueous monovalent electrolyte of bulk concentration ϱ as a function of surface-to-surface separation

ψ_s, see Fig. 12.24. Expansion of the sinh term is (12.97a) and insertion of (12.95b) via (12.97b) yields the particularly simple result

$$F(d) = 128\pi R \left(e^2 \varrho_b^2 \lambda_D^3 / \varepsilon\varepsilon_0\right) \tanh^2(i \psi_0 / 2V_0) \exp(-d/\lambda_D) \;. \tag{12.98}$$

Especially, if one has low surface potentials, $\psi \lesssim -iV_0/2$, application of (12.94) together with the Debye–Hückel approximation, (12.95c), yields via second-order expansion of (12.97a)

$$F(d) = 4\pi R \left(\sigma^2 \lambda_D / \varepsilon\varepsilon_0\right) \exp(-d/\lambda_D) \;. \tag{12.99}$$

The above results show that the ionic double layer forces in an electrolytic environment drop exponentially with probe–sample separation. The decay length is given by the Debye screening length which only depends on the bulk electrolytic concentration. The behavior is in strong contrast to ionic forces in non-electrolytic immersions exhibiting a logarithmic to $1/d$ force law as discussed in the previous section. However, as for pure liquids, the total probe–sample force also has to include the VdW component. The interference of ionic and VdW forces is well known from the classical Derjaguin–Landau–Verwey–Overbeck (DLVO) theory of lyophobic colloid stability [12.29]. Figure 12.26 shows representative curves of the total probe–sample interaction that may occur if SFM is performed under electrolytic immersion. The results, shown for two different surface charges and various bulk electrolytic concentrations, may be generalized as follows: For highly charged surfaces and dilute electrolytes, there is strong repulsion more or less throughout the whole regime relevant to non-contacting SFM. For electrolytes of higher concentration the total probe–sample interaction exhibits a minimum (attractive), preferably at probe–sample separations of a few nanometers. If surface charging is relatively low, the force exhibits a broad maximum (repulsive) some nanometers away from the substrate surface, and approaches the pure VdW curve for increasing electrolyte concentration.

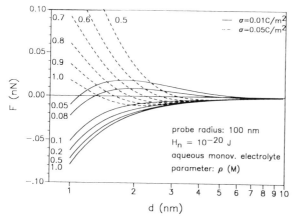

Fig. 12.26. Total probe–sample force as a function of working distance if the force microscope is operated under electrolytic immersion. The curves correspond to two different surface charge densities σ and to various bulk electrolytic concentrations ϱ

The above analysis has important bearings on VdW and magnetic force microscopy performed under liquid immersion. Unwanted ionic forces due to surface dissociation processes of probe and sample can largely be suppressed by adding an appropriate amount of inorganic salts to the immersion fluid. For magnetic measurements performed at ultra-low working distances, the total non-magnetic force may be reduced by approximately compensating repulsive ionic and attractive VdW forces for a given average probe–sample spacing. Equation (12.99) can immediately be extended to the situation of electrolytic immersions containing divalent ions or a mixture of ions of arbitrary valency n_j and bulk particle densities ϱ_{bj}. This only requires the use of a generalized Debye screening length [12.8], now given by

$$\lambda_D = \frac{1}{e}\sqrt{\frac{\varepsilon\varepsilon_0\,kT}{\sum_j n_j^2\,\varrho_{bj}}}\;. \tag{12.100}$$

However, it must be emphasized, that the weak overlap approximation used to derive the probe–sample ionic interaction is implicitly based on the assumption that the working distance is well beyond λ_D. For smaller separations one must resort to numerical solutions of the Poisson–Boltzmann equation [12.29]. For example, pure water at pH 7 exhibits a room temperature value of $\lambda_D \approx 950$ nm, while $\lambda_D \approx 0.8$ nm is found for ocean water, and $\lambda_D \approx 0.7$ nm for many biological samples [12.8]. However, for almost all SFM applications it is satisfactory to use the simple analytical results obtained in Sect. 12.3.1 if $\lambda_D \gtrsim 1$ nm, while the above results are preferred for smaller Debye lengths. Finally it may be instructive to address at least one of the open problems in the field of collective

VdW-ionic interaction in SFM. The above treatment was consequently based on the assumption that both contributions can be evaluated separately, where the total force exerted on the probe is then given by a linear superposition. However, fundamental statistical mechanics do not provide any firm basis for this treatment [12.5]. A rigorous ab initio ansatz would involve a Poisson–Boltzmann equation,

$$\nabla^2(\psi + \phi) = -\frac{e}{\varepsilon\varepsilon_0}\sum_j n_j\,\varrho_{bj}\exp[-n_j e\,(\psi + \phi)/kT]\ , \tag{12.101a}$$

which contains the sum of the equilibrium ionic potential ψ and the fluctuating VdW potential ϕ. Linearization yields

$$\nabla^2\phi \approx -\frac{e^2}{\varepsilon\varepsilon_0\,kT}\left[\sum_j n_j\,\varrho_{bj}\exp(-n_j e\,\psi/kT)\right]\phi\ , \tag{12.101b}$$

For a sharp probe in close proximity to a substrate, ψ may be a complicated function of position and may show deviations from the simple behavior assumed in the above treatment; (12.101b) in general is extremely difficult to solve in a self-consistent way. However, it confirms that, at least for high ionic mobility, VdW and ionic forces are not completely independent. Detailed SFM experiments on ionic forces may help in future to further clarify this point. If the detailed nature of the interaction is understood, SFM of ionic forces would be particularly valuable to measure surface charge densities at high spatial resolution. This may also include externally superimposed electrostatic potential differences between probe and sample.

12.4 Squeezing of Individual Molecules: Solvation Forces

The theories of VdW and ionic probe–sample interactions discussed so far are pure continuum theories in which immersion liquids present in the intervening gap between probe and sample are solely treated in terms of bulk properties, such as dielectric permittivity and average ionic concentration. This treatment breaks down when the probe–sample separation is decreased to some molecular diameters. In this regime the discrete molecular nature of immersion media can no longer be ignored since the effective intermolecular pair potentials in the liquids become a sensitive, anisotropic function of the distance between probe and sample. This phenomenon may cause quite long-range ordering effects of the liquid molecules [12.30], as shown in Fig. 12.27. Attractive interaction between the trapped molecules and the surfaces of probe and sample together with the geometric constraining effect give rise to density oscillations which may extend over several molecular diameters [12.30]. Forces related to these ordering phenomena are known as "solvation forces" [12.8].

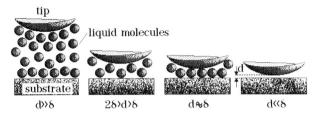

Fig. 12.27. Squeezing of individual liquid molecules between probe apex and substrate leads to long-range ordering phenomena. The resulting molecular density oscillations exhibit a periodicity roughly equal to the molecular diameter δ

An excess near-surface molecular density is, according to the contact value theorem, (12.84), related to a repulsive pressure between two slabs in close proximity. Modeling of probe–sample solvation forces thus consists of calculating the excess surface molecular density for the two slabs separated by a certain distance with respect to that of a free surface. Over the past years there has been much study of the liquid structure near constraining walls [12.31]. Different theoretical approaches to the problem include linear theories, non-linear density-functional theories and Monte Carlo simulations. However, the somewhat controversial results indicate that the field is not yet fully exploited. Thus, the present treatment is devoted to a derivation of a representative order of magnitude of the effects and to an analysis of the physics behind solvation interactions in SFM.

As intuitively expected from the simple-minded model in Fig. 12.27, all theoretical work [12.30, 31] as well as some experimental observations [12.8] have invariably confirmed an oscillating near-surface excess molecular density which may roughly be modeled by

$$\varrho(z) - \varrho(\infty) = \varrho_0 \cos(2\pi z/\delta) \exp(-z/d) \,, \tag{12.102}$$

where z denotes the separation of the two slabs and δ the effective molecular diameter of the intervening fluid. The empirical ansatz describes an exponentially damped oscillatory variation of the surface excess molecular density. ϱ_0 determines the excess density if the gap between the slabs just equals one molecular diameter. The solvation force acting on a typical SFM probe is then given, according to (12.34b, 84), by integrating (12.102):

$$F(d) = F(\delta) \left[\cos(2\pi d/\delta) - 2\pi \sin(2\pi d/\delta)\right] \exp(1 - d/\delta) \,, \tag{12.103a}$$

where one approximately has

$$F(\delta) = kT \varrho_0 \delta R/[2\pi \exp(1)] \,, \tag{12.103b}$$

with an effective probe radius according to (12.32). The problem of estimating a somewhat realistic order of magnitude of the force is now reduced to an estimation of ϱ_0, i.e., of the molecular density if the probe–sample spacing is just one molecular diameter δ. This problem is of course hard to solve in general, since ϱ_0 is expected to be sensitive to the geometry of the opposing surfaces [12.31]. However, a rough estimate may be obtained by considering an upper limit of the total order–disorder difference of an ideal hard sphere liquid. In the total ordering limit, i.e., solidification of the hard sphere molecules between probe and sample in a close-packed lattice, the maximum number density would be $\varrho(\delta) = \sqrt{2}/\delta^3$. If it is further assumed that the excess near-surface molecular density of a free surface is almost negligible with respect to this value, i.e., $\varrho(\infty) \ll \varrho(\delta)$, (12.102) yields $\varrho_0 = \sqrt{2}\exp(1)/\delta^3$. Thus one obtains from (12.103b)

$$F(\delta)/R = (1/\sqrt{2\pi})\,kT/\delta^2 \ . \tag{12.104}$$

This particularly simple relationship represents an upper limit of the solvation force per unit probe radius measured at a probe–sample separation of one molecular diameter for an ideal hard sphere VdW immersion liquid. For $\delta = 1$ nm one obtains a value of about 1 mN/m. This is ten times smaller than the typical VdW magnitude mentioned is Sect. 12.2.10. The oscillating solvation force according to (12.103a) is shown in Fig. 12.28 in comparison with a small attractive VdW interaction. While the empirical two-slab pressure according to (12.102) exhibits a maximum when the gap width d corresponds to multiples of the molecular diameter δ, the force measured with a paraboloidal or ellipsoidal probe exhibits, according to (12.103a), a shift of the molecular peaks by about 25% of the molecular diameter toward lower gap values. The amplitude of the force oscillations increases with the square of the reciprocal of the molecular diameter, while the latter also determines the characteristic decay length with increasing probe–sample separation.

The total probe–sample interaction at molecular probe–sample separations is of course composed by both solvation and VdW interactions. The solvation

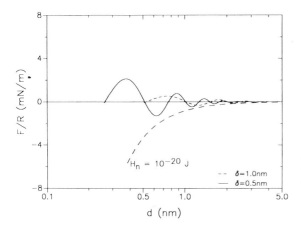

Fig. 12.28. Oscillatory solvation force per unit probe radius as a function of probe–sample separation for two different molecular diameters. A weak attractive van der Waals force is shown for reference

forces result from density fluctuations of the molecules trapped between probe
and substrate which are due to long-range molecular ordering processes. The
VdW force between probe and sample is sensitive to the dielectric permittivity of
the intervening gap, which in turn depends on the actual molecular density of
the immersion fluid. Thus it is clear that solvation and VdW forces cannot be
treated separately to obtain the total interaction by linear superposition. Both
components act collectively rather than simply additively. Since the VdW theory
developed in Sect. 12.2 is a pure continuum theory it is convenient to treat the
molecular ordering processes in a quasi-macroscopic way. This can be done via
the Clausius–Mossotti equation

$$\alpha(iv) = \varepsilon_0 \left[\varepsilon(iv) - 1\right]/\left[\varepsilon(iv) + 2\right] \varrho , \tag{12.105}$$

which relates the effective molecular polarizability of hard sphere molecules in
the gas phase to the dielectric permittivity and average molecular density which
would be measured for a macroscopic ensemble of the molecules. Since the
effective polarizability of non-interacting hard sphere molecules is invariant to
density fluctuations, a certain average molecular density $\bar{\varrho}$, deviating from the
bulk liquid density ϱ_b, transforms into a modification of the dielectric permittivity through

$$\varepsilon(\bar{\varrho}) = \frac{2 + \varepsilon_b + 2\bar{\varrho}/\varrho_b\,(\varepsilon_b - 1)}{2 + \varepsilon_b - \bar{\varrho}/\varrho_b\,(\varepsilon_b - 1)} , \tag{12.106}$$

where $\varepsilon_b = \varepsilon(\varrho_b)$ corresponds to the permittivity of the bulk liquid. Under the
assumed boundary conditions, (12.106) holds for any spectral contribution, i.e.,
for static orientational as well as for higher-frequency electronic permittivities.
The consequence is, that, according to (12.8b, 12b), the entropic and non-
retarded Hamaker constants exhibit a density-induced modulation.

   If one assumes that the average molecular density $\bar{\varrho}(d)$ within the volume
between probe and sample is approximately equal to the surface molecular
density $\varrho(d)$, (12.102) yields

$$\bar{\varrho}(d)/\varrho_b = \varrho(\infty)/\varrho_b \left[1 + \cos(2\pi d/\delta)\exp(\eta - d/\delta)\right] , \tag{12.107a}$$

with

$$\eta = 1 + \ln\left[\varrho(\delta)/\varrho(\infty) - 1\right] . \tag{12.107b}$$

At this point some heuristic assumptions have to be made concerning the ratio
of the free surface density $\varrho(\infty)$ to the bulk density of the liquid ϱ_b as well as the
ratio of the gap's excess surface density $\varrho(\delta)$ to $\varrho(\infty)$. A reasonable assumption
is, that the permittivity of the small gap between probe and sample approaches
its vacuum value somewhere between $d = 3\delta/4$ and $d = \delta/2$, when there is
no space left to trap any liquid molecules (Fig. 12.27). According to (12.106),
$\varepsilon = 1$ requires $\bar{\varrho} = 0$, and thus, according to (12.107a), $3/2 > \eta \geq 1/2$ which

in turn yields for the gap-induced increase of molecular ordering $2.6 > \varrho(\delta)/\varrho(\infty) \geq 1.6$. In other words, the packing fraction of molecules on probe and sample surface increases by a factor of 1.6 to 2.6 when the probe approaches the sample surface and finally reaches a separation corresponding to only one molecular diameter [12.32]. The second free parameter left in (12.107a) is the effective excess molecular packing fraction $\varrho(\infty)/\varrho_b$ which is simply not known for a system consisting of a sharp tip opposite to an arbitrarily shaped sample surface. Information on this quantity can only be obtained by performing Monte Carlo simulations under realistic boundary conditions. However, first results obtained for the structure of hard spheres near flat or spherical walls [12.31] suggest that the packing fraction is a complicated function of molecular diameter and constraining wall geometry and may by far exceed unity. Due to these uncertainties it is convenient to choose a somewhat pragmatic way: (12.107a) is strictly valid only for probe sample separations of a few molecular diameters since it was assumed $\bar{\varrho}(d) = \varrho(d)$. However, to ensure bulk convergence for large probe–sample separation one has to fulfill $\bar{\varrho}(\infty) = \varrho_b$. This in turn formally requires $\varrho(\infty) = \varrho_b$. This pragmatic approach permits at least an order of magnitude estimate of oscillatory VdW forces without too much ambigious parameters. A typical result for a metal–dielectric combination of probe and sample immersed in an ideal hard sphere liquid with $\varrho(\delta)/\delta(\infty) = 2.1$ (which can be considered as a somewhat typical value according to the above analysis) is shown in Fig. 12.29. The oscillating Hamaker constant has been obtained according to (12.54) with $n_3(d) = \sqrt{\varepsilon_3(d)}$, and $\varepsilon_3[\varrho(d)]$ according to

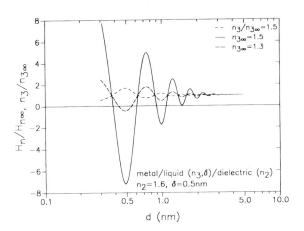

Fig. 12.29. Periodic molecular ordering of molecules trapped between probe and sample causes oscillations of the effective optical refractive index n_3 of the immersion fluid, the bulk index of which is given by $n_{3\infty}$. These oscillations transform into a huge periodic variation of the non-retarded Hamaker constant H_n with respect to its bulk value $H_{n\infty}$. δ denotes the effective molecular diameter of the immersion medium and n_2 the refractive index of the dielectric sample. The probing tip is metallic

(12.106, 107). The oscillating refractive index n_3 of the immersion liquid transforms into a huge "overshoot" of the non-retarded Hamaker constant with respect to its bulk value. If the bulk index of the immersion fluid is close to that of the dielectric (sample), the originally purely attractive interaction may become repulsive for certain probe–sample separations, while it is solely attractive but oscillating if probe and sample are made from the same material. Refractive index and Hamaker constant both exhibit the exponential damping ultimately resulting from the decrease of the molecular excess osmotic pressure according to (12.102). However, they are completely out of phase, but both show the molecular periodicity.

The total VdW solvation force exerted on the probe is now obtained by a linear superposition of the osmotic contribution according to (12.103a) and the VdW contribution according to (12.40a) using a density-modulated Hamaker constant according to (12.6, 12b, 106, 107). A typical result, again for a metal–dielectric combination of probe and sample, is shown in Fig. 12.30. The total interaction still shows the molecular periodicity δ. However, since osmotic and VdW contributions are mutually phase-shifted in a complicated way, the oscillating curve does generally not peak when the probe–sample separation exactly equals a multiple of half the molecular diameter. The damping at small probe–sample distances is stronger than that of the excess osmotic pressure in (12.102), and approaches the latter a few molecular diameters away from the sample surface. At very small probe–sample separations, i.e., just before interatomic repulsion occurs, the total interaction approaches the VdW continuum expected for a vacuum interaction between probe and sample. However, if probe and substrate are separated by more than about one molecular dia-

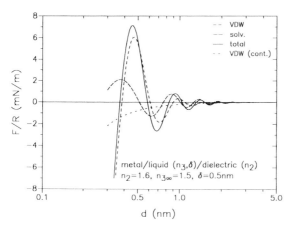

Fig. 12.30. Force per unit probe radius as a function of probe–sample separation for a metal/dielectric (optical refractive index n_2) configuration of probe and sample, immersed in a hard sphere liquid with an effective molecular diameter δ and a bulk optical refractive index $n_{3\infty}$. Superposition of the oscillatory van der Waals and osmotic contributions yields the total force exerted on the probing tip. For reference, the van der Waals curve resulting from the pure continuum theory is also shown

meter, the giant oscillations of the VdW solvation force exceed by far the continuum of VdW forces.

Finally it should be emphasized that the field of solvation force phenomena in SFM is completely open and, to the author's knowledge, no detailed observation of an oscillating attractive/repulsive interaction at molecular working distances has ever been reported. However, the present theoretical analysis confirms that, at least for some model configurations, oscillatory solvation forces should be detectable. Quite promising inert immersion liquids, which contain fairly rigid spherical or quasi-spherical molecules, are, e.g., OctoMethyl-CycloTetraSiloxane (OMCTS, non-polar, $\delta \approx 0.9$ nm), carbontetrachloride (non-polar, $\delta \approx 0.28$ nm), cyclohexane (non-polar, $\delta \approx 0.29$ nm), and propyl-enecarbonate (highly polar hydrogen-bonding, $\delta = 0.5$ nm) [12.8]. SFM measurements on these and other immersion liquids could help to obtain a deeper insight into molecular ordering processes near surfaces and in small cavities. As already emphasized with respect to VdW and ionic interactions, solvation forces certainly have to be accounted for as unwanted contributions, if electric or magnetic force microscopy is performed at ultra-low working distances and under liquid immersion. In general, the situation is complicated by the fact that VdW, ionic, and solvation forces may contribute to the total probe–sample interaction in a non-additive way. Unfortunately this is only part of the whole story. If SFM experiments are performed under aqueous immersion, or if only trace amounts of water are present, and this is the case for almost all experiments under ambient conditions, hydrophilic and hydrophobic interactions have often additionally to be taken into account [12.8]. The phenomena are mainly of entropic origin and result from the rearrangement of water molecules if probe and sample come into close contact. In this sense hydrophilic and hydrophobic forces clearly belong to the general field of solvation forces, however, macroscopic experiments [12.8] confirm, that they are generally not well characterized by the simple theory presented above. Hydration forces result, whenever water molecules strongly bind to hydrophilic surface groups of probe and sample. A strong repulsion results, which exhibits an exponential decay over a few molecular diameters [12.8]. In the opposite situation, for hydrophobic probe and sample, the rearrangement of water molecules in the overlapping solvation zones results in a strong attractive interaction. These phenomena once again show that water is one of the most complicated liquids that we know. However, its importance in SFM experiments under ambient conditions must not be emphasized and more detailed information on its microscopic behavior is of great importance.

12.5 Capillary Forces

Under humid conditions, a liquid bridge between probe and sample can be formed in two different ways: by spontaneous capillary condensation of vapours and by directly dipping the tip into a wetting film which is present on top of the substrate surface. Capillary condensation is a first-order phase-transition whereby the undersaturated vapour condenses in the small cavity between probe apex and sample surface. Due to surface tension a liquid bridge between probe and sample results in a mutual attraction. At thermodynamic equilibrium the meniscus radii according to Fig. 12.31 are related to the relative vapour pressure by the well-known Kelvin equation [12.33],

$$\frac{1}{r_1} + \frac{1}{r_2} = \frac{CT\varrho}{2\gamma M}\ln(p/p_s) \ , \tag{12.108}$$

where C denotes the universal gas constant and ϱ, M, γ are the mass density, the molar mass, and the specific surface free energy or surface tension of the liquid forming the capillary. Since $p < p_s$, the Kelvin mean radius, $|r_K| = r_1 r_2/(r_1 + r_2)$, for a concave meniscus as in Fig. 12.31 is negative. Figure 12.32 shows the equilibrium Kelvin radius for a water capillary between probe and sample as a function of relative humidity of the experimental environment. For $r_K \to -\infty$, i.e., for a relative humidity approaching 100%, the swelling capillary degenerates to a wetting film. In the opposite extreme, at a relative humidity of a few percent no capillary is formed, or a pre-existing capillary evaporates, since the Kelvin radius approaches molecular dimensions.

The mutual attraction of probe and sample results from the Laplace pressure,

$$\pi = \gamma/r_k \ , \tag{12.109a}$$

within the liquid bridge. The total capillary force exerted on the probe is thus

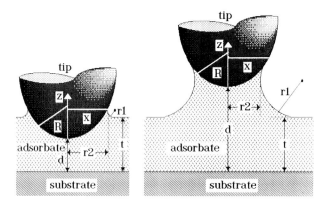

Fig. 12.31. Capillary interaction between the probe and a substrate which has a surface covered with a liquid adsorbate. When the probe is dipped into the adsorbate the liquid surface exhibits curvature near the probe's surface (*left side*). Withdrawal of the probe or spontaneous capillary condensation before the probe contacts the liquid surface results in an elongated liquid bridge (*right side*)

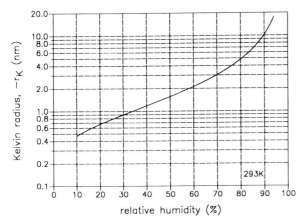

Fig. 12.32. Equilibrium dimension of the Kelvin radius for a water capillary between probe and sample

given by

$$F(d) = \pi x^2(d)\, \gamma/r_k \ , \tag{12.109b}$$

where, as shown in Fig. 12.31, x is the radius of the area, where the meniscus is in contact with the probe's surface. The problem is thus to determine this radius as a function of the probe–substrate separation, since the Kelvin radius is known at thermodynamic equilibrium from (12.108). One first considers the situation sketched in the left part of Fig. 12.31, i.e., the probe–substrate separation d is less than or equal to the adsorbate thickness t. For simplicity, further an ideally wetting liquid with vanishing contact angle at the probe is considered. From (12.31) one obtains the relation $x^2 = 2Rz$, where the effective probe radius is determined by (12.32). From geometrical considerations one then immediately obtains

$$z \approx t - d + r_1[1 + R/(R + r_1)] \ , \tag{12.110}$$

which is valid for thin adsorbate films with $t \ll R$. Since $r_1 \ll r_2$ one has to a good approximation $r_1 \approx -r_k$. The force according to (12.109b) is thus given by

$$F(d) = -2\pi R\, \gamma\, [1 + R/(R - r_k) - (t - d)/r_k] \ . \tag{12.111}$$

Force-versus-distance curves according to the above relation contain complete information about an adsorbate layer. At low partial vapour pressure leading to $-r_k \ll R$, the force measured for a virgin probe–adsorbate contact, $F/R = -4\pi\gamma$, permits a measurement of the adsorbate's surface tension. The Kelvin radius is directly obtained from the slope $\partial F/\partial d = -2\pi\gamma R/r_k$. Finally the adsorbate thickness may be obtained by a simple dipping experiment, whereby $F(d)$ is detected for $0 \le d \le t$. The maximum capillary force is obtained just before the tip touches the substrate: $F(0) = -2\pi\gamma(2r_k - t)R/r_K$ for

$-r_k \ll R$ and $F(0) = 2\pi\gamma tR/r_k$ for $-r_k \ll t$. For a water film the specific surface free energy is 73 mJ/m^2 [12.8]. The capillary force acting on a probe which dips into a water film on top of the sample is thus $|F|/R > 0.9$ N/m which is about 90 times the typical VdW magnitude mentioned in Sect. 12.2.10.

When the probe is withdrawn after it had contact with the liquid adsorbate, an elongated capillary is formed as shown in the right part of Fig. 12.31. Since for $d \geq t$ both meniscus radii r_1 and r_2 now vary over a considerable range, the calculation of the probe–sample capillary force is slightly more complicated than in the previous situation. Simple geometrical arguments lead to

$$x = R(r_2 + r_1)/(R + r_1) , \tag{12.112a}$$

and

$$x = R\sqrt{1 - (R + d - t - r_1)/(R + r_1)} , \tag{12.112b}$$

where, at thermodynamic equilibrium, r_1 and r_2 are additionally related to each other by (12.108). After a little algebra, the radius of the probe–capillary contact area is determined by the solution of the following cubic equation:

$$x^3 - (d - t)x^2 + 2R(2r_k + d - t)x - R(d - t)^2 = 0 , \tag{12.113}$$

which is valid for $d - t \ll R$. The result can of course be obtained analytically but is then a little bit unwieldy. For $d - t \gg -r_k$, $r_1 \approx (d - t)/2$ and $r_2 \approx -r_k \ll R$ lead, according to (12.112a, 109b), to the asymptotic force

$$F(d) = \pi\gamma R^2(d - t)^2/r_k (2R + d - t)^2 , \tag{12.114}$$

from which an upper limit of $F(\infty) = \pi\gamma R^2/r_k$ can be deduced for the capillary force. For example, this upper limit amounts for a 100 nm probe, interacting with the substrate via a water meniscus, to 1.5 μN at an ambient humidity of 50–60%. This order of magnitude shows that capillary forces in SFM are generally much stronger than any aforementioned interaction.

Figure 12.33 shows some force-versus-distance curves obtained according to (12.111, 109b, 113). When the approaching probe first touches the water film of 5 nm thickness it experiences a sudden attractive force of magnitude $4\pi\gamma R$ which is for a 100 nm probe about 90 nN. The probe then penetrates the adsorbate layer and exhibits a linear increase in attractive force according to (12.111). The maximum value achieved just before touching the substrate depends on the ambient humidity, where the highest values are obtained in a relatively dry atmosphere. If the probe is withdrawn before making contact with the substrate, the attractive force decreases reversibly until the adsorbate–air interface is reached. From then on an elongated capillary is formed leading to a pronounced hystereris effect. The curves are now described by (12.109b, 113). Upon further withdrawal of the probe the force first exhibits a further decrease until some minimum value close to zero is reached. From this point it increases again approaching the asymptotic behavior according to (12.114) (not shown). If

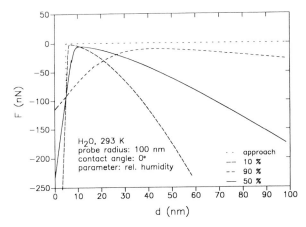

Fig. 12.33. Capillary force as a function of probe–substrate separation. The substrate is covered with a 5 nm water film of fixed thickness, while the probe–sample interaction is shown for three different values of the ambient relative humidity

thermodynamic equilibrium conditions would be present throughout the complete measurement the capillary between probe and sample would assume an arbitrary length, while the smaller meniscus radius r_2 becomes equal to the Kelvin radius $-r_K$, i.e., the circumference of the meniscus becomes stable. However, since the adsorbate film has a finite thickness, material transport into the growing capillary is disrupted at some time leading to an irreversibility, whereby the force suddenly vanishes. The same occurs if the capillary is destroyed by external perturbations (vibration, air currents).

The above results are only precise for nearly spherical probes and vanishing contact angle. However, the analysis can immediately be extended to paraboloidal or ellipsoidal probes, even of low aspect ratios, $R_</R_>$ according to (12.32) and to arbitrary contact angles. The main results predicted by the above treatment remain unchanged. A particularly interesting feature is related to (12.113). A careful analytical examination shows that, for a certain regime of the substrate–sample separation, the cubic equation involves three real roots, all leading to stable solutions for the force according to (12.109b). This implies the possibility of discontinuous transitions between different force curves upon variation of the probe–sample separation. It should further be noted, that the Kelvin equation (12.108) as well as the Laplace relation (12.109a) are strictly macroscopic equations, i.e., to ensure validity, the system has to be in thermodynamic equilibrium and the Kelvin radius must well exceed the molecular diameter. For very small Kelvin radii, the liquid's surface tension is no longer a constant. While for simple Lennard–Jones liquids, such as cyclohexane or benzene, a macroscopic behavior is already manifest at molecular Kelvin radii, for water a value of about 5 nm is assumed [12.8], which corresponds, according to Fig. 12.32, to a relative humidity of 80%.

The other important question is, whether an ideal thermodynamic equilibrium can be assumed for an arbitrary adsorbate. In the extreme situation, where an adsorbate exhibits a nearly vanishing evaporation rate since it forms a stable film on top of the substrate, the simple free-liquid equilibrium conditions considered above are no longer valid. In this case, the Kelvin equation (12.108), which controls the interplay of the meniscus radii r_1 and r_2 has to be replaced by a relation representing the condition of zero material transport into the capillary. According to Fig. 12.31 the meniscus volume for $d \geq t$ is given by

$$V = \pi \left(\sum_{j=1}^{2} \left\{ [(r_1 + r_2)^2 + r_1^2] z_j - (r_1 + r_2) \sqrt{r_1^2 - z_j^2}\, z_j - r_1^2 (r_1 + r_2) \right. \right.$$

$$\left. \left. \times \arcsin \frac{z_j}{r_1} - \frac{z_j^3}{3} \right\} - \frac{z_1^2}{3} (3R - z_1) \right), \quad (12.115a)$$

with

$$\left. \begin{aligned} z_1 &= R + d - t - r_1 - \sqrt{R^2 - x^2} \\ z_2 &= r_1 \end{aligned} \right\}. \quad (12.115b)$$

Zero material transport is then ensured by the constraint $\partial V/\partial d = 0$. This latter condition then relates $r_2(d)$ to $r_1(d)$. The additional use of (12.112) then leads to a first-order non-linear differential equation for the meniscus–probe contact radius x. The numerical solution permits, together with (12.109b), a calculation of the capillary force for any probe–sample separation d. However, in the most interesting regime $t \leq d \ll R$, the differential equation can be considerably simplified which leads to an analytical solution for x. $z_1 \approx z_2 = r_1$ and $r_1 = (x^2 + 2R[d - t])/4R$ inserted into (12.115a) yields the simple equilibrium condition

$$\left[2(d - t) + \frac{x^2}{R} \right] \frac{\partial x}{\partial d} + x = 0, \quad (12.116a)$$

with the solution

$$x^2 = \frac{4R\, r_K^2}{d - t - r_K}. \quad (12.116b)$$

The capillary force according to (12.109b) is thus

$$F(d) = \frac{r_K}{d - t - r_K}, \quad (12.117)$$

which matches the result of (12.111) for $d = t$. On the other hand, if $d - t \gg -r_K$, $r_1 \approx (d - t)/2 - r_K$ directly inserted into (12.112b) yields via (12.109b) the asymptotic behavior

$$F(d) = 2\pi\, \gamma \frac{R^3}{r_K(2R + d + t)}. \quad (12.118)$$

This latter relation clearly shows that the meniscus between probe and substrate may extend over probe–sample separations several times exceeding the adsorbate thickness, and even exceeding the probe radius. The capillary instability point can be estimated by considering the decrease of the meniscus radius r_2 (Fig. 12.31) with increasing probe–sample separation d. Using $r_1 \approx (d - t)/2$ the combination of (12.112) yields

$$r_2 = \sqrt{2R^2 + R(d - t)} - (d - t)/2 , \qquad (12.119a)$$

where the capillary becomes instable if $r_2 = -r_K$. This gives a critical probe–substrate separation of

$$d = 2(1 + \sqrt{3})\, R + t \approx t + 5.5R . \qquad (12.119b)$$

Thus, an ideal, externally unperturbed capillary may extend over more than a hundred nanometers.

The force obtained according to the constant–volume equilibrium via (12.117) exhibits a behavior completely different from that shown in Fig. 12.33. The result obtained for a PerFluoroPolyEther (PFPE) polymer liquid film adsorbed on a substrate is shown in Fig. 12.34. The surface tension and Kelvin radius values were taken from [12.27]. Upon approach to the sample, a sudden attractive force is exerted on the probe, when it first touches the adsorbate film. The linear behavior upon dipping the probe into the adsorbate film is again described by (12.111). Upon withdrawal, a considerable hysteresis again occurs, since an elongated liquid bridge is now formed between probe and adsorbate surface. This leads to a monotonic decrease of the force with increasing probe–sample separation; initially, according to (12.117), and then, in the asymptotic regime, according to (12.118). The theoretical result shown in Fig. 12.34 is in good quantitative agreement with experimental results on PFPE polymer liquid films presented by *Mate* et al. [12.27].

The existence of long-range capillary forces has been demonstrated by several experimental results. Detailed measurements for water were presented

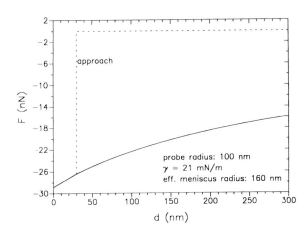

Fig. 12.34. Capillary force as a function of probe–substrate separation. The adsorbate thickness is assumed to be 30 nm. The model calculation actually applies to an adsorbed perfluoropolyether polymer liquid film

by *Weisenhorn* et al. [12.34]. However, a detection of the pure capillary forces appears to be difficult in some cases since the intermediate contact of the probe with the substrate yields additional adhesion forces which may considerably modify the curves shown in Figs. 12.33, 34. Yet, not enough experimental data is available to strictly decide whether liquid adsorbates in general exhibit capillary forces according to Fig. 12.33 or according to Fig. 12.34, or if they in general exhibit a more complex intermediate behavior. However, especially for very thin films showing clear capillary formation at highly undersaturated vapour pressure, it is likely that Eqs. (12.117, 118) are valid. It is interesting, that the force according to these relations does not involve the multistabilities which occur if the free-liquid thermodynamic equilibrium equation (12.113) is used to derive the probe–meniscus contact radius.

In spite of these uncertainties the above analysis has an important implication for force measurements in the presence of liquids: If the SFM probe is dipped into a liquid adsorbate of a few nanometer thickness, the force always exhibits the linear dependence on the probe–sample separation d given by (12.111). Since all forces dealt with before involve a non-linear dependence on the probe–sample separation, they can be measured in the presence of capillary forces. A complete immersion of the whole SFM can thus be reduced to dipping only the tip into an immersion film of sufficient thickness, while ultimate corrections of the measured force-versus-distance curves are simply linear. Additionally, capillary forces may be used to lock-in the non-contacting probe at a certain separation from the substrate surface, instead of using electrostatic servo forces. However, in general, much more experimental information on capillary phenomena is needed to further elucidate this complex but interesting area [12.35]. A more detailed examination should also include the effect of electrostatic fields on capillary equilibrium, since voltages between probe and substrate are used quite often in scanned probe microscopy.

12.6 Conclusions

In non-contact scanning force microscopy, various kinds of long-range probe–sample interactions may be present. Apart from electro- and magneto-static forces, the detection of which is the basis of the well-established techniques of electric and magnetic force microscopy, four other basic types of forces are identified as relevant interactions in non-contact scanning force microscopy. Ultimately, these forces are of course all of electromagnetic origin, however, their actual magnitudes and ranges are quite different. Van der Waals forces provide an ever-present contribution to the probe–sample interaction. Further theoretical and experimental work is needed here to clarify in particular the microscopic dielectric behavior of sharp probes, especially of metal probes, and of the near-surface regime of the samples.

Under liquid immersions ionic forces generally provide an additional contribution to the total probe–sample interaction. In this area, progress in solving the Poisson–Boltzmann equation under largely realistic boundary conditions may lead to the discovery of specific phenomena related to the unique electrostatic field configuration near the apices of sharp tips.

The area of solvation forces which may be present at very small probe–sample separations under liquid immersion, appears to be completely open with respect to both improved theoretical treatments and detailed experimental investigations.

Capillary forces are unwanted in most experiments. However, more detailed theoretical and experimental examinations may provide a deeper insight into the microscopic thermodynamics of liquids.

The present analysis predicts concrete orders of magnitude for all mentioned types of probe–sample interaction. All discussed phenomena are accessible to present-day experimental set-ups. It is thus hoped that this theoretical analysis stimulates some experimental approaches to open new avenues in the application of non-contact scanning force microscopy.

Acknowledgements. Thanks are due to Prof. C. Heiden (KFA Jülich/University of Gießen) for his continuous support of the present work. The author is indebted to Dr. N.A. Burnham (KFA Jülich) for numerous clarifying discussions and for the careful proof-read of the manuscript.

References

12.1   E.M. Lifshitz: J. Exper. Theoret. Phys. USSR **29**, 94 (1955) [Sov. Phys. JETP 2, 73 (1956)]

12.2   R. Eisenschitz, F. London: Z. Phys. **60**, 491 (1930)

12.3   H.B.G. Casimir, D. Polder: Phys. Rev. **73**, 360 (1948); H.B.G. Casimir: Proc. Kon. Ned. Akad. Wetensch. **51**, 793 (1948)

12.4   I.E. Dzyaloshinskii, E.M. Lifshitz, L.P. Pitaevskii: Adv. Phys. **10**, 165 (1961)

12.5   A comprehensive survey of the basic theory of van der Waals forces between macroscopic bodies is, e.g., given by J. Mahanty and B.W. Ninham: *Dispersion Forces* (Academic, London 1976)

12.6   H.C. Hamaker: Physica **4**, 1058 (1937)

12.7   U. Hartmann: Phys. Rev. B **42**, 1541 (1990)

12.8   An excellent review on various micro- and macroscopic aspects of molecular interactions is given by J.N. Israelachvili: *Intermolecular and Surface Forces with Applications to Colloidal and Biological Systems* (Academic, London 1985)

12.9   U. Hartmann: Phys. Rev. B **43**, 2404 (1991)

12.10 See for example: J.N. Israelachvili: Proc. R. Soc. Lond. A **331**, 39 (1972)

12.11 B.V. Derjaguin: Koll. Z. **69**, 155 (1934)

12.12 U. Hartmann: J. Vac. Sci. Technol. B **9**, 465 (1991)

12.13 Yu.N. Moiseev, V.M. Mostepanenko, V.I. Panov, I.Yu. Sokolov: Phys. Lett. A **132**, 354 (1988)

12.14 M. Anders (unpublished result)

12.15 L.D. Landau and E.M. Lifshitz: *Electrodynamics of Continuous Media* (Addison-Wesley, Reading/MA 1960)

12.16 U. Hartmann: Adv. Mat. **2**, 594 (1991)

12.17 J.N. Israelachvili: Proc. R. Soc. Lond. A **331**, 19 (1972)

12.18 See standard textbooks, e.g., J.D. Jackson: *Classical Electrodynamics* (Wiley, New York 1975)

12.19 G. Feinberg and S. Sucher: Phys. Rev. A **2**, 2395 (1970); G. Feinberg: Phys. Rev. B **9**, 2490 (1974)

12.20 T. Datta, L.H. Ford: Phys. Lett. A **83**, 314 (1981)

12.21 About the same conclusions have previously been drawn by V.M. Mostepanenko, I.Yu. Sokolov: Dokl. Akad. Nauk SSSR **298**, 1380 (1988) [Sov. Phys. Dokl. **33**, 140 (1988)]

12.22 E. Zaremba, W. Kohn: Phys. Rev. B **13**, 2270 (1976)

12.23 C. Girad: Phys. Rev. B **43**, 8822 (1991)

12.24 D.M. Eigler, E.K. Schweizer: Nature **334**, 524 (1990)

12.25 K.E. Drexler: J. Vac. Sci. Technol. B **9**, 1394 (1991)

12.26 H. Lemke, T. Göddenhenrich, H.P. Bochem, U. Hartmann, C. Heiden: Rev. Sci. Instrum. **61**, 2538 (1990)

12.27 C.M. Mate, M.R. Lorenz, V.J. Novotny: J. Chem. Phys. **90**, 7550 (1989)

12.28 See, e.g., H. Räther: *Surface Plasmons on Smooth and Rough Surfaces and on Gratings* (Springer, Berlin, Heildelberg 1988), as well as several articles on plasmon observation by scanning tunneling microscopy

12.29 For a fuller discussion of this issue see, e.g., P.C. Hiemenz: *Principles of Colloid and Surface Chemistry* (Dekker, New York 1977)

12.30 For some extensive reviews on this subject see, e.g., D. Nicholson and N.D. Personage: *Computer Simulations and the Statistical Mechanics of Adsorption* (Academic, New York 1982); G. Rickayzen and P. Richmond: in *Thin Liquid Films*, ed. by I.B. Ivanov (Dekker, New York 1985)

12.31 S.T. Chui: Phys. Rev. B **43**, 10654 (1991), and references therein

12.32 The upper limit is additionally constrained by the fact that $\varepsilon(\delta)$ must of course be finite. Convergence of (12.106) requires $\varrho(\delta) < \varrho_b(\varepsilon_b + 2)/(\varepsilon_b - 1)$. However, this criterion only becomes relevant if the excess surface density for the gap between probe and sample is almost the same as for the free surfaces, and if this free surface molecular density is much higher than the bulk liquid density. For $\varrho(\infty) \approx \varrho_b$, as used in the following, $\varrho(\delta)/\varrho(\infty) < 2.6$ can be considered as the relevant criterion for all immersion liquids (with $\varepsilon_b < 2.9$).

12.33 See, e.g., A.W. Adamson: *Physical Chemistry of Surfaces* (Wiley, New York 1976)

12.34 A.L. Weisenhorn, P.K. Hamsma, T.R. Albrecht, C.F. Quate: Appl. Phys. Lett. **54**, 2651 (1989)

12.35 See R. Evans, U.M.B. Marconi, P. Tarazona: J. Chem. Phys. **84**, 2376 (1986), and references therein

Subject Index

Contents of

Scanning Tunneling Microscopy I

(Springer Series in Surface Sciences, Vol. 20)

Contents of

Scanning Tunneling Microscopy II

(Springer Series in Surface Sciences, Vol. 28)

Printing: Mercedesdruck, Berlin
Binding: Buchbinderei Lüderitz & Bauer, Berlin